Six-Minute Solutions

for Civil PE Exam Problems

Water Resources

R. Wane Schneiter, PhD, PE, DEE

Professional Publications, Inc. • Belmont, CA

How to Locate Errata and Other Updates for This Book

At Professional Publications, we do our best to bring you error-free books. But when errors do occur, we want to make sure that you know about them so they cause as little confusion as possible.

A current list of known errata and other updates for this book is available on the PPI website at **www.ppi2pass.com/errata**. We update the errata page as often as necessary, so check in regularly. You will also find instructions for submitting suspected errata. We are grateful to every reader who takes the time to help us improve the quality of our books by pointing out an error.

SIX-MINUTE SOLUTIONS FOR CIVIL PE EXAM PROBLEMS: WATER RESOURCES

Current printing of this edition: 3

Printing History

edition number	printing number	update
1	1	New book.
1	2	Minor corrections.
1	3	Minor corrections.

Printed in the United States of America

Professional Publications, Inc.
1250 Fifth Avenue, Belmont, CA 94002
(650) 593-9119
www.ppi2pass.com

Library of Congress Cataloging-in-Publication Data
Schneiter, R. W.
    Six-minute solutions for civil PE exam problems: water resources / R. Wane Schneiter.
        p. cm.
    ISBN 1-888577-90-8
    1. Hydraulic engineering--Examinations, questions, etc. 2. Hydrology--Examinations, questions, etc. 3. Water--Purification--Examinations, questions, etc. 4. National Council of Examiners for Engineering and Surveying--Examinations--Study guides. I. Title.

TC157.5.S35 2003
627'.0076--dc21
                                                2003047123

Table of Contents

About the Author

R. Wane Schneiter, PhD, PE, DEE, is the Benjamin H. Powell, Jr., '36 Professor of Engineering in the Civil and Environmental Engineering Department at the Virginia Military Institute where he is responsible for the environmental engineering program. Prior to VMI, Dr. Schneiter was a principal engineer with an environmental consulting firm in California. He has been an environmental consultant to numerous industrial, commercial, and local governmental clients and nongovernmental organizations, and has served as an expert witness in litigated disputes involving a variety of environmental issues.

Preface and Acknowledgments

The Principles and Practice of Engineering examination (PE exam) for civil engineering, prepared by the National Council of Examiners for Engineering and Surveying (NCEES), is developed from sample problems submitted by educators and professional engineers representing consulting, government, and industry. PE exams are designed to test examinees' understanding of both conceptual and practical engineering concepts. Problems from past exams are not available from NCEES nor any other source. However, NCEES does identify the general subject areas covered on the exam.

The topics covered in *Six-Minute Solutions for Civil PE Exam Problems: Water Resources* coincide with those subject areas identified by NCEES for the water resources engineering depth module of the civil PE exam. These problem topics are hydraulics, hydrology, and water treatment.

The problems presented in this book are representative of the type and difficulty of problems you will encounter on the PE exam. The book's problems are both conceptual and practical, and they are written to provide varying levels of difficulty. Though you probably won't encounter problems on the exam exactly like those presented here, reviewing these problems and solutions will increase your familiarity with the exam problems' form, content, and solution methods. This preparation will help you considerably during the exam.

Problems and solutions have been carefully prepared and reviewed to ensure that they are appropriate and understandable, and that they were solved correctly. If you find errors or discover a more efficient way to solve a problem, please bring it to PPI's attention so your suggestions can be incorporated into future editions. You can report errors and keep up with the changes made to this book, as well changes to the exam, by logging on to PPI's website at www.ppi2pass.com and clicking on "Errata."

Thank you to the many persons in the editorial and production departments at Professional Publications who contributed to the successful publication of this book. They are an enjoyable group to work with, are thorough and professional, and are dedicated to providing the best possible publication. Thanks are also due to James R. Sheetz, PE, DEE, for his thorough technical review.

R. Wane Schneiter, PhD, PE, DEE

Introduction

EXAM FORMAT

The Principles and Practice of Engineering examination (PE exam) in civil engineering is an 8 hr exam divided into a morning and an afternoon session. The morning session is known as the *breadth* exam, and the afternoon is known as the *depth* exam.

The morning session consists of 40 problems from all of the five civil engineering subdisciplines (environmental, geotechnical, structural, transportation, and water resources), with each subdiscipline representing about 20% of the problems. As the "breadth" designation implies, morning session problems are general in nature and wide-ranging in scope.

The afternoon session allows the examinee to select a depth exam module from one of the five subdisciplines. The 40 problems in the afternoon session require more specialized knowledge than those in the morning session.

All problems from the morning and afternoon sessions are multiple choice. They include a problem statement with all required defining information, followed by four logical choices. Only one of the four options is correct. Nearly every problem is independent of the others, so an incorrect choice on one problem typically will not carry over to subsequent problems.

Topics and the approximate distribution of problems on the afternoon session of the civil water resources exam are as follows.

Water Resources: approximately 65% of exam problems

- Hydraulics
- Hydrology
- Water Treatment

Environmental: approximately 25% of exam problems

- Wastewater Treatment
- Aquatic Biology and Microbiology
- Solid and Hazardous Waste
- Groundwater and Well Fields

Geotechnical: approximately 10% of exam problems

- Subsurface Exploration and Sampling
- Engineering Properties of Soils
- Soil Mechanics Analysis

For further information and tips on how to prepare for the civil water resources PE exam, consult the *Civil Engineering Reference Manual* or Professional Publications' website, www.ppi2pass.com.

THIS BOOK'S ORGANIZATION

Six-Minute Solutions for Civil PE Exam Problems: Water Resources is organized into two sections. The first section, Breadth Problems, presents 20 water resources engineering problems of the type that would be expected in the morning part of the civil engineering PE exam. The second section, Depth Problems, presents 80 problems representative of the afternoon part of this exam. The two sections of the book are further subdivided into the topic areas covered by the water resources exam.

Most of the problems are quantitative, requiring calculations to arrive at a correct solution. A few are nonquantitative. Some problems will require a little more than six minutes to answer and others a little less. On average, you should expect to complete 80 problems in 480 minutes (eight hours), or spend six minutes per problem.

Six-Minute Solutions for Civil PE Exam Problems: Water Resource does not include problems related directly to environmental and geotechnical engineering, although problems from these subdisciplines will be included in the civil water resources exam. *Six-Minute Solutions for Civil PE Exam Environmental Problems* and *Six-Minute Solutions for Civil PE Exam Geotechnical Problems* provide problems for review in these areas of civil engineering.

HOW TO USE THIS BOOK

In *Six-Minute Solutions for Civil PE Exam Problems: Water Resources*, each problem statement, with its supporting information and answer choices, is presented in the same format as the problems encountered on the

PE exam. The solutions are presented in a step-by-step sequence to help you follow the logical development of the correct solution and to provide examples of how you may want to approach your solutions as you take the PE exam.

Each problem includes a hint to provide direction in solving the problem. In addition to the correct solution, you will find an explanation of the faulty solutions leading to the three incorrect answer choices. The incorrect solutions are intended to represent common mistakes made when solving each type of problem. These may be simple mathematical errors, such as failing to square a term in an equation, or more serious errors, such as using the wrong equation.

To optimize your study time and obtain the maximum benefit from the practice problems, consider the following suggestions.

1. Complete an overall review of the problems and identify the subjects that you are least familiar with. Work a few of these problems to assess your general understanding of the subjects and to identify your strengths and weaknesses.

2. Locate and organize relevant resource materials. As you work problems, some of these resources will emerge as more useful to you than others. These are what you will want to have on hand when taking the PE exam.

3. Work the problems in one subject area at a time, starting with the subject areas that you have the most difficulty with.

4. When possible, work problems without utilizing the hint. Always attempt your own solution before looking at the solutions provided in the book. Use the solutions to check your work or to provide guidance in finding solutions to the more difficult problems. Use the incorrect solutions to help identify pitfalls and to develop strategies to avoid them.

5. Use each subject area's solutions as a guide to understanding general problem-solving approaches. Although problems identical to those presented in *Six-Minute Solutions for Civil PE Exam Problems: Water Resources* will not be encountered on the PE exam, the approach to solving problems will be the same.

Solutions presented for each problem may represent only one of several methods for obtaining a correct answer. Although we have tried to prepare problems with unique solutions, alternative problem-solving methods may produce a different, but nonetheless appropriate, answer.

References

The minimum recommended library for the civil exam consists of PPI's *Civil Engineering Reference Manual*. You may also find the following references helpful in completing some of the problems in *Six-Minute Solutions for Civil PE Exam Problems: Water Resources.*

Aisenbrey, A.J., Jr. et al. *Design of Small Canal Structures*. Denver, Colo.: U.S. Department of Interior, Bureau of Reclamation, 1978.

Fetter, C.W. *Applied Hydrogeology*. 3rd ed. New York, N.Y.: Macmillan, 1994.

Linsley, R.K., et al. *Hydrology for Engineers*. 3rd ed. New York, N.Y.: McGraw Hill, 1982.

Linsley, R.K., et al. *Water-Resources Engineering*. 4th ed. New York, N.Y.: McGraw Hill, 1991.

Luthin, J.N. *Drainage Engineering*. Huntington, N.Y.: RE Krieger, 1978.

McGhee, T.J. *Water Supply and Sewerage Engineering*. 6th Ed. New York, N.Y.: McGraw Hill, 1991.

Munson, B.R., et al. *Fundamentals of Fluid Mechanics*. 3rd ed. New York, N.Y.: Wiley, 1999.

Peavy, H.S., et al. *Environmental Engineering*. New York, N.Y. McGraw Hill, 1985.

Viessman, W., Jr., et al. *Introduction to Hydrology*. 4th ed. Menlo Park, Calif.: Addison Wesley, 1997.

Viessman, W., Jr. and M.J. Hammer. *Water Supply and Pollution Control*. 6th ed. Menlo Park, Calif.: Addison Wesley, 1998.

Breadth Problems

HYDRAULICS

PROBLEM 1

A radial gate is used to control flow into a wasteway turnout and prevent erosion of a section of downstream channel. The gate is operated partially open. The normal flow in the wasteway channel is 80 ft^3/sec with 4.3 ft of available head at the gate. The gate discharge coefficient is 0.72. What is the required area of the gate opening?

    (A) 0.40 ft^2
    (B) 0.83 ft^2
    (C) 6.7 ft^2
    (D) 54 ft^2

Hint: A partially open gate has the characteristics of a submerged orifice.

PROBLEM 2

Water flows at 4.2 ft/sec in a 2 in diameter pipe that is connected through a reducer to a 1.5 in diameter pipe. What is the flow velocity in the 1.5 in diameter pipe?

    (A) 2.4 ft/sec
    (B) 5.6 ft/sec
    (C) 7.5 ft/sec
    (D) 28 ft/sec

Hint: This is a continuity equation problem.

PROBLEM 3

For the following illustration, what is most nearly the pressure in pipe B at point P? Pipe A diameter is 15 cm, pipe B diameter is 10 cm, and pipe C diameter is 8 cm.

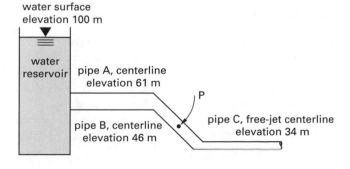

PROBLEM 4

Water flows in an open rectangular channel that is 5 ft wide with a normal water depth of 8 in. The channel is concrete lined along its entire length and has a constant slope of 0.2%. What is the flow rate of the water in the channel?

    (A) 8.7 ft^3/sec
    (B) 11 ft^3/sec
    (C) 130 ft^3/sec
    (D) 820 ft^3/sec

Hint: With the normal depth given, is uniform flow implied? What equation fits these conditions?

PROBLEM 5

What is the head loss per unit of length in a smooth earthen trapezoidal channel with a base width of 2 m, a water depth of 0.5 m, a water flow rate of 3.2 m/s, and 1-to-1 side slopes?

    (A) 0.00065 m/m
    (B) 0.0040 m/m
    (C) 0.013 m/m
    (D) 0.024 m/m

Hint: Can any simplifying assumptions be made regarding head loss? What equations apply to head loss in open channel flow?

PROBLEM 6

What is the flow rate through a 24 in Cipoletti weir when the water height above the notch is 8.6 in?

    (A) 3.8 ft^3/sec
    (B) 4.1 ft^3/sec
    (C) 5.4 ft^3/sec
    (D) 14 ft^3/sec

Hint: Find the discharge equation for a Cipoletti weir.

    (A) 88 kN/m^2
    (B) 120 kN/m^2
    (C) 260 kN/m^2
    (D) 1700 kN/m^2

Hint: This is an energy equation problem.

PROBLEM 7

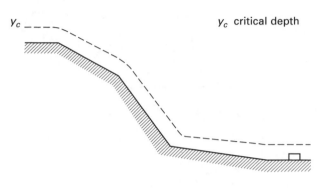

y_c critical depth

channel section

Which figure most likely represents the flow profile over the channel section shown?

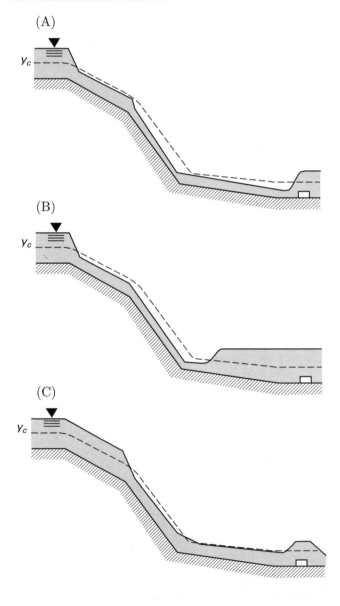

(D)

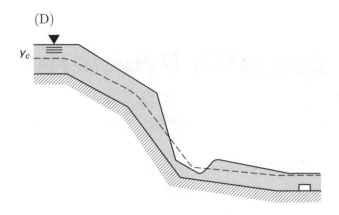

Hint: How is critical flow influenced by slope?

PROBLEM 8

Water flows by gravity between two tanks through 250 ft of 2 in diameter steel pipeline with screwed fittings. The pipeline includes 7 regular 90° elbows, 16 couples, 4 unions, 5 tees (flow-through line), and 2 gate valves. What is the total equivalent length of the pipeline?

    (A) 110 ft
    (B) 260 ft
    (C) 350 ft
    (D) 360 ft

Hint: How is equivalent length defined?

PROBLEM 9

A gate that is hinged at the top is used to prevent back-flow of tidal water into a 0.5 m circular storm water drain line. At high tide, the hinge at the top of the gate is 2.3 m below the water surface. What force is required to open the gate at high tide?

    (A) 3.0 kN
    (B) 3.8 kN
    (C) 12 kN
    (D) 19 kN

Hint: Find the resultant force on the gate.

PROBLEM 10

A concrete-lined open channel is used to convey storm water runoff along a roadway. The roadway and channel make an abrupt transition from 12% slope to 2% slope, which causes a hydraulic jump to occur. The channel is a triangular cross section and 1-to-1 side slopes. The upstream water depth is 10 cm. What is the water depth in the downstream channel section with the 2% slope?

(A) 0.20 m
(B) 0.25 m
(C) 0.46 m
(D) 1.2 m

Hint: Be careful of the triangular channel. Use the momentum equation.

PROBLEM 11

A 1 m high, 27 m long box culvert placed at 2% slope sees an upstream water surface elevation of 1202.83 m and a downstream water surface elevation of 1202.38 m. The invert elevation of the culvert outlet is 1201.17 m. What is the culvert flow classification?

(A) type 3
(B) type 4
(C) type 5
(D) type 6

Hint: Sketch the culvert.

HYDROLOGY

PROBLEM 12

The cross section of a river channel is shown in the illustration. Velocity measurements for each section of river channel at the indicated depths are summarized in the table. What is most nearly the total flow in the river?

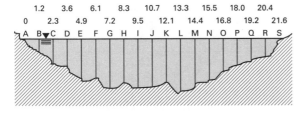

distance from left bank (m) ⟶

section	velocity at 0.2 depth (m/s)	velocity at 0.8 depth (m/s)	average depth (m)
AB	–	–	0.7
BC	0.41	0.32	1.9
CD	0.44	0.32	2.3
DE	0.48	0.34	2.7
EF	0.48	0.33	2.9
FG	0.49	0.36	3.0
GH	0.49	0.35	3.1
HI	0.51	0.37	2.9
IJ	0.50	0.36	2.9
JK	0.52	0.37	2.8
KL	0.50	0.38	3.1
LM	0.49	0.35	2.8
MN	0.50	0.36	2.7
NO	0.47	0.34	2.5
OP	0.43	0.31	2.0
PQ	0.41	0.32	1.8
QR	0.39	0.30	1.6
RS	–	–	0.5

(A) 1.3 m³/s
(B) 20 m³/s
(C) 40 m³/s
(D) 330 m³/s

Hint: Use section areas and average velocities.

PROBLEM 13

The 25 yr return period rainfall frequency-depth-duration curves for a coastal region is shown in the illustration. For a mean annual precipitation (P_{ma}) of 27 in, what is the rainfall intensity for a 2.5 hr storm?

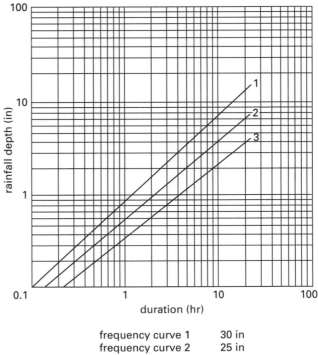

frequency curve 1	30 in
frequency curve 2	25 in
frequency curve 3	20 in

(A) 0.40 in/hr
(B) 0.60 in/hr
(C) 1.5 in/hr
(D) 11 in/hr

Hint: How do the units of the answer choices relate to the illustration?

PROBLEM 14

Frequency-intensity-duration curves for a watershed are presented in the illustration. What is the frequency of the 60 min duration and 5.8 in/hr intensity storm?

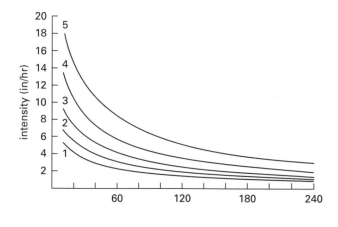

duration (min)

frequency curve 1	50 %
frequency curve 2	10 %
frequency curve 3	4 %
frequency curve 4	2 %
frequency curve 5	1 %

(A) 2 yr
(B) 10 yr
(C) 25 yr
(D) 50 yr

Hint: What information can be obtained from the illustration?

PROBLEM 15

A unit hydrograph for a drainage area is shown in the illustration. Twelve hours after the beginning of runoff, the discharge measured at a gaging station at the outlet of the drainage area is 150 m³/s. What is the peak discharge from the drainage area?

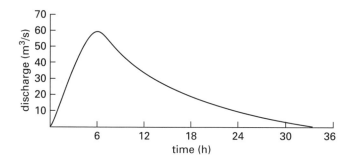

(A) 59 m³/s
(B) 180 m³/s
(C) 210 m³/s
(D) 260 m³/s

Hint: How does the unit hydrograph relate to the actual runoff from the drainage area?

PROBLEM 16

A manufacturing facility is willing to accept only 1% risk of flooding during its 50 yr design life. What is the annual probability that flooding will occur during the facility design life?

(A) 0.00020%
(B) 0.020%
(C) 0.50%
(D) 1.0%

Hint: How are design life and risk related?

PROBLEM 17

A 131 ac drainage area has the following characteristics and 10-year storm frequency-intensity-duration curve.

What is the peak runoff from the drainage area for the 30 min duration, 10-year storm?

land use	area %
apartments	30
landscaped open space (park)	25
light industrial	45

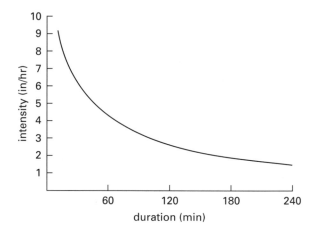

(A) 24 ac-ft/hr
(B) 26 ac-ft/hr
(C) 72 ac-ft/hr
(D) 310 ac-ft/hr

Hint: Account for the variation in runoff coefficients for different land uses.

WATER TREATMENT

PROBLEM 18

A water treatment plant serves a population of 65,000 people. Treatment costs are $0.12/(10)³ gal with 20% attributable to electrical power. Assuming continuous operation, what is the approximate annual electric bill for the plant?

(A) $58,000/yr
(B) $67,000/yr
(C) $94,000/yr
(D) $2,300,000/yr

Hint: What is the typical average annual daily per capita water demand?

PROBLEM 19

A filter gallery of five multimedia sand filters treats 28,500 m^3/d of water. Typically, each filter is backwashed twice during every 24 h period with backwashing occurring at 36 $m^3/m^2 \cdot h$ for 25 min, followed by a conditioning period of 8 min. The filter loading rate is 225 $m^3/m^2 \cdot d$. What is the approximate net daily production per filter?

(A) 4900 m^3/d
(B) 5200 m^3/d
(C) 5300 m^3/d
(D) 5500 m^3/d

Hint: What flow rates do the remaining filters see when one filter is backwashing?

PROBLEM 20

A mass diagram for a municipal water storage tank is shown. What is the minimum required capacity of the storage tank?

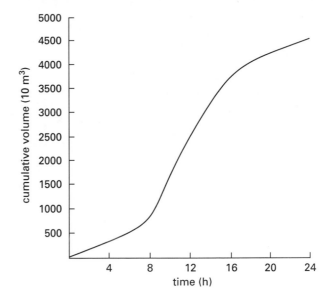

(A) 1300 m^3
(B) 6500 m^3
(C) 13 000 m^3
(D) 45 000 m^3

Hint: Determine the average demand.

Depth Problems

HYDRAULICS

PROBLEM 21

A hydraulic jump with a stilling pool is selected to dissipate energy over a spillway prior to the water entering a natural river channel. The spillway is 2.5 m wide and the hydraulic jump occurs when the water depth at the toe of the spillway is 0.15 m for a flow of 2.93 m³/s. What is most nearly the total head dissipated?

(A) 1.4 m
(B) 1.8 m
(C) 1.9 m
(D) 4.2 m

Hint: The solution involves the calculation of conjugate depths.

PROBLEM 22

A baffled outlet is used for energy dissipation at a drainage discharge into a basin from a square channel flowing full. The flow from the drainage channel is 120 ft³/sec with a head of 32 ft. The basin is shown in the illustration. What is the minimum required depth (D_b) of the baffled outlet basin if its width is fixed at 6 ft?

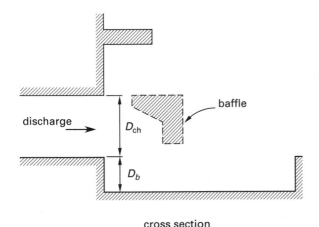

cross section

(A) 0.44 ft
(B) 0.68 ft
(C) 0.85 ft
(D) 0.91 ft

Hint: Use the Froude number.

PROBLEM 23

A low-head siphon spillway with the following characteristics and as shown in the illustration is used to divert irrigation flow from a canal. What is the approximate diameter required for the siphon throat?

discharge flow	90 ft³/sec
operating head	4 ft
atmospheric pressure head	34 ft
radius to throat centerline	2 ft
siphon entrance coefficient	0.6

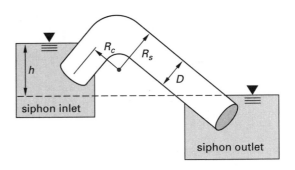

(A) 0.86 ft
(B) 1.2 ft
(C) 2.0 ft
(D) 3.5 ft

Hint: This is a special case of a discharge through an orifice.

PROBLEM 24

A square baffled outlet is used to dissipate energy at the bottom of a rectangular chute with a 3 ft base width. The head and flow are 42 ft and 200 ft³/sec, respectively. What is the minimum required width of the baffled outlet basin? The Froude number and the basin width to channel outlet depth ratio are related by $w_b/D_c = 0.875 \, \text{Fr} + 2.7$.

(A) 1.3 ft
(B) 2.6 ft
(C) 12 ft
(D) 26 ft

Hint: Do not confuse the outlet width with the basin width.

PROBLEM 25

Water flowing in a schedule-40 steel pipe divides at a tee as shown in the illustration. The characteristics of each pipe are summarized in the table. The flow rate in pipe A is 0.76 ft³/sec. What is the flow rate in pipe C?

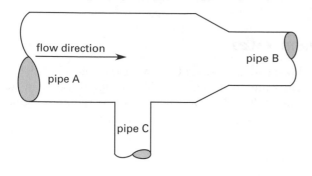

pipe	nominal diameter (in)	pressure (psig)
A	2.5	100
B	1.5	90
C	1	–

(A) 0.13 ft³/sec
(B) 0.22 ft³/sec
(C) 0.35 ft³/sec
(D) 0.76 ft³/sec

Hint: Use both the continuity and the energy equations.

PROBLEM 26

Water flows at a depth of 1.5 m and a velocity of 2.5 m/s in an 8 m wide rectangular open channel. The channel transitions to a circular culvert to cross 50 m under a highway. Both the channel and the culvert are constructed of concrete at a constant slope of 0.002 m/m. What is the required culvert diameter if the culvert is to be constructed of standard concrete pipe and is to flow half full?

(A) two pipes, 9 ft diameter
(B) three pipes, 9 ft diameter
(C) four pipes, 9 ft diameter
(D) eight pipes, 9 ft diameter

Hint: Does the continuity equation apply?

PROBLEM 27

Water is pumped from a tank through a pipe to a point 340 m above. The pump is located at the same elevation as the tank water surface. The pipe inside diameter is 2.54 cm for most of its length, but decreases to discharge from a 0.5 cm diameter nozzle at 7000 kPa. The flow

in the pipe is 0.002 m³/s, and the water temperature is 20°C. What approximate pressure is required at the discharge side of the pump?

(A) 1600 kPa
(B) 7500 kPa
(C) 10 000 kPa
(D) 15 000 kPa

Hint: Does the energy equation apply?

PROBLEM 28

Water is pumped from an elevation of 3457 ft to an elevation of 3503 ft through 500 ft of 2 in schedule-80 steel pipe at a velocity of 5 ft/sec. The pipe includes 15 couples, 8 90° regular elbows, 4 45° regular elbows, 6 tees (straight flow), and 2 globe valves. All fittings are standard pipe thread. The water temperature is 60°F. What is the total head loss in the pipe from all sources?

(A) 70 ft
(B) 86 ft
(C) 120 ft
(D) 270 ft

Hint: Head losses in this problem occur from three sources.

PROBLEM 29

An existing nominal 6 in steel water line is unable to meet the projected demand of 2.4 ft³/sec for a growing residential development. A nominal 4 in steel line will be placed parallel to the existing line. Both pipes begin and end at the same point and are 1400 ft long. What will be the approximate flow in the 4 in pipe?

(A) 0.51 ft³/sec
(B) 0.61 ft³/sec
(C) 0.64 ft³/sec
(D) 0.74 ft³/sec

Hint: Assume an initial value for the Reynolds number.

PROBLEM 30

A run of 3 in schedule-40 pipe with long radius 90° elbows is fixed in an overhead rack. The elbows are in the horizontal plane with no change in inlet and outlet elevations. The flow in the pipe is 0.55 ft³/sec at 50 psig. What is the resultant force exerted by the water on the pipe rack at each elbow?

(A) 516 lbf
(B) 523 lbf
(C) 649 lbf
(D) 727 lbf

Hint: The pressure and area are constant.

PROBLEM 31

A trapezoidal channel has concrete walls placed on a 1-to-1 slope and a gravel bottom. The channel slope is 0.0005 and the normal water depth is 1 m. Because of the gravel bottom, the maximum flow velocity in the channel is limited to 0.75 m/s. What is the width of the channel base?

(A) 0.62 m
(B) 0.87 m
(C) 1.8 m
(D) 3.6 m

Hint: Define the hydraulic radius in terms of the channel base.

PROBLEM 32

The height of a submerged barrier placed across the entire width of a 15 ft wide rectangular channel can be increased or decreased according to varying flow conditions. At a flow velocity of 3 ft/sec and a normal depth of 8 ft, how high can the barrier be raised before creating a change in the water depth upstream of the barrier?

(A) 1.5 ft
(B) 2.6 ft
(C) 3.0 ft
(D) 4.2 ft

Hint: This is a critical depth problem.

PROBLEM 33

The water height above the crest of a dam spillway is 2 m and the spillway crest is 21 m above the bottom of the reservoir. Most nearly what force is exerted on the dam by the water?

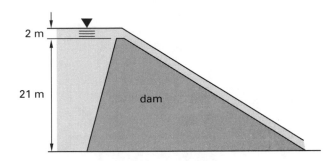

(A) 17 kN/m
(B) 340 kN/m
(C) 2300 kN/m
(D) 2500 kN/m

Hint: Be careful of energy and depth relationships.

PROBLEM 34

Two reservoirs differ in elevation by 20 m and are connected by a 9 km long earth-lined channel with a constant slope. The channel bottom is 2 m wide with a normal water depth of 1 m and the channel has 3-to-1 horizontal-to-vertical sides. What is the flow rate of water in the channel?

(A) 4.2 m³/s
(B) 9.3 m³/s
(C) 37 m³/s
(D) 49 m³/s

Hint: Use the Manning equation.

PROBLEM 35

Monitoring results from a sewer inflow and infiltration (I/I) evaluation are presented in the table.

section	total infiltration to section (m³/d)	pipe diameter (mm)	pipe length (km)
1	2315	100	13.5
		200	6.8
		300	6.2
2	958	100	9.2
		200	7.1
		300	4.4
3	3996	100	24.9
		200	12.1
		300	11.9
4	1867	100	21.3
		200	11.0
		300	4.7

Which section of city sewer should receive first priority for rehabilitation?

(A) section 1
(B) section 2
(C) section 3
(D) section 4

Hint: Reduce the data for each section to a unit value that allows for direct comparison.

PROBLEM 36

A storm-water detention pond uses a submerged orifice to control discharge to an open channel. A standpipe is used to prevent the pool elevation from exceeding 100 ft. The centerline of the orifice is at an elevation of 92.6 ft and the orifice opening is sharp edged. What diameter of the orifice is required to limit the discharge through it to 20 ft³/sec?

(A) 0.77 ft
(B) 0.85 ft
(C) 1.1 ft
(D) 1.4 ft

Hint: This is similar to discharge from an open tank.

PROBLEM 37

Pressure is maintained at 100 psig in a fire hose with a nozzle diameter of 0.5 in. The nozzle coefficient is 0.98 and the water temperature is 70°F. What is the flow rate at the nozzle outlet?

(A) 13 gal/min
(B) 75 gal/min
(C) 240 gal/min
(D) 600 gal/min

Hint: This problem resembles a problem involving the discharge from an orifice at the bottom of a tank.

PROBLEM 38

A real estate developer has proposed a project that will include 346 single-family houses each located on a ⅛ ac cleared lot. The houses will be about equally divided between one- and two-story wood-frame structures. The one-story houses will average 1400 ft² and the two-story houses will average 2300 ft². The average occupancy for each house is expected to be four people. Most nearly what design flow should the water line servicing the development be sized to carry?

(A) 990 gal/min
(B) 1300 gal/min
(C) 1400 gal/min
(D) 1500 gal/min

Hint: Consider fire demand by alternative criteria.

PROBLEM 39

A commercial sod farm maintains 500 ha in current production year-round. The sod requires about 260 cm of water annually, and is applied to subdivided plots by a fixed sprinkler irrigation system at an application rate of 2 cm/h three times weekly. What is the maximum subdivided plot size if irrigation at the sod farm is to occur for 12 h/d and 7 d/wk?

(A) 5 ha
(B) 15 ha
(C) 21 ha
(D) 45 ha

Hint: Use time ratios.

PROBLEM 40

The water surface in a reservoir is maintained at the spillway elevation by sluiceways except during periods of heavy rainfall. During dry weather conditions the average flow into the reservoir is 1100 ft³/sec, but during the 10-year storm, inflow increases to 1550 ft³/sec after 8 hr and the water level rises 6 ft above the spillway crest. The reservoir storage capacity when the pool elevation is at the spillway crest is 160,000 ac-ft and the spillway is 30 ft wide. By approximately how much is the reservoir storage increased during the 10-year storm?

(A) 200 ac-ft
(B) 430 ac-ft
(C) 560 ac-ft
(D) 160,000 ac-ft

Hint: Use critical depth and energy.

PROBLEM 41

For the pipe network shown in the illustration, what is the flow velocity in pipe BC? The Darcy friction factor for all pipes in the network is 0.023 and all nodes are at equal elevation. Pipe lengths and nominal diameters are shown on the illustration.

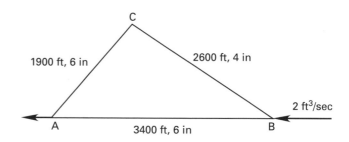

(A) 0.56 ft/sec
(B) 2.9 ft/sec
(C) 6.5 ft/sec
(D) 17 ft/sec

Hint: The solution should converge quickly. If it does not, an error has been made.

PROBLEM 42

The illustration shows three water storage tanks that are connected at a common node. The flow between the node and tank D is 4 ft³/sec. The tank elevations, pipe lengths, and nominal diameters are shown on the illustration. The pipe is welded steel. What is most nearly the water surface elevation in tank B?

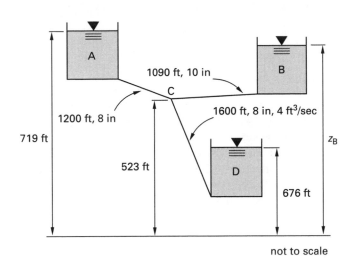

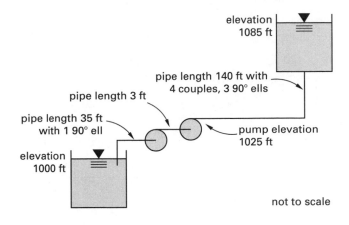

(A) 420 ft
(B) 500 ft
(C) 540 ft
(D) 550 ft

Hint: Use the energy and continuity equations and head-loss tables.

PROBLEM 43

The head-capacity curves for two pumps and the pump operating conditions are shown in the following illustrations. The pipe is nominal 2 in schedule-40 steel. What is the total discharge at the combined operating point of the two pumps?

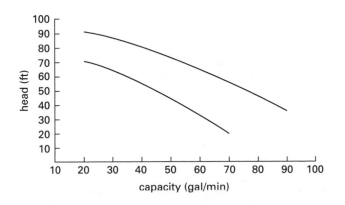

(A) 56 gal/min
(B) 60 gal/min
(C) 65 gal/min
(D) 120 gal/min

Hint: How does series or parallel operation influence pumping capacity?

PROBLEM 44

A centrifugal pump operates at a speed of 1750 rpm and is rated at 850 gal/min for 78% efficiency and 180 ft of head. For constant efficiency and head, what is most nearly the flow rate if the pump is operated at 2200 rpm?

(A) 530 gal/min
(B) 680 gal/min
(C) 830 gal/min
(D) 1100 gal/min

Hint: Apply the principles of affinity.

PROBLEM 45

A submersible pump operated at 1750 rpm is needed to deliver 800 gal/min of water from a 350 ft deep well. At 93% efficiency, the average specific speed of the pump is 2300. How many stages are required for the pump?

(A) 1 stage
(B) 3 stages
(C) 6 stages
(D) 35 stages

Hint: Specific speed is derived from dynamic similarity.

PROBLEM 46

In some parts of its length, the storm sewer shown in the illustration flows full during extreme rainfall events. At what manhole, if any, does the storm water overflow through the manhole onto the street surface?

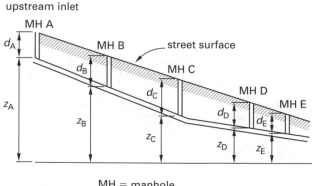

upstream inlet

MH = manhole
d = depth below grade
z = elevation

not to scale

location	pipe centerline depth (ft)	pipe centerline elevation (ft)	water pressure (lbf/in²)
MH A	8.2	100	0
MH B	10.5	86	4.2
MH C	12.4	71	6.8
MH D	6.8	54	3.1
MH E	6.2	43	1.2

(A) none
(B) manholes A, B, and E
(C) manholes C and D
(D) manholes B, C, and D

Hint: Where is the hydraulic grade line?

PROBLEM 47

A concrete-lined trapezoidal channel with an 8 m bottom width and 1-to-1 side slope will be used for storm water drainage. The maximum anticipated flow in the channel is 22 m³/s. What uniform slope of the channel bottom is required to just produce critical flow at the maximum flow rate?

(A) 0.000060 m/m
(B) 0.00054 m/m
(C) 0.0019 m/m
(D) 0.25 m/m

Hint: Use equations for Froude number and channel cross-sectional area.

PROBLEM 48

The equivalent length of a 12 in diameter welded steel pipe (assume standard pipe) is 1 mi. The water flow in the pipe varies between the two extremes of 2.20 ft³/sec and 4.40 ft³/sec. If head loss is given by $h_f = kQ^n$, what is the value of k?

(A) 0.18
(B) 2.5
(C) 19
(D) 25

Hint: Find two equations to solve for the two unknowns.

PROBLEM 49

The net positive suction head (NPSH) characteristics of a pump are shown in the following illustration. The pump was selected to deliver 75 gal/min of water at 48°F. The friction losses between the water surface and the pump inlet are 6.7 ft. If the water surface elevation is 4573 ft above mean sea level (MSL), what is the maximum permissible elevation of the pump?

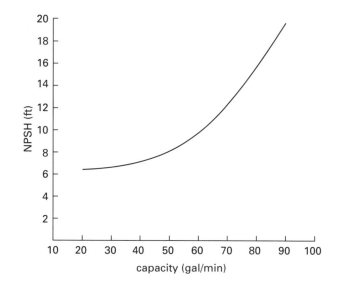

(A) 4566 ft
(B) 4580 ft
(C) 4586 ft
(D) 4608 ft

Hint: Find an equation that relates net positive suction head to elevation.

PROBLEM 50

Water at 10°C is flowing 43 mm deep in a 100 mm diameter PVC pipe. The pipe is placed on a uniform 3% slope. What is most nearly the numeric value of the friction factor?

(A) 0.014
(B) 0.017
(C) 0.070
(D) 0.37

Hint: Find or develop an equation that relates the friction factor to known parameters.

PROBLEM 51

A pneumatically operated valve at the end of 30 m of 200 mm schedule-40 steel pipe closes abruptly. The pipe is anchored against any axial movement. With the valve fully open, the water velocity in the pipe is 2.4 m/s. What is most nearly the maximum head produced by the valve closure?

(A) 9.8 m
(B) 310 m
(C) 330 m
(D) 590 m

Hint: This is a water hammer problem.

PROBLEM 52

Alternative $^1/_{10}$ scale model diving sleds are being tested in a deep, freshwater column at 15°C. The ideal prototype sled will fall at 0.5 m/s in 8°C seawater. What is the required fall rate for the model sled if it is to duplicate the desired prototype?

(A) 0.041 m/s
(B) 0.34 m/s
(C) 4.1 m/s
(D) 6.0 m/s

Hint: This is a similarity problem.

PROBLEM 53

A cyclist riding on a flat, newly paved road has a frontal area of 0.47 m^2. The wind speed with the cyclist at rest is zero and the air temperature is 39°C. The mass of the cyclist and bicycle is 79 kg. The drag coefficient is 0.24 and rolling resistance is 0.05% of weight. The Reynolds number is less than 10^5. How many calories does the cyclist expend to overcome drag and rolling resistance if he travels 100 km at a constant velocity of 11.6 m/s?

(A) 0.21 kcal
(B) 34 kcal
(C) 200 kcal
(D) 210 kcal

Hint: The basic parameters in this problem are force, distance, and work.

PROBLEM 54

What is most nearly the maximum allowable suction head for a pump with a cavitation constant of 0.26 that is to operate under the following conditions?

elevation	1370 m above mean sea level (MSL)
water temperature	13°C
total dynamic head	38 m

(A) 1.2 m
(B) 1.5 m
(C) 8.5 m
(D) 15 m

Hint: Define the cavitation constant.

PROBLEM 55

A new 5 km long pipeline connects two reservoirs. The water surface elevation in the upper reservoir is 1100 m and in the lower reservoir is 835 m. The pipeline is 400 mm nominal diameter welded steel with a square mouth inlet and includes the following flanged fittings.

10 gate valves
19 standard radius 90° ells
37 standard radius 45° ells
8 straight tees

What is the maximum water flow rate between the reservoirs?

(A) 0.49 m^3/s
(B) 0.51 m^3/s
(C) 0.64 m^3/s
(D) 0.83 m^3/s

Hint: Use the energy equation and the Hazen-Williams equation.

PROBLEM 56

Water flowing 3 ft deep in a trapezoidal channel enters a box culvert through an abrupt transition. The culvert extends for 1000 ft and then makes an abrupt transition back to a trapezoidal channel. The channel and culvert are constructed of reinforced concrete. The channel base width is 8 ft with 1-to-1 side slopes, and the culvert width is 8 ft. The channel and culvert slope are constant at 2%. What is most nearly the head loss through the culvert?

(A) 12 ft
(B) 20 ft
(C) 26 ft
(D) 67 ft

Hint: Use the definition for hydraulic radius.

PROBLEM 57

Manometers are installed on the upstream and downstream sides of a straight pipe 20 ft long with an inside diameter 0.75 in. The flow rate through the pipe is 8 gal/min. The upstream manometer reading is 37.1 in and the downstream manometer reading is 9.2 in. What is the specific roughness of the pipe?

(A) 0.000014 ft
(B) 0.00022 ft
(C) 0.014 ft
(D) 0.019 ft

Hint: Use the Darcy equation.

PROBLEM 58

A pipe friction apparatus has two sets of water manometers. The first set of manometers is separated by 10 cm of straight pipe and includes the pipe only, with no fittings or valves. The second set of manometers includes a fully open gate valve with 20 cm of straight pipe at either end. The manometer readings for a water velocity of 3.2 m/s are shown in the following table. What is the loss coefficient for a fully open gate valve placed between the second set of manometers?

manometer	reading (cm)
upstream, no valve	14.2
downstream, no valve	12.9
upstream, with valve	18
downstream, with valve	3.7

(A) 0.17
(B) 0.22
(C) 0.25
(D) 0.27

Hint: Find an equation for minor losses that includes the loss coefficient. What is the relationship between the two sets of manometers?

PROBLEM 59

A hydraulic jump occurs in a trapezoidal channel with a 4.2 m bottom width and 1-to-1 side slopes. The flow in the channel is 38 m^3/s and the water depth upstream of the jump is 0.74 m. What is the water depth downstream of the jump?

(A) 0.69 m
(B) 0.74 m
(C) 2.3 m
(D) 3.3 m

Hint: Use either conjugate depths or the specific force equation.

PROBLEM 60

A sluice gate is placed across a rectangular channel as an energy dissipation device as shown in the illustration. The water velocity in the channel just downstream of

the sluice gate is 8 m/s and the water depth is 0.65 m. Approximately how much energy is dissipated by the hydraulic jump that forms downstream of the sluice gate?

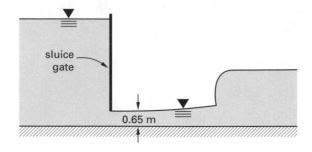

(A) 38 kW/m
(B) 57 kW/m
(C) 210 kW/m
(D) 230 kW/m

Hint: The power dissipated is related to the change in specific energy.

PROBLEM 61

An ordinary firm loam-lined ditch is used to convey clear irrigation water from a reservoir to fields below. The first 24 m section of the ditch follows a slope of 5%, with discharge occurring over a weir that can be raised or lowered to control flow. The ditch has a 90° V-shaped cross section and 20 cm diameter logs are secured in the ditch channel perpendicular to the flow direction at 4.0 m spacing along the steep section. The scour velocity for clear water flowing over ordinary firm loam is 0.76 m/s. What is the approximate maximum flow in the ditch if the scour velocity should not be exceeded?

(A) 0.0014 m^3/s
(B) 0.033 m^3/s
(C) 0.064 m^3/s
(D) 0.28 m^3/s

Hint: When does the maximum water velocity occur?

PROBLEM 62

What is the flow rate through the venturi meter shown in the following illustration? The upstream pipe diameter is 16 in, and the venturi throat diameter is 10 in. The fluid in the manometer is mercury, and the water and mercury temperatures are equal at 70°F.

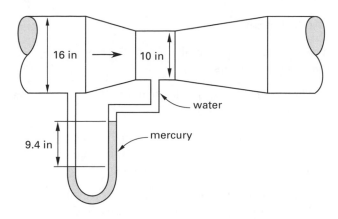

(A) 5.9 ft³/sec
(B) 15 ft³/sec
(C) 23 ft³/sec
(D) 38 ft³/sec

Hint: Find an equation that relates the manometer fluid reading to the venturi meter velocity.

PROBLEM 63

A dye tracer is used to measure the flow velocity in a 16 in sewer line. The dye is introduced into the sewer at a manhole located 1076 ft upstream of the observation manhole. The depth of flow in the upstream sewer is 3.8 in and in the observation manhole is 5.2 in. The dye is first observed in the observation manhole 398 sec after introduction into the sewer and the last trace of the dye is observed after 512 sec. What is the approximate average flow rate between the two manholes?

(A) 0.021 ft³/sec
(B) 0.76 ft³/sec
(C) 0.82 ft³/sec
(D) 0.93 ft³/sec

Hint: Use average depths and times.

PROBLEM 64

Determine the maximum discharge for the culvert under the conditions shown in the following illustration. The culvert entrance and discharge structures are square flush and constructed of reinforced concrete. The culvert barrel is round reinforced concrete pipe.

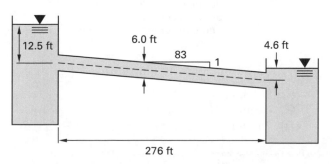

(A) 350 ft³/sec
(B) 410 ft³/sec
(C) 470 ft³/sec
(D) 560 ft³/sec

Hint: Determine the culvert flow classification.

HYDROLOGY

PROBLEM 65

An isohyetal map for a defined region is shown in the illustration and the gross area enclosed by each isohyet within the region boundary is summarized in the table. What is the areal average precipitation for the region?

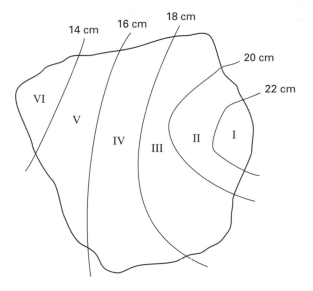

area	isohyet (cm)	enclosed area (km²)
I	>22	84
II	20	252
III	18	578
IV	16	892
V	14	1136
VI	<14	1294

(A) 15.0 cm
(B) 16.8 cm
(C) 17.5 cm
(D) 52.0 cm

Hint: What computation method does the form of the data suggest?

PROBLEM 66

A histograph of flood peak flows representing 112 events over a 94 year period is presented in the following illustration. If the largest 18 floods resulted in significant economic impact, what is the minimum peak flow of these floods?

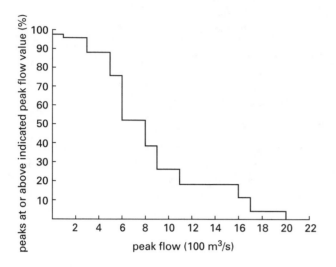

(A) 500 m³/s
(B) 1100 m³/s
(C) 1600 m³/s
(D) 16 000 m³/s

Hint: What format does the figure require of the data?

PROBLEM 67

The illustration presents a histograph of peak flows for a 112 yr period. What is the recurrence interval for the event corresponding to a flow of 40,000 ft³/sec?

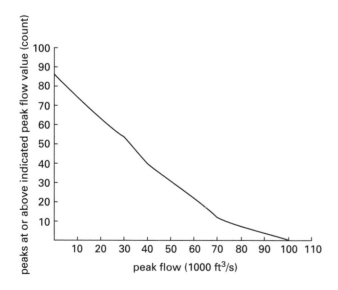

(A) 0.35 yr
(B) 2.2 yr
(C) 2.8 yr
(D) 3.4 yr

Hint: How can the figure be used to determine recurrence interval?

PROBLEM 68

Annual flood data compiled for a 97 yr period provide the following results.

The log of the arithmetic mean of all floods (for flow units of ft³/sec) is 3.571.

The sum of the squared difference of the log of the flood magnitude for each probability and the log of the arithmetic mean of all floods is 3.894.

The sum of the cubed difference of the log of the flood magnitude for each probability and the log of the arithmetic mean of all floods is 0.181.

What is most nearly the flow magnitude of the 25-year flood?

(A) 4100 ft³/sec
(B) 8300 ft³/sec
(C) 8700 ft³/sec
(D) 10,000 ft³/sec

Hint: What method of calculation is suggested by the given information?

PROBLEM 69

A rainfall frequency-depth-duration curve for a watershed is shown in the following illustration. The values for the Steel formula constants K and b are 180 in-min/hr and 25 min. What is the time of concentration for the 2 h duration 50-year storm?

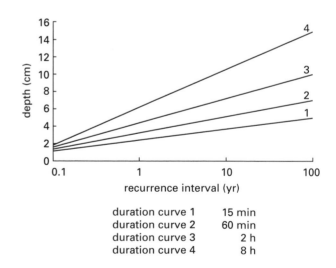

duration curve 1	15 min
duration curve 2	60 min
duration curve 3	2 h
duration curve 4	8 h

(A) 15 min
(B) 26 min
(C) 42 min
(D) 75 min

Hint: How does the Steel formula apply?

PROBLEM 70

Rainfall records for four precipitation stations are summarized in the table. Stations B, C, and D are those located in closest proximity to station A. What is the estimated precipitation at station C for 1991?

station	normal annual precipitation (cm)	annual precipitation for year indicated (cm)			
		1985	1986	1987	1989
A	39	32	37	42	41
B	31	27	30	37	34
C	42	34	39	45	46
D	37	34	32	–	39

station	normal annual precipitation (cm)	annual precipitation for year indicated (cm)			
		1990	1991	1992	1993
A	39	44	34	36	40
B	31	37	28	30	30
C	42	43	–	39	40
D	37	37	32	35	36

(A) 31 cm
(B) 37 cm
(C) 40 cm
(D) 48 cm

Hint: The solution is not a simple average.

PROBLEM 71

City parking lots A and B slope to join at a common gutter that discharges to a storm sewer inlet located at one end of the lots, as shown. Both lots are 100 ft wide, with lot A being 150 ft deep and lot B being 90 ft deep. Lot A slopes at 0.001 and lot B slopes at 0.0006. Both lots are newly paved with smooth asphalt. For a rainfall intensity of 0.89 in/hr over a 1 hr period, what is the time of concentration for flow to the gutter?

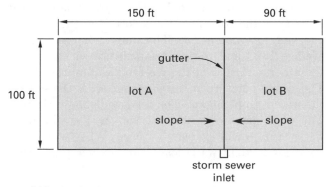

(A) 0.14 min
(B) 18 min
(C) 28 min
(D) 36 min

Hint: The small drainage area and the form of the rainfall data dictate selection of the appropriate equation.

PROBLEM 72

A unit hydrograph developed for a 3 h storm and a 420 km^2 drainage area is shown in the illustration. What is the peak flow of three successive 3 h periods of rainfall that produce 1.6 cm, 3.1 cm, and 2.7 cm of runoff?

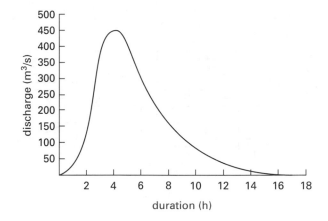

(A) 1400 m^3/s
(B) 2100 m^3/s
(C) 2900 m^3/s
(D) 3300 m^3/s

Hint: Add the hydrographs.

PROBLEM 73

The elemental hydrograph for a small urban drainage area is presented in the illustration. What is the volume of rainfall in surface detention?

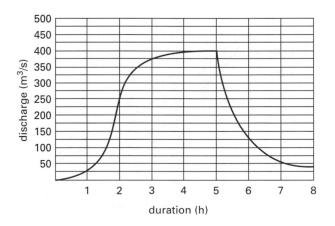

(A) 1.2 × 10^6 m^3
(B) 2.7 × 10^6 m^3
(C) 4.5 × 10^6 m^3
(D) 5.7 × 10^6 m^3

Hint: What are the characteristics of an elemental hydrograph?

PROBLEM 74

What is the approximate direct runoff from the storm characterized by the hydrograph shown in the illustration? The basin drainage area is 5480 mi^2.

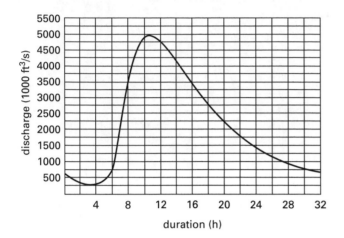

(A) 0.016 in
(B) 2.6 in
(C) 13 in
(D) 16 in

Hint: Separate groundwater from direct runoff and integrate.

PROBLEM 75

The illustration presents a moisture-tension curve for a soil. What is the moisture content of the soil that corresponds to the wilting point?

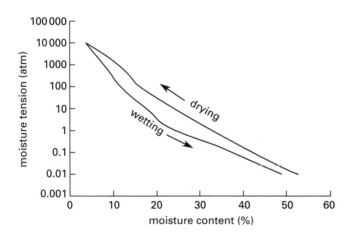

(A) 17%
(B) 20%
(C) 23%
(D) 32%

Hint: How is the wilting point defined?

PROBLEM 76

The area distribution by elevation in a mountainous region is shown. The snow line is at 1370 m and the weather station where temperature is measured is at 2036 m. The ambient lapse rate is $-0.016°C/m$ and the degree-day factor is 3 mm/°C·d. What is most nearly the snowmelt volume for a day when the average temperature at the weather station is 12°C?

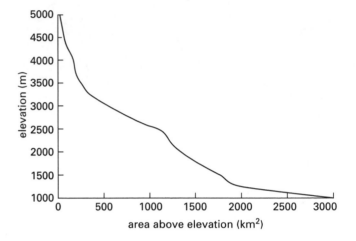

(A) 2.0×10^7 m^3
(B) 3.7×10^7 m^3
(C) 6.3×10^7 m^3
(D) 4.2×10^{10} m^3

Hint: Between which elevations does snow melt occur?

PROBLEM 77

The feasibility of a dam for impounding irrigation water in the southwestern United States is to be determined, in part, by the volume of water lost from the resulting reservoir to evaporation. The water temperature will average 18°C and the air temperature within 2 m of the water surface over the reservoir will average 24°C. The air temperature at the water surface is the same as the water temperature. The relative humidity is 16%. The wind speed over the reservoir at 4 m above the water surface will average 3.6 m/s. What will be the approximate daily water loss from the reservoir to evaporation?

(A) 0.17 mm/d
(B) 4.0 mm/d
(C) 5.2 mm/d
(D) 7.0 mm/d

Hint: What estimation method is suggested by the data provided? Be careful when defining vapor pressure.

PROBLEM 78

The average annual sediment load to a reservoir is presented in the following illustration. The sediment is composed of 64% clay and 36% silt and will always be submerged. The reservoir original volume was 3600 ac-ft. For an average annual stream flow of 20 ft³/sec, what is most nearly the useful life of the reservoir if its minimum useful volume is 75% of its original volume?

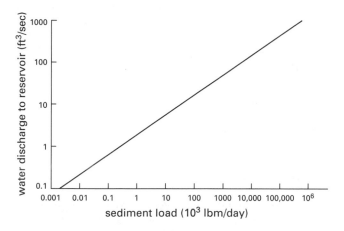

(A) 14 yr
(B) 59 yr
(C) 240 yr
(D) 960 yr

Hint: Be careful of how the specific weights of the sediments are determined.

PROBLEM 79

Land use classification for a watershed in the western United States is summarized in the table. The longest flow path within the watershed to the storm sewer is 337 yd and the average ground slope over the watershed is 0.013. What is the flow to the storm sewer resulting from a 10-year storm event?

land use	area (ac)
shingle roof	0.8
concrete surface	1.1
asphalt surface	2.6
poorly drained lawn	10

(A) 1.7 ft³/sec
(B) 6.3 ft³/sec
(C) 9.2 ft³/sec
(D) 11 ft³/sec

Hint: The solution requires multiple steps to find time of concentration, intensity, and runoff.

PROBLEM 80

A small, developed area covered with manicured sod and well-defined drainage channels has a flow distance of 83 m and an average surface slope of 0.011. The rainfall intensity for the area from the 15 min duration, 10-year storm is 5.3 cm/h. What is most nearly the time of concentration?

(A) 2.7 min
(B) 26 min
(C) 29 min
(D) 38 min

Hint: Use the given information to select the most appropriate formula or equation.

PROBLEM 81

The inflow, outflow, and water surface elevation for a deep reservoir with an uncontrolled discharge is summarized in the table. The spillway crest is at an elevation of 371 m. The reservoir volume in m³ can be estimated by $9.4(\text{elevation, m})^2 + 3854(\text{elevation, m}) + 14\,470$. How long is required for the reservoir to reach its peak storage coinciding with a maximum water surface elevation of 377.8 m?

inflow (m³/s)	outflow (m³/s)	water surface elevation (m)
0.67	0.67	371.1
1.2	0.70	373.2
2.8	2.9	379.1
2.5	3.8	377.8

(A) 3.9 h
(B) 7.6 h
(C) 30 h
(D) 130 h

Hint: Use mass balance or find an equation for routing through an uncontrolled reservoir.

PROBLEM 82

Frequency-intensity-duration curves and a sketch of a watershed that will be developed into a theme park are presented in the illustrations. All land within the watershed will be paved with an average ground slope of 1.2%. What is the change in the time of concentration from the 10-year to the 25-year storm for a storm duration of 45 min?

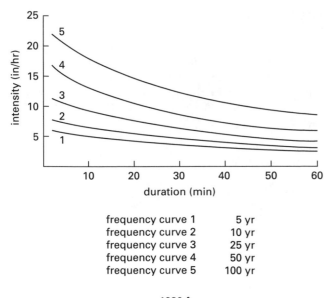

frequency curve 1	5 yr
frequency curve 2	10 yr
frequency curve 3	25 yr
frequency curve 4	50 yr
frequency curve 5	100 yr

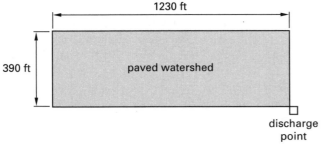

(A) 0.70 min
(B) 1.4 min
(C) 2.3 min
(D) 30 min

Hint: Find an equation appropriate for paved surfaces that includes rainfall intensity as a parameter.

PROBLEM 83

A slope-stage-discharge rating curve and observed fall correction curve for a stream gage station are presented in the illustrations. What is the stream flow for a 2.3 m stage with an observed fall of 0.3 m?

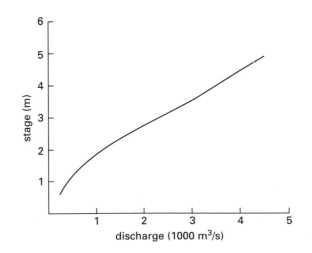

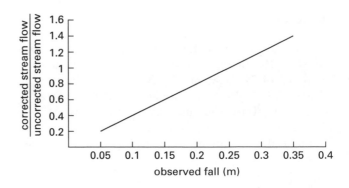

(A) $1.8 \text{ m}^3/\text{s}$
(B) $300 \text{ m}^3/\text{s}$
(C) $1200 \text{ m}^3/\text{s}$
(D) $1800 \text{ m}^3/\text{s}$

Hint: Define the terms shown on the axes of the illustrations.

PROBLEM 84

An average annual discharge record for a river covers a period of 83 yr. The average flow during the entire period of record was $1947 \text{ ft}^3/\text{sec}$ with a standard deviation of $613 \text{ ft}^3/\text{sec}$. What is the average flow that will produce the 50-year flood?

(A) $1200 \text{ ft}^3/\text{sec}$
(B) $3700 \text{ ft}^3/\text{sec}$
(C) $4000 \text{ ft}^3/\text{sec}$
(D) $6100 \text{ ft}^3/\text{sec}$

Hint: The information given suggests an analysis method.

WATER TREATMENT

PROBLEM 85

A planned community will have a population of 3200 people. What is the likely maximum daily water demand during summer months, including fire demand?

(A) 790 gal/min
(B) 1900 gal/min
(C) 2200 gal/min
(D) 2600 gal/min

Hint: Are seasonal and peak multipliers applicable?

PROBLEM 86

Census records for a city population and average annual per capita water use are presented in the table.

year	population	per capita water use (gal/day)
1920	20,800	83
1930	23,400	89
1940	25,100	94
1950	27,900	97
1960	29,800	137
1970	32,600	159
1980	35,200	148
1990	37,700	144
2000	40,100	142

What is the projected approximate average water demand for the city in 2020?

- (A) 5.4×10^6 gal/day
- (B) 6.3×10^6 gal/day
- (C) 1.2×10^7 gal/day
- (D) 3.6×10^8 gal/day

Hint: Check for an arithmetic or a geometric population growth trend.

PROBLEM 87

Average diurnal water demand for a municipality is summarized in the table. A single water storage reservoir serves the municipality's population. Water is pumped to the reservoir continuously at a constant rate equal to the average daily demand. During what period of the day will the net water flow be into the reservoir?

time period	average demand (gal/min)
0000–0200	6900
0200–0400	6500
0400–0600	8100
0600–0800	12,100
0800–1000	14,300
1000–1200	15,600
1200–1400	13,800
1400–1600	12,500
1600–1800	9900
1800–2000	8900
2000–2200	8300
2200–2400	7200

- (A) 6:00 a.m.–4:00 p.m.
- (B) 4:00 p.m.–6:00 a.m.
- (C) 4:20 p.m.–7:10 a.m.
- (D) 4:40 p.m.–6:10 a.m.

Hint: Determine how the average daily flow relates to the average flow per period.

PROBLEM 88

A reservoir is proposed for storing irrigation water. To meet irrigation demand, the required minimum annual yield of the reservoir is 90,000 ac-ft. The reservoir will receive water through inflow from several small streams. The mass inflow from the streams over the preceding 3 yr period is presented in the figure. What reservoir capacity is required to meet the minimum yield?

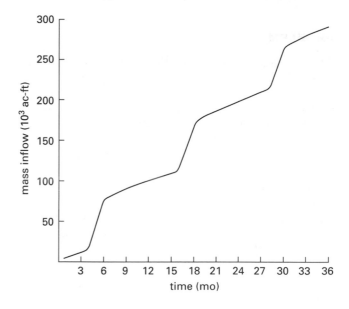

- (A) 40,000 ac-ft
- (B) 78,000 ac-ft
- (C) 90,000 ac-ft
- (D) 270,000 ac-ft

Hint: What is the relationship between the information provided in the figure and the average annual demand?

PROBLEM 89

A water supply contains a total hardness of 382 mg/L as $CaCO_3$. The water treatment plant uses ion exchange to provide water to its customers with a total hardness to 100 mg/L as $CaCO_3$ at a flow rate of 130 000 m^3/d. Reactor vessels with a bed volume of 4 m^3 are available and the resin capacity is 95 kg/m^3. What is the total number of reactor vessels required if regeneration occurs once daily?

- (A) 34
- (B) 71
- (C) 97
- (D) 130

Hint: How is the desired hardness retained in the water supplied to the customers?

PROBLEM 90

What standard motor size is required for a tank-impellor flash mixer sized to treat 5.0×10^6 gal/day at a water temperature of 60°F? The design velocity gradient and residence time are 700 sec^{-1} and 2 min, respectively, and the motor efficiency is 88%.

- (A) 15 hp
- (B) 20 hp
- (C) 22 hp
- (D) 25 hp

Hint: How do efficiency and standard motor size influence the solution?

PROBLEM 91

The design parameters for a flocculation basin are summarized in the table.

design flow rate
    per flocculator   18 000 m³/d
average time-velocity
    gradient 4.5×10^4, unitless
average velocity gradient   40 s^{-1}
number of sections   3
depth   3.5 m
paddle configuration   horizontal paddle wheel, axis perpendicular to flow

What are the length and width dimensions of the flocculation basin?

- (A) width = 4.0 m, length = 16 m
- (B) width = 5.0 m, length = 15 m
- (C) width = 6.5 m, length = 10.5 m
- (D) width = 10 m, length = 10.5 m

Hint: The horizontally mounted paddles influence the basin dimensions.

PROBLEM 92

A flocculation basin consists of two sections each with an equal length and depth of 4.0 m and a width of 8 m. The average velocity gradient is 45 s^{-1} and the paddle speed in both sections is 3 rpm with flat paddles. The paddles are horizontally mounted with the axis perpendicular to flow. The water temperature is 20°C. What is the required paddle area in the first section of the flocculation basin?

- (A) 0.72 m²
- (B) 1.9 m²
- (C) 2.8 m²
- (D) 4.5 m²

Hint: Consider paddle-wall clearance and paddle slip.

PROBLEM 93

A sedimentation basin with a settling zone surface area of 5700 ft² accepts a flow of 2.7×10^6 gal/day. For a

particle-settling velocity of 0.008 in/sec, what is the approximate removal efficiency of the sedimentation basin?

- (A) 1.5%
- (B) 38%
- (C) 46%
- (D) 92%

Hint: Is there a relationship between overflow rate and particle-settling velocity?

PROBLEM 94

A reverse osmosis (RO) system is required to treat a drinking water source that is subject to saltwater intrusion. The water and RO system have the following characteristics.

desired fresh water flow rate   30 000 m³/d
permeate recovery   77%
membrane flux rate   0.93 m³/m²·d
membrane packing density   800 m²/m³
membrane module volume   0.028 m³
pressure vessel capacity   12 modules

How many pressure vessels are required to treat the water?

- (A) 92
- (B) 120
- (C) 160
- (D) 1900

Hint: Determine the flow rate that requires treatment.

PROBLEM 95

What is the head loss through a clean single media sand filter defined by the following characteristics?

water temperature   20°C
clean filtering velocity   0.003 m³/m²·s
media bed depth   0.75 m
media mesh 12×16

- (A) 2.9 cm
- (B) 7.5 cm
- (C) 19 cm
- (D) 25 cm

Hint: Start with the Reynolds number, but be careful of how the particle diameter is determined.

PROBLEM 96

For a flow rate of 10 000 m³/d, what is the approximate overflow rate for a settling basin to achieve 80% efficiency? Results from a settling column test for the Type I suspension are shown in the illustration.

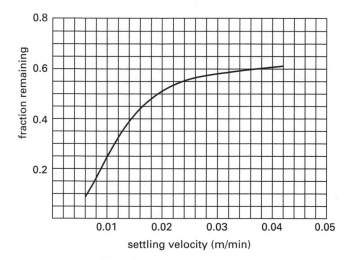

(A) 0.0085 m/min
(B) 0.011 m/min
(C) 0.015 m/min
(D) 0.02 m/min

Hint: Integrate the curve.

PROBLEM 97

What is the monthly caustic soda mass required for softening the water represented by the following analysis? The flow rate requiring treatment is 30 000 m^3/d and the total hardness is to be removed to 100 mg/L as $CaCO_3$. The caustic soda is available at 83% purity. A tonne is 1000 kg.

ion	concentration (mg/L)
Ca^{2+}	187
Mg^{2+}	49
HCO_3^-	618

(A) 460 tonne/mo
(B) 560 tonne/mo
(C) 610 tonne/mo
(D) 650 tonne/mo

Hint: Use molar concentrations.

PROBLEM 98

A water supply experiences taste problems that have been traced to disinfection byproducts (DBP) at a cumulative concentration of 138 μg/L. What is the powdered activated carbon (PAC) dose required to reduce the DBP concentration to 5 μg/L? For the Freundlich model, the intercept of the adsorption isotherm is 21 mg/g and the slope is 0.54.

(A) 110 mg/L
(B) 320 mg/L
(C) 1200 mg/L
(D) 2700 mg/L

Hint: Determine which isotherm equation applies and be careful to use the proper units required by the equation.

PROBLEM 99

The Safe Drinking Water Act (SDWA) includes all of the following EXCEPT

(A) regulation of all public drinking water systems in the U.S.
(B) regulation of both naturally occurring and man-made contaminants in drinking water.
(C) regulation of private drinking water wells serving less than 25 people.
(D) regulation of waste disposal through injection wells.

Hint: Consider the scope and purpose of the SDWA.

PROBLEM 100

A 250 mL water sample with an initial pH of 9.7 is titrated with 0.03 N H_2SO_4. A pH of 8.3 is reached after 6 mL of acid are added and a pH of 4.5 is reached after another 12 mL of acid are added. Which alkalinity specie dominates and what is its concentration?

(A) carbonate dominates at 43 mg/L as $CaCO_3$
(B) carbonate dominates at 67 mg/L as $CaCO_3$
(C) carbonate dominates at 72 mg/L as $CaCO_3$
(D) bicarbonate dominates at 72 mg/L as $CaCO_3$

Hint: How does the initial pH influence the distribution of the alkalinity species?

Breadth Solutions

HYDRAULICS

SOLUTION 1

A_o	area of gate opening	ft^2
C_d	gate discharge coefficient	–
g	gravitational acceleration	$32.2 \ ft/sec^2$
h	available head at the gate	ft
Q	discharge through the gate	ft^3/sec

Because a partially open gate has the characteristics of a submerged orifice, the following equation applies.

$$A_o = \frac{Q}{C_d\sqrt{2gh}} = \frac{80 \ \frac{ft^3}{sec}}{(0.72)\sqrt{(2)\left(32.2 \ \frac{ft}{sec^2}\right)(4.3 \ ft)}}$$

$$= 6.7 \ ft^2$$

The answer is (C).

Why Other Options Are Wrong

(A) This incorrect solution fails to take the square root of the term in the denominator. Definitions are unchanged from the correct solution.

$$A_o = \frac{Q}{C_d(2gh)} = \frac{80 \ \frac{ft^3}{sec}}{(0.72)(2)\left(32.2 \ \frac{ft}{sec^2}\right)(4.3 \ ft)}$$

$$= 0.40 \ ft\text{-}sec$$

The units do not work.

(B) This incorrect solution applies the square root to the head only, instead of the head and gravitational acceleration term. Definitions are unchanged from the correct solution.

$$A_o = \frac{Q}{C_d 2g\sqrt{h}} = \frac{80 \ \frac{ft^3}{sec}}{(0.72)(2)\left(32.2 \ \frac{ft}{sec^2}\right)\sqrt{(4.3 \ ft)}}$$

$$= 0.83 \ ft^{3/2} \ sec$$

The units do not work.

(D) This incorrect solution uses the equation for flow over a weir instead of through a submerged orifice. This requires an erroneous assumption that the gate area is equal to the product of the upstream head and the gate width. Definitions are unchanged from the correct solution.

A	gate opening area	ft^2
b	gate length	ft
H	upstream head	ft

$$Q = C_d b H^{3/2}$$

$$b = \frac{Q}{C_d H^{3/2}} = \frac{80 \ \frac{ft^3}{sec}}{(0.72)(4.3 \ ft)^{3/2}} = 12.5 \ ft$$

The units do not work.

$$A = bH = (12.5 \ ft)(4.3 \ ft) = 54 \ ft^2$$

SOLUTION 2

Because specific information regarding the type of pipe is not given, assume pipe diameters given are actual inside diameters.

A_1, A_2	cross-sectional area of upstream and downstream pipe, respectively	in^2
d_1, d_2	inside diameter of upstream and downstream pipe, respectively	in

$$A_1 = \frac{\pi d_1^2}{4} = \frac{\pi (2 \ in)^2}{4} = 3.14 \ in^2$$

$$A_2 = \frac{\pi d_2^2}{4} = \frac{\pi (1.5 \ in)^2}{4} = 1.77 \ in^2$$

v_1	water velocity in upstream pipe	ft/sec
v_2	water velocity in downstream pipe	ft/sec

$$v_2 = \frac{A_1 v_1}{A_2} = \frac{(3.14 \ in^2)\left(4.2 \ \frac{ft}{sec}\right)}{1.77 \ in^2} = 7.5 \ ft/sec$$

The answer is (C).

Why Other Options Are Wrong

(A) This incorrect solution reverses the values for area in the velocity equation. Other definitions and equations are the same as used in the correct solution.

$$A_1 = \frac{\pi d_1^2}{4} = \frac{\pi (2 \text{ in})^2}{4} = 3.14 \text{ in}^2$$

$$A_2 = \pi \frac{d_2^2}{4} = \frac{\pi (1.5 \text{ in})^2}{4} = 1.77 \text{ in}^2$$

$$v_2 = \frac{A_2 v_1}{A_1} = \frac{(1.77 \text{ in}^2)\left(4.2 \dfrac{\text{ft}}{\text{sec}}\right)}{3.14 \text{ in}^2} = 2.4 \text{ ft/sec}$$

(B) This incorrect solution calculates circumference instead of area in the area equations. Other definitions and equations are the same as used in the correct solution.

$$A_1 = \pi d_1 = \pi(2 \text{ in}) = 6.28 \text{ in}$$

The units do not work.

$$A_2 = \pi d_2 = \pi(1.5 \text{ in}) = 4.71 \text{ in}$$

The units do not work.

$$v_2 = \frac{(6.28 \text{ in})\left(4.2 \dfrac{\text{ft}}{\text{sec}}\right)}{4.71 \text{ in}} = 5.6 \text{ ft/sec}$$

(D) This incorrect solution makes a unit conversion error in the velocity equation and inverts the areas. Other definitions and equations are the same as used in the correct solution.

$$A_1 = \frac{\pi d_1^2}{4} = \frac{\pi (2 \text{ in})^2}{4} = 3.14 \text{ in}^2$$

$$A_2 = \frac{\pi d_2^2}{4} = \frac{\pi (1.5 \text{ in})^2}{4} = 1.77 \text{ in}^2$$

$$v_2 = \frac{A_2 v_1}{A_1} = \frac{(1.77 \text{ in}^2)\left(4.2 \dfrac{\text{ft}}{\text{sec}}\right)\left(12 \dfrac{\text{in}}{\text{ft}}\right)}{3.14 \text{ in}^2}$$
$$= 28 \text{ ft/sec}$$

The units do not work.

SOLUTION 3

Apply the energy equation.

g	gravitational acceleration	9.81 m/s^2
p	pressure	kN/m^2
v	flow velocity	m/s
z	elevation	m
z_o	water surface elevation	m
ρ	water density	kg/m^3

$$\frac{p_A}{\rho g} + z_A + \frac{v_A^2}{2g} = \frac{p_B}{\rho g} + z_B + \frac{v_B^2}{2g}$$
$$= \frac{p_C}{\rho g} + z_C + \frac{v_C^2}{2g}$$
$$= z_o$$

$$p_C = 0 \ \frac{\text{kN}}{\text{m}^2} \quad \text{[free jet]}$$

$$\frac{p_C}{\rho g} + z_C + \frac{v_C^2}{2g} = 0 + 34 \text{ m} + \frac{v_C^2}{(2)\left(9.81 \dfrac{\text{m}}{\text{s}^2}\right)}$$

$$z_o = 100 \text{ m}$$
$$v_C = 36 \text{ m/s}$$

A	pipe cross-sectional area	m^2
D	pipe diameter	m

$$A = \pi \frac{D^2}{4}$$

$$A_C = \frac{\pi (8 \text{ cm})^2 \left(\dfrac{1 \text{ m}^2}{10^4 \text{ cm}^2}\right)}{4}$$
$$= 0.0050 \text{ m}^2$$

Q	flow rate	m^3/s

$$Q = Av$$
$$= (0.0050 \text{ m}^2)\left(36 \ \frac{\text{m}}{\text{s}}\right)$$
$$= 0.18 \text{ m}^3/\text{s}$$

$$A_B = \frac{\pi (10 \text{ cm})^2 \left(\dfrac{1 \text{ m}^2}{10^4 \text{ cm}^2}\right)}{4} = 0.0079 \text{ m}^2$$

$$v_B = \frac{Q}{A_B} = \frac{0.18 \dfrac{\text{m}^3}{\text{s}}}{0.0079 \text{ m}^2} = 23 \text{ m/s}$$

Assume a temperature of 20°C for determining the density of water (998.23 kg/m³).

$$\frac{p_B}{\rho g} + z_B + \frac{v_B^2}{2g} = z_o$$

$$\overline{\left(\begin{array}{c}\left(998.23 \ \dfrac{\text{kg}}{\text{m}^3}\right)\left(9.81 \ \dfrac{\text{m}}{\text{s}^2}\right) \\ \times \left(\dfrac{\text{N·s}^2}{\text{kg·m}}\right)\left(\dfrac{1 \text{ kN}}{1000 \text{ N}}\right)\end{array}\right)}$$

$$+ 46 \text{ m} + \frac{\left(23 \ \dfrac{\text{m}}{\text{s}}\right)^2}{(2)\left(9.81 \ \dfrac{\text{m}}{\text{s}^2}\right)} = 100 \text{ m}$$

$$p_B = 265 \text{ kN/m}^2$$
$$(260 \text{ kN/m}^2)$$

The answer is (C).

Why Other Options Are Wrong

(A) This incorrect solution transposes the elevation value for pipe B. Other assumptions, definitions, and equations are the same as used in the correct solution.

$$\frac{p_A}{\rho g} + z_A + \frac{v_A^2}{2g} = \frac{p_B}{\rho g} + z_B + \frac{v_B^2}{2g}$$

$$= \frac{p_C}{\rho g} + z_C + \frac{v_C^2}{2g}$$

$$= z_o$$

$$p_C = 0 \ \frac{\text{kN}}{\text{m}^2} \quad \text{[free jet]}$$

$$\frac{p_C}{\rho g} + z_C + \frac{v_C^2}{2g} = 0 \ \frac{\text{kN}}{\text{m}^2} + 34 \text{ m} + \frac{v_C^2}{(2)\left(9.81 \ \frac{\text{m}}{\text{s}^2}\right)}$$

$$= 100 \text{ m}$$

$$v_C = 36 \text{ m/s}$$

$$A_C = \frac{\pi (8 \text{ cm})^2 \left(\frac{1 \text{ m}^2}{(10)^4 \text{ cm}^2}\right)}{4} = 0.0050 \text{ m}^2$$

$$Q = (0.0050 \text{ m}^2)\left(36 \ \frac{\text{m}}{\text{s}}\right) = 0.18 \text{ m}^3/\text{s}$$

$$A_B = \frac{\pi (10 \text{ cm})^2 \left(\frac{1 \text{ m}^2}{(10)^4 \text{ cm}^2}\right)}{4} = 0.0079 \text{ m}^2$$

$$v_B = \frac{0.18 \ \frac{\text{m}^3}{\text{s}}}{0.0079 \text{ m}^2} = 23 \text{ m/s}$$

$$\frac{p_B}{\left(\left(998.23 \ \frac{\text{kg}}{\text{m}^3}\right)\left(9.81 \ \frac{\text{m}}{\text{s}^2}\right)\right)}{\times \left(\frac{\text{N·s}^2}{\text{kg·m}}\right)\left(\frac{1 \text{ kN}}{1000 \text{ N}}\right)}$$

$$+ 64 \text{ m} + \frac{\left(23 \ \frac{\text{m}}{\text{s}}\right)^2}{(2)\left(9.81 \ \frac{\text{m}}{\text{s}^2}\right)} = 100 \text{ m}$$

$$p_B = 88 \text{ kN/m}^2$$

(B) This incorrect solution fails to take the square root when determining the velocity in pipe C. Other assumptions, definitions, and equations are the same as used in the correct solution.

$$\frac{p_A}{\rho g} + z_A + \frac{v_A^2}{2g} = \frac{p_B}{\rho g} + z_B + \frac{v_B^2}{2g}$$

$$= \frac{p_C}{\rho g} + z_C + \frac{v_C^2}{2g}$$

$$= z_o$$

$$p_C = 0 \ \frac{\text{kN}}{\text{m}^2} \quad \text{[free jet]}$$

$$\frac{p_C}{\rho g} + z_C + \frac{v_C^2}{2g} = 0 \ \frac{\text{kN}}{\text{m}^2} + 34 \text{ m} + \frac{v_C^2}{(2)\left(9.81 \ \frac{\text{m}}{\text{s}^2}\right)}$$

$$= 100 \text{ m}$$

$$v_C = 1294.9 \text{ m/s}$$

$$A_C = \frac{\pi (8 \text{ cm})^2 \left(\frac{1 \text{ m}^2}{(10)^4 \text{ cm}^2}\right)}{4}$$

$$= 0.0050 \text{ m}^2$$

$$Q = (0.0050 \text{ m}^2)\left(1294.9 \ \frac{\text{m}}{\text{s}}\right)$$

$$= 6.475 \text{ m}^3/\text{s}$$

$$A_B = \frac{\pi (10 \text{ cm})^2 \left(\frac{1 \text{ m}^2}{(10)^4 \text{ cm}^2}\right)}{4}$$

$$= 0.0079 \text{ m}^2$$

$$v_B = \frac{6.475 \ \frac{\text{m}^3}{\text{s}}}{0.0079 \text{ m}^2} = 819.6 \text{ m/s}$$

$$\frac{p_B}{\left(\left(998.23 \ \frac{\text{kg}}{\text{m}^3}\right)\left(9.81 \ \frac{\text{m}}{\text{s}^2}\right)\right)}{\times \left(\frac{\text{N·s}^2}{\text{kg·m}}\right)\left(\frac{1 \text{ kN}}{1000 \text{ N}}\right)}$$

$$+ 46 \text{ m} + \frac{\left(819.6 \ \frac{\text{m}}{\text{s}}\right)}{(2)\left(9.81 \ \frac{\text{m}}{\text{s}^2}\right)} = 100 \text{ m}$$

$$p_B = (100 \text{ m} - 41.77 \text{ s} - 46 \text{ m})\left(9.793 \ \frac{\text{kN}}{\text{m}^3}\right)$$

The units in the difference term do not work and are ignored.

$$p_B = 120 \text{ kN/m}^2$$

(D) This incorrect solution uses the value of specific weight for customary U.S. units instead of density and gravitational acceleration for SI units. Other assumptions, definitions, and equations are the same as used in the correct solution.

$$\frac{p_A}{\rho g} + z_A + \frac{v_A^2}{2g} = \frac{p_B}{\rho g} + z_B + \frac{v_B^2}{2g}$$

$$= \frac{p_C}{\rho g} + z_C + \frac{v_C^2}{2g}$$

$$= z_o$$

γ specific weight   kN/m^3

$$\rho g = \gamma$$

$$= 62.3 \text{ kN/m}^3 \text{ at } 20°\text{C}$$

$$\frac{p_A}{\gamma} + z_A + \frac{v_A^2}{2g} = \frac{p_B}{\gamma} + z_B + \frac{v_B^2}{2g}$$

$$= \frac{p_C}{\gamma} + z_C + \frac{v_C^2}{2g}$$

$$= z_o$$

$$p_C = 0 \ \frac{\text{kN}}{\text{m}^2} \quad \text{[free jet]}$$

$$\frac{p_C}{\gamma} + z_C + \frac{v_C^2}{2g} = 0 + 34 \text{ m} + \frac{v_C^2}{(2)\left(9.81 \ \frac{\text{m}}{\text{s}^2}\right)}$$

$$= 100 \text{ m}$$

$$v_C = 36 \text{ m/s}$$

$$A_C = \frac{\pi(8 \text{ cm})^2 \left(\frac{1 \text{ m}^2}{(10)^4 \text{ cm}^2}\right)}{4}$$

$$= 0.0050 \text{ m}^2$$

$$Q = (0.0050 \text{ m}^2)\left(36 \ \frac{\text{m}}{\text{s}}\right)$$

$$= 0.18 \text{ m}^3/\text{s}$$

$$A_B = \frac{\pi(10 \text{ cm})^2 \left(\frac{1 \text{ m}^2}{(10)^4 \text{ cm}^2}\right)}{4}$$

$$= 0.0079 \text{ m}^2$$

$$v_B = \frac{0.18 \ \frac{\text{m}^3}{\text{s}}}{0.0079 \text{ m}^2}$$

$$= 23 \text{ m/s}$$

$$\frac{p_B}{62.3 \ \frac{\text{kN}}{\text{m}^3}} + 46 \text{ m} + \frac{\left(23 \ \frac{\text{m}}{\text{s}^2}\right)}{(2)\left(9.81 \ \frac{\text{m}}{\text{s}}\right)^2}$$

$$= 100 \text{ m}$$

$$p_B = 1684 \text{ kN/m}^2 \quad (1700 \text{ kN/m}^2)$$

SOLUTION 4

When the channel width is much greater than the normal depth, the simplifying assumption that the hydraulic radius is equal to the normal depth applies. This is not the case for this problem. Use the Manning equation.

A	area of flow	ft^2
d_n	normal depth	ft
n	Manning roughness coefficient	–
Q	flow rate	ft^3/sec
S	channel slope	%/100%
R	hydraulic radius	ft
w	channel width	ft

The Manning roughness coefficient for concrete is 0.013.

$$A = d_n w = (8 \text{ in})\left(\frac{1 \text{ ft}}{12 \text{ in}}\right)(5 \text{ ft})$$

$$= 3.33 \text{ ft}^2$$

$$R = \frac{d_n w}{w + 2d_n} = \frac{(8 \text{ in})\left(\frac{1 \text{ ft}}{12 \text{ in}}\right)(5 \text{ ft})}{5 \text{ ft} + (2)(8 \text{ in})\left(\frac{1 \text{ ft}}{12 \text{ in}}\right)}$$

$$= 0.526 \text{ ft}$$

$$Q = \frac{1.49}{n} A R^{2/3} \sqrt{S}$$

$$= \left(\frac{1.49}{0.013}\right)(3.33 \text{ ft}^2)(0.526 \text{ ft})^{2/3}\sqrt{\frac{0.2\%}{100\%}}$$

$$= 11.12 \text{ ft}^3/\text{sec} \quad (11 \text{ ft}^3/\text{sec})$$

The answer is (B).

Why Other Options Are Wrong

(A) This incorrect solution uses the Manning equation for SI units instead of customary U.S. units and makes the simplifying assumption that the hydraulic radius is equal to the normal depth. Assumptions and definitions are unchanged from the correct solution.

$$Q = \frac{d_n^{5/3} w \sqrt{S}}{n} = \frac{\left(8 \text{ in}\left(\frac{1 \text{ ft}}{12 \text{ in}}\right)\right)^{5/3}(5 \text{ ft})\sqrt{\frac{0.2\%}{100\%}}}{0.013}$$

$$= 8.7 \text{ ft}^3/\text{sec}$$

(C) This incorrect solution uses the slope as percent instead of as a fraction and makes the simplifying assumption that the hydraulic radius is equal to the normal depth. Assumptions, definitions, and equations are unchanged from the correct solution.

$$Q = \frac{1.49 d_n^{5/3} w \sqrt{S}}{n}$$

$$= \frac{(1.49)\left(8 \text{ in}\left(\frac{1 \text{ ft}}{12 \text{ in}}\right)\right)^{5/3}(5 \text{ ft})\sqrt{0.2\%}}{0.013}$$

$$= 130 \text{ ft}^3/\text{sec}$$

(D) This incorrect solution fails to convert the normal depth from inches to feet. This requires an assumption that the units for flow rate are in ft^3/sec for normal depth units in inches. This solution also makes the simplifying assumption that the hydraulic radius is equal to the normal depth. Assumptions, definitions, and equations are unchanged from the correct solution.

$$Q = \frac{1.49 d_n^{5/3} w \sqrt{S}}{n} = \frac{(1.49)(8 \text{ in})^{5/3}(5 \text{ ft})\sqrt{\frac{0.2\%}{100\%}}}{0.013}$$

$$= 820 \text{ ft}^3/\text{sec}$$

SOLUTION 5

For a unit length of smooth earthen channel, it is reasonable to assume that flow is uniform and that all head loss is due to friction. With these assumptions, the head loss is the product of the channel length and slope, with the slope determined using the Manning equation.

b channel base width   m
d water depth   m
R hydraulic radius   m
θ angle from the horizontal
     to the side wall   degree

For 1-to-1 side slopes, the angle from the horizontal to the side wall is 45°.

$$
\begin{aligned}
R &= \frac{bd\sin\theta + d^2\cos\theta}{b\sin\theta + 2d} \\
&= \frac{(2\text{ m})(0.5\text{ m})(\sin 45°) + (0.5\text{ m})^2(\cos 45°)}{(2\text{ m})(\sin 45°) + (2)(0.5\text{ m})} \\
&= 0.366\text{ m}
\end{aligned}
$$

h_f head loss from friction   m/unit length
                     of channel
L channel length   m
n Manning roughness coefficient –
v water-flow velocity   m/s

Use 1 m for the unit length of the channel. The Manning roughness coefficient for a smooth earthen lining is 0.018.

$$
\begin{aligned}
h_f &= \frac{Ln^2 v^2}{R^{4/3}} = \frac{(1\text{ m})(0.018)^2\left(3.2\,\frac{\text{m}}{\text{s}}\right)^2}{(0.366\text{ m})^{4/3}} \\
&= 0.013\text{ m/m length of channel}
\end{aligned}
$$

The answer is (C).

Why Other Options Are Wrong

(A) This incorrect solution calculates the wetted perimeter instead of the hydraulic radius for the channel. Other assumptions, definitions, and equations are unchanged from the correct solution.

$$
\begin{aligned}
R &= b + 2\left(\frac{d}{\sin\theta}\right) = 2\text{ m} + (2)\left(\frac{0.5\text{ m}}{\sin 45°}\right) \\
&= 3.4\text{ m}
\end{aligned}
$$

$$
\begin{aligned}
h_f &= \frac{Ln^2 v^2}{R^{4/3}} = \frac{(1\text{ m})(0.018)^2\left(3.2\,\frac{\text{m}}{\text{s}}\right)^2}{(3.4\text{ m})^{4/3}} \\
&= 0.00065\text{ m/m}
\end{aligned}
$$

(B) This incorrect solution fails to square the velocity term in the head loss equation. Other assumptions, definitions, and equations are unchanged from the correct solution.

$$
\begin{aligned}
R &= \frac{bd\sin\theta + d^2\cos\theta}{b\sin\theta + 2d} \\
&= \frac{(2\text{ m})(0.5\text{ m})(\sin 45°) + (0.5\text{ m})^2(\cos 45°)}{(2\text{ m})(\sin 45°) + (2)(0.5\text{ m})} \\
&= 0.366\text{ m}
\end{aligned}
$$

$$
\begin{aligned}
h_f &= \frac{Ln^2 v}{R^{4/3}} = \frac{(1\text{ m})(0.018)^2\left(3.2\,\frac{\text{m}}{\text{s}}\right)}{(0.366\text{ m})^{4/3}} \\
&= 0.0040\text{ m/m}
\end{aligned}
$$

(D) This incorrect solution uses the Manning roughness coefficient for natural channels instead of smooth earth. Other assumptions, definitions, and equations are unchanged from the correct solution.

$$
\begin{aligned}
R &= \frac{bd\sin\theta + d^2\cos\theta}{b\sin\theta + 2d} \\
&= \frac{(2\text{ m})(0.5\text{ m})(\sin 45°) + (0.5\text{ m})^2(\cos 45°)}{(2\text{ m})(\sin 45°) + (2)(0.5\text{ m})} \\
&= 0.366\text{ m}
\end{aligned}
$$

Assume the Manning Roughness coefficient for a natural channel is 0.025.

$$
\begin{aligned}
h_f &= \frac{Ln^2 v}{R^{4/3}} = \frac{(1\text{ m})(0.025)^2\left(3.2\,\frac{\text{m}}{\text{s}}\right)^2}{(0.366\text{ m})^{4/3}} \\
&= 0.024\text{ m/m}
\end{aligned}
$$

SOLUTION 6

b weir length   ft
H head above notch   ft
Q flow rate   ft³/sec

The discharge equation giving flow rate in ft³/sec for a Cipoletti weir when b and H are in ft is

$$
\begin{aligned}
Q &= 3.367bH^{3/2} \\
&= (3.367)(24\text{ in})\left(\frac{1\text{ ft}}{12\text{ in}}\right)\left((8.6\text{ in})\left(\frac{1\text{ ft}}{12\text{ in}}\right)\right)^{3/2} \\
&= 4.1\text{ ft}^3/\text{sec}
\end{aligned}
$$

The answer is (B).

Why Other Options Are Wrong

(A) This incorrect solution uses the discharge equation for a standard contracted rectangular weir. Definitions are unchanged from the correct solution.

$$Q = (3.33)(b - 0.2H)H^{3/2}$$

$$= (3.33)\left(\begin{array}{c}(24 \text{ in})\left(\dfrac{1 \text{ ft}}{12 \text{ in}}\right) - (0.2) \\[2mm] \times (8.6 \text{ in})\left(\dfrac{1 \text{ ft}}{12 \text{ in}}\right)\end{array}\right)$$

$$\times \left((8.6 \text{ in})\left(\dfrac{1 \text{ ft}}{12 \text{ in}}\right)\right)^{3/2}$$

$$= 3.8 \text{ ft}^3/\text{sec}$$

(C) This incorrect solution makes a mathematical error with the exponent in the head term of the discharge equation. Definitions and the equation are unchanged from the correct solution.

$$Q = 3.367bH^{2/3}$$

$$= (3.367)(24 \text{ in})\left(\dfrac{1 \text{ ft}}{12 \text{ in}}\right)\left((8.6 \text{ in})\left(\dfrac{1 \text{ ft}}{12 \text{ in}}\right)\right)^{2/3}$$

$$= 5.4 \text{ ft}^3/\text{sec}$$

(D) This incorrect solution makes an error in converting from inches to feet. Definitions and the equation are unchanged from the correct solution.

$$Q = 3.367bH^{3/2}$$

$$= (3.367)(24 \text{ in})(8.6 \text{ in})^{3/2}\left(\dfrac{1 \text{ ft}^2}{144 \text{ in}^2}\right)$$

$$= 14 \text{ ft}^3/\text{sec}$$

SOLUTION 7

As the water flows over the initial break from horizontal to a relatively steep slope and the flow transitions from subcritical to supercritical, the water depth will likely be at critical depth at the break in slope. Water will flow at less than critical depth (supercritical velocity) through both steeply sloped sections. At the bottom of the second steep section where the slope changes from relatively steep to relatively flat, a hydraulic jump will form as flow transitions from supercritical back to subcritical. The hydraulic jump will form close to the slope transition. The hump in the channel bottom could influence flow in a variety of ways, but its relative distance from the steep to shallow slope change and occurrence after the hydraulic jump where the water is deeper will limit this influence.

The answer is (B).

Why Other Options Are Wrong

(A) The flow profile generally follows what would be expected until it reaches the transition from the steep to nearly horizontal section. The hydraulic jump would likely form near the transition, some distance upstream of the hump instead of, as the illustration shows, at the hump.

(C) The flow profile would likely show critical flow beginning at the top of the first steep section and a hydraulic jump would likely occur at the bottom of the second steep section. This illustration does not show either of these features.

(D) Because of the slope change from relatively steep to relatively flat, it is unlikely that the flow would be subcritical in the upstream horizontal section and then continue as subcritical flow after the slope transition as shown in the illustration.

SOLUTION 8

The equivalent lengths for the fittings are summarized in the table. The equivalent lengths were taken from standard reference tables.

fitting	quantity	unit fitting equivalent length (ft)	total fitting equivalent length (ft)
90° ell	7	8.5	59.5
couple	16	0.45	7.2
union	4	0.45	1.8
tee	5	7.7	38.5
valve	2	1.5	3.0
			110

L length of pipe   ft
L_e total equivalent length   ft
L_f equivalent length of fittings   ft

$$L_e = L + \sum L_f$$

$$= 250 \text{ ft} + 110 \text{ ft}$$

$$= 360 \text{ ft}$$

The answer is (D).

Why Other Options Are Wrong

(A) This incorrect solution ignores the pipe length in the equivalent length calculation. Other definitions are unchanged from the correct solution.

fitting	quantity	unit fitting equivalent length (ft)	total fitting equivalent length (ft)
90° ell	7	8.5	59.5
couple	16	0.45	7.2
union	4	0.45	1.8
tee	5	7.7	38.5
valve	2	1.5	3.0
			110

$$L_e = \sum L_f = 110 \text{ ft}$$

(B) This incorrect solution uses inches instead of feet as the units for fitting equivalent lengths. Other definitions are unchanged from the correct solution.

fitting	quantity	unit fitting equivalent length (in)	total fitting equivalent length (in)
90° ell	7	8.5	59.5
couple	16	0.45	7.2
union	4	0.45	1.8
tee	5	7.7	38.5
valve	2	1.5	3.0
			110

$$L_e = \sum L_f = 250 \text{ ft} + (110 \text{ in}) \left(\frac{1 \text{ ft}}{12 \text{ in}} \right)$$
$$= 259 \text{ ft} \quad (260 \text{ ft})$$

(C) This incorrect solution uses equivalent lengths for branched flow tees and long radius elbows. Other definitions are unchanged from the correct solution.

fitting	quantity	unit fitting equivalent length (ft)	total fitting equivalent length (ft)
90° ell	7	3.6	25.2
couple	16	0.45	7.2
union	4	0.45	1.8
tee	5	12	60
valve	2	1.5	3.0
			97.2

$$L_e = L + \sum L_f = 250 \text{ ft} + 97.2 \text{ ft}$$
$$= 347 \text{ ft} \quad (350 \text{ ft})$$

SOLUTION 9

A gate surface area m^2
D gate diameter m

$$A = \frac{\pi D^2}{4} = \frac{\pi (0.5 \text{ m})^2}{4} = 0.20 \text{ m}^2$$

g gravitational acceleration 9.81 m/s^2
h_1 water depth above the gate top m
h_2 water depth above the gate bottom m
R resultant force on the gate N
ρ water density kg/m^3

$$h_2 = 2.3 \text{ m} + 0.5 \text{ m} = 2.8 \text{ m}$$

Assume the density of water to be 1000 kg/m^3 since the water temperature was not given.

$$R = A\rho g \left(\frac{h_1 + h_2}{2} \right)$$
$$= (0.20 \text{ m}^2) \left(1000 \ \frac{\text{kg}}{\text{m}^3} \right) \left(9.81 \ \frac{\text{m}}{\text{s}^2} \right)$$
$$\times \left(\frac{2.3 \text{ m} + 2.8 \text{ m}}{2} \right) \left(\frac{\text{N} \cdot \text{s}^2}{\text{kg} \cdot \text{m}} \right)$$
$$= 5003 \text{ N}$$

d_R distance below water surface where the resultant force is applied m
I_c centroid area moment of inertia m

For a circle,

$$I_c = \frac{\pi D^4}{64} = \frac{\pi (0.5 \text{ m})^4}{64}$$
$$= 0.0031 \text{ m}^4$$

$$d_R = \frac{(h_1 + h_2)}{2} + \frac{I_c}{A \dfrac{(h_1 + h_2)}{2}}$$
$$= \frac{(2.3 \text{ m} + 2.8 \text{ m})}{2} + \frac{0.0031 \text{ m}^4}{(0.20 \text{ m}^2) \dfrac{(2.3 \text{ m} + 2.8 \text{ m})}{2}}$$
$$= 2.6 \text{ m}$$

F_G force at the non-hinged side of the gate N

$$F_G = \frac{R(d_R - h_1)}{D}$$
$$= \left(\frac{(5003 \text{ N})(2.6 \text{ m} - 2.3 \text{ m})}{0.5 \text{ m}} \right) \left(\frac{1 \text{ kN}}{1000 \text{ N}} \right)$$
$$= 3.0 \text{ kN}$$

The answer is (A).

Why Other Options Are Wrong

(B) This incorrect solution erroneously defines the water depths used to calculate the resultant force and its location. The negative sign in the final answer is ignored. Other assumptions, definitions, and equations are unchanged from the correct solution.

$$A = \frac{\pi D^2}{4} = \frac{\pi (0.5 \text{ m})^2}{4} = 0.20 \text{ m}^2$$

h_1 water depth above the gate top m
h_2 water depth above the gate bottom m

$$R = A\rho g\left(\frac{h_1 + h_2}{2}\right)$$

$$= (0.20 \text{ m}^2)\left(1000 \ \frac{\text{kg}}{\text{m}^3}\right)\left(9.81 \ \frac{\text{m}}{\text{s}^2}\right)$$

$$\times \left(\frac{2.3 \text{ m} + 0.5 \text{ m}}{2}\right)\left(\frac{\text{N·s}^2}{\text{kg·m}}\right)$$

$$= 2747 \text{ N}$$

$$d_R = \left(\frac{2}{3}\right)\left(h_1 + h_2 - \frac{h_1 h_2}{h_1 + h_2}\right)$$

$$= \left(\frac{2}{3}\right)\left(2.3 \text{ m} + 0.5 \text{ m} - \frac{(2.3 \text{ m})(0.5 \text{ m})}{2.3 \text{ m} + 0.5 \text{ m}}\right)$$

$$= 1.6 \text{ m}$$

$$F_G = \frac{R(d_R - h_1)}{D}$$

$$= \left(\frac{(2747 \text{ N})(1.6 \text{ m} - 2.3 \text{ m})}{(0.5 \text{ m})}\right)\left(\frac{1 \text{ kN}}{1000 \text{ N}}\right)$$

$$= -3.8 \text{ kN}$$

The negative sign is ignored.

(C) This incorrect solution miscalculates the area of the gate. Other assumptions, definitions, and equations are unchanged from the correct solution.

$$A = \pi r^2 = \pi (0.5 \text{ m})^2 = 0.79 \text{ m}^2$$

$$h_2 = 2.3 \text{ m} + 0.5 \text{ m} = 2.8 \text{ m}$$

$$R = A\rho g\left(\frac{h_1 + h_2}{2}\right)$$

$$= (0.79 \text{ m}^2)\left(1000 \ \frac{\text{kg}}{\text{m}^3}\right)\left(9.81 \ \frac{\text{m}}{\text{s}^2}\right)$$

$$\times \left(\frac{2.3 \text{ m} + 2.8 \text{ m}}{2}\right)\left(\frac{\text{N·s}^2}{\text{kg·m}}\right)$$

$$= 19\,762 \text{ N}$$

$$d_R = \left(\frac{2}{3}\right)\left(h_1 + h_2 - \frac{h_1 h_2}{h_1 + h_2}\right)$$

$$= \left(\frac{2}{3}\right)\left(2.3 \text{ m} + 2.8 \text{ m} - \frac{(2.3 \text{ m})(2.8 \text{ m})}{2.3 \text{ m} + 2.8 \text{ m}}\right)$$

$$= 2.6 \text{ m}$$

$$F_G = \frac{R(d_R - h_1)}{D}$$

$$= \left(\frac{(19\,762 \text{ N})(2.6 \text{ m} - 2.3 \text{ m})}{(0.5 \text{ m})}\right)\left(\frac{1 \text{ kN}}{1000 \text{ N}}\right)$$

$$= 12 \text{ kN}$$

(D) This incorrect solution miscalculates the location of the resultant force. Other assumptions, definitions, and equations are unchanged from the correct solution.

$$A = \frac{\pi D^2}{4} = \frac{\pi (0.5 \text{ m})^2}{4} = 0.20 \text{ m}^2$$

$$h_2 = 2.3 \text{ m} + 0.5 \text{ m} = 2.8 \text{ m}$$

$$R = A\rho g\left(\frac{h_1 + h_2}{2}\right)$$

$$= (0.20 \text{ m}^2)\left(1000 \ \frac{\text{kg}}{\text{m}^3}\right)\left(9.81 \ \frac{\text{m}}{\text{s}^2}\right)$$

$$\times \left(\frac{2.3 \text{ m} + 2.8 \text{ m}}{2}\right)\left(\frac{\text{N·s}^2}{\text{kg·m}}\right)$$

$$= 5003 \text{ N}$$

$$d_R = \left(\frac{2}{3}\right)\left(h_1 + h_2 + \frac{h_1 h_2}{h_1 + h_2}\right)$$

$$= \left(\frac{2}{3}\right)\left(2.3 \text{ m} + 2.8 \text{ m} + \frac{(2.3 \text{ m})(2.8 \text{ m})}{2.3 \text{ m} + 2.8 \text{ m}}\right)$$

$$= 4.2 \text{ m}$$

$$F_G = \frac{R(d_R - h_1)}{D}$$

$$= \left(\frac{(5003 \text{ N})(4.2 \text{ m} - 2.3 \text{ m})}{(0.5 \text{ m})}\right)\left(\frac{1 \text{ kN}}{1000 \text{ N}}\right)$$

$$= 19 \text{ kN}$$

SOLUTION 10

d_1 upstream section water depth   m
R_1 upstream section hydraulic radius   m
θ channel side slope angle measured
              from the horizontal   degree

For 1-to-1 side slopes, the channel side slope angle measured from the horizontal is 45°.

$$R_1 = \frac{d_1 \cos\theta}{2} = \frac{(10 \text{ cm})\left(\frac{1 \text{ m}}{100 \text{ cm}}\right)(\cos 45°)}{2}$$

$$= 0.035 \text{ m}$$

n Manning roughness coefficient   –
S_1 upstream section channel slope   m/m
v_1 upstream section flow velocity   m/s

For concrete-lined channels, the Manning roughness coefficient is 0.013.

$$v_1 = \frac{R_1^{2/3}\sqrt{S_1}}{n} = \frac{(0.035 \text{ m})^{2/3}\sqrt{0.12}}{0.013}$$

$$= 2.85 \text{ m/s}$$

$$d_1 = (10 \text{ cm})\left(\frac{1 \text{ m}}{100 \text{ cm}}\right) = 0.1 \text{ m}$$

A area m^2
d_2 downstream section water depth   m
F force   N
g gravitational acceleration 9.81 m/s^2
Q flow rate m^3/s
v_2 downstream section flow velocity   m/s
z depth to centroid   m
γ specific weight kN/m^3
ρ water density kg/m^3

Because the problem uses a triangular channel, the hydraulic jump equation must be derived from the momentum equation.

$$F_1 - F_2 = \rho Q v_2 - \rho Q v_1$$
$$\rho = \frac{\gamma}{g}$$
$$F = \gamma z A$$
$$Q = A v$$

For 1:1 triangular channels,

$$A = d_2$$
$$z = d/2$$

Make substitutions into the momentum equation and solve for d_2.

$$F_1 - F_2 = \rho Q v_2 - \rho Q v_1$$
$$\frac{d_1^3}{3} - \frac{d_2^3}{3} = \frac{Q^2}{g d_2^2} - \frac{Q^2}{g d_1^2}$$
$$Q = A_1 v_1 = d_1^2 v_1$$
$$= (0.1 \text{ m})^2 \left(2.85 \ \frac{\text{m}}{\text{s}} \right)$$
$$= 0.0285 \text{ m}^3/\text{s}$$

$$\frac{0.1 \text{ m}^3}{3} - \frac{d_2^3}{3} = \frac{\left(0.0285 \ \frac{\text{m}^3}{\text{s}} \right)^2}{\left(9.81 \ \frac{\text{m}}{\text{s}^2} \right) d_2^2} - \frac{\left(0.0285 \ \frac{\text{m}^3}{\text{s}} \right)^2}{\left(9.81 \ \frac{\text{m}}{\text{s}^2} \right) (0.1 \text{ m})^2}$$
$$d_2 = 0.28 \text{ m}$$

The answer is (B).

Why Other Options Are Wrong

(A) This incorrect solution fails to square the velocity term in the conjugate depth equation and uses the wrong equation. Other assumptions, definitions, and equations are unchanged from the correct solution.

$$R_1 = \frac{d_1 \cos \theta}{2}$$
$$= \frac{(10 \text{ cm}) \left(\frac{1 \text{ m}}{100 \text{ cm}} \right) (\cos 45^\circ)}{2}$$
$$= 0.035 \text{ m}$$
$$v_1 = \frac{R_1^{2/3} \sqrt{S_1}}{n} = \frac{(0.035 \text{ m})^{2/3} \sqrt{0.12}}{0.013}$$
$$= 2.85 \text{ m/s}$$
$$d_1 = (10 \text{ cm}) \left(\frac{1 \text{ m}}{100 \text{ cm}} \right) = 0.1 \text{ m}$$

$$d_2 = -0.5 d_1 + \sqrt{\frac{2 v_1 d_1}{g} + \frac{d_1^2}{4}}$$

$$= (-0.5)(0.1 \text{ m}) + \sqrt{\left(\frac{(2) \left(2.85 \ \frac{\text{m}}{\text{s}} \right)^2 (0.1 \text{ m})}{9.81 \ \frac{\text{m}}{\text{s}^2}} \right) + \frac{(0.1 \text{ m})^2}{4}}$$

$$= 0.20 \text{ m}$$

(C) This incorrect solution adds instead of subtracts the first term in the conjugate depth equation and uses the wrong equation. Other assumptions, definitions, and equations are unchanged from the correct solution.

$$R_1 = \frac{d_1 \cos \theta}{2}$$
$$= \frac{(10 \text{ cm}) \left(\frac{1 \text{ m}}{100 \text{ cm}} \right) (\cos 45^\circ)}{2}$$
$$= 0.035 \text{ m}$$
$$v_1 = \frac{R_1^{2/3} \sqrt{S_1}}{n} = \frac{(0.035 \text{ m})^{2/3} \sqrt{0.12}}{0.013}$$
$$= 2.85 \text{ m/s}$$
$$d_1 = (10 \text{ cm}) \left(\frac{1 \text{ m}}{100 \text{ cm}} \right) = 0.1 \text{ m}$$
$$d_2 = 0.5 d_1 + \sqrt{\frac{2 v_1^2 d_1}{g} + \frac{d_1^2}{4}}$$

$$= (0.5)(0.1 \text{ m}) + \sqrt{\left(\frac{(2) \left(2.85 \ \frac{\text{m}}{\text{s}} \right)^2 (0.1 \text{ m})}{9.81 \ \frac{\text{m}}{\text{s}^2}} \right) + \frac{(0.1 \text{ m})^2}{4}}$$

$$= 0.46 \text{ m}$$

(D) This incorrect solution makes a math error involving the exponents in the velocity equation and uses the wrong conjugate depth equation. Other assumptions, definitions, and equations are unchanged from the correct solution.

$$R_1 = \frac{d_1 \cos \theta}{2} = \frac{(10 \text{ cm}) \left(\frac{1 \text{ m}}{100 \text{ cm}} \right) (\cos 45^\circ)}{2}$$
$$= 0.035 \text{ m}$$
$$v_1 = \frac{R_1^{2/3} \sqrt{S_1}}{n} = \frac{(0.035 \text{ m})^{1/3} \sqrt{0.12}}{0.013}$$
$$= 8.7 \text{ m/s}$$
$$d_1 = (10 \text{ cm}) \left(\frac{1 \text{ m}}{100 \text{ cm}} \right)$$
$$= 0.1 \text{ m}$$

$$d_2 = -0.5d_1 + \sqrt{\frac{2v_1^2 d_1}{g} + \frac{d_1^2}{4}}$$

$$= (-0.5)(0.1 \text{ m}) + \sqrt{\left(\frac{(2)\left(8.7\ \frac{\text{m}}{\text{s}}\right)^2 (0.1 \text{ m})}{9.81\ \frac{\text{m}}{\text{s}^2}}\right) + \frac{(0.1 \text{ m})^2}{4}}$$

$$= 1.2 \text{ m}$$

SOLUTION 11

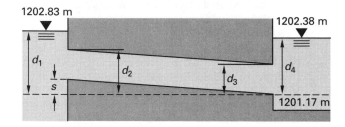

d_3 culvert diameter   m
L culvert length   m
S slope   m/m

Other terms are defined in the illustration. All have units of m.

$$d_1 = 1202.83 \text{ m} - 1201.17 \text{ m} = 1.66 \text{ m}$$

$$s = SL = \left(\frac{2 \text{ m}}{100 \text{ m}}\right)(27 \text{ m})$$

$$= 0.54 \text{ m}$$

$$d_2 = d_3 + s = 1 \text{ m} + 0.54 \text{ m}$$

$$= 1.54 \text{ m}$$

Because d_1 is greater than d_2, the culvert inlet is submerged.

$$d_4 = 1202.38 \text{ m} - 1201.17 \text{ m} = 1.21 \text{ m}$$

$$d_3 = 1 \text{ m}$$

Because d_4 is greater than d_3, the culvert outlet is submerged.

Submerged inlet and outlet classify the culvert flow as type-4, submerged outlet.

The answer is (B).

Why Other Options Are Wrong

(A) This incorrect solution confuses elevations used when calculating the inlet and outlet water depths, resulting in an exposed (not submerged) inlet and free

flow outlet, which classify the culvert flow as type-3, tranquil flow. Definitions are the same as used in the correct solution.

$$d_1 = 1202.38 \text{ m} - 1201.17 \text{ m} = 1.21 \text{ m}$$

$$s = SL = \left(\frac{2 \text{ m}}{100 \text{ m}}\right)(27 \text{ m})$$

$$= 0.54 \text{ m}$$

$$d_2 = D + s = 1 \text{ m} + 0.54 \text{ m}$$

$$= 1.54 \text{ m}$$

Because d_1 is less than d_2, the culvert inlet is not submerged.

$$d_4 = 1202.83 \text{ m} - 1202.38 \text{ m}$$

$$= 0.45 \text{ m}$$

$$d_3 = D$$

$$= 1 \text{ m}$$

Because d_4 is less than d_3, the culvert outlet is free flow.

(C) This incorrect solution checks the inlet conditions only. Submerged inlet classifies the culvert flow as type-5, rapid flow, at inlet. Definitions are the same as used in the correct solution.

$$d_1 = 1202.83 \text{ m} - 1201.17 \text{ m} = 1.66 \text{ m}$$

$$s = SL = \left(\frac{2 \text{ m}}{100 \text{ m}}\right)(27 \text{ m})$$

$$= 0.54 \text{ m}$$

$$d_2 = D + s = 1 \text{ m} + 0.54 \text{ m}$$

$$= 1.54 \text{ m}$$

Because d_1 is greater than d_2, the culvert inlet is submerged.

(D) This incorrect solution confuses the elevation values when calculating the outlet water depth, resulting in a submerged inlet and free-flow outlet, which classify the culvert flow as type-6, full flow, free outlet. Definitions are the same as used in the correct solution.

$$d_1 = 1202.83 \text{ m} - 1201.17 \text{ m} = 1.66 \text{ m}$$

$$s = SL = \left(\frac{2 \text{ m}}{100 \text{ m}}\right)(27 \text{ m})$$

$$= 0.54 \text{ m}$$

$$d_2 = D + s = 1 \text{ m} + 0.54 \text{ m}$$

$$= 1.54 \text{ m}$$

Because d_1 is greater than d_2, the culvert inlet is submerged.

$$d_4 = 1202.83 \text{ m} - 1201.38 \text{ m} = 0.45 \text{ m}$$

$$d_3 = D$$

$$= 1 \text{ m}$$

Because d_4 is less than d_3, the culvert outlet is free flow.

HYDROLOGY
SOLUTION 12

section	velocity at 0.2 depth (m/s)	velocity at 0.8 depth (m/s)	average velocity (m/s)
AB	–	–	–
BC	0.41	0.32	0.365
CD	0.44	0.32	0.38
DE	0.48	0.34	0.41
EF	0.48	0.33	0.405
FG	0.49	0.36	0.425
GH	0.49	0.35	0.42
HI	0.51	0.37	0.44
IJ	0.50	0.36	0.43
JK	0.52	0.37	0.445
KL	0.50	0.38	0.44
LM	0.49	0.35	0.42
MN	0.50	0.36	0.43
NO	0.47	0.34	0.405
OP	0.43	0.31	0.37
PQ	0.41	0.32	0.365
QR	0.39	0.30	0.345
RS	–	–	–

section	average depth (m)	section width (m)	flow area (m²)	flow rate (m³/s)
AB	0.7	1.2	0.84	–
BC	1.9	1.1	2.09	0.76
CD	2.3	1.3	2.99	1.14
DE	2.7	1.3	3.51	1.44
EF	2.9	1.2	3.48	1.41
FG	3.0	1.1	3.30	1.40
GH	3.1	1.1	3.41	1.43
HI	2.9	1.2	3.48	1.53
IJ	2.9	1.2	3.48	1.50
JK	2.8	1.4	3.92	1.74
KL	3.1	1.2	3.72	1.64
LM	2.8	1.1	3.08	1.29
MN	2.7	1.1	2.97	1.28
NO	2.5	1.3	3.25	1.32
OP	2.0	1.2	2.40	0.89
PQ	1.8	1.2	2.16	0.79
QR	1.6	1.2	1.92	0.66
RS	0.5	1.2	0.60	–
				$\overline{20.22}$

The tabulated calculations are from the following equations.

v section average flow velocity   m/s
v_2 section velocity at 0.2 depth   m/s
v_8 section velocity at 0.8 depth   m/s

$$v = \frac{v_2 + v_8}{2}$$

A section flow area m^2
d section average flow depth   m
w section flow width   m

$$A = dw$$

q section flow rate m^3/s

$$q = Av$$

Q total flow rate m^3/s

$$Q = \sum q$$
$$= 20.22 \ m^3/s \quad (20 \ m^3/s)$$

The answer is (B).

Why Other Options Are Wrong

(A) This incorrect solution calculates the average flow in each section instead of the total flow. The table and other definitions and equations are unchanged from the correct solution.

$$Q = \frac{20.22 \ \frac{m^3}{s}}{16}$$
$$= 1.26 \ m^3/s \quad (1.3 \ m^3/s)$$

(C) This incorrect solution uses the sum of the velocities at each depth fraction instead of calculating the average velocity. Other definitions and equations are unchanged from the correct solution.

section	velocity at 0.2 depth (m/s)	velocity at 0.8 depth (m/s)	average velocity (m/s)
AB	–	–	–
BC	0.41	0.32	0.73
CD	0.44	0.32	0.76
DE	0.48	0.34	0.82
EF	0.48	0.33	0.81
FG	0.49	0.36	0.85
GH	0.49	0.35	0.84
HI	0.51	0.37	0.88
IJ	0.50	0.36	0.86
JK	0.52	0.37	0.89
KL	0.50	0.38	0.88
LM	0.49	0.35	0.84
MN	0.50	0.36	0.86
NO	0.47	0.34	0.81
OP	0.43	0.31	0.74
PQ	0.41	0.32	0.73
QR	0.39	0.30	0.69
RS	–	–	–

section	average depth (m)	section width (m)	flow area (m²)	flow rate (m³/s)
AB	0.7	1.2	0.84	–
BC	1.9	1.1	2.09	1.53
CD	2.3	1.3	2.99	2.27
DE	2.7	1.3	3.51	2.88
EF	2.9	1.2	3.48	2.82
FG	3.0	1.1	3.30	2.81
GH	3.1	1.1	3.41	2.86
HI	2.9	1.2	3.48	3.06
IJ	2.9	1.2	3.48	2.99
JK	2.8	1.4	3.92	3.49
KL	3.1	1.2	3.72	3.27
LM	2.8	1.1	3.08	2.59
MN	2.7	1.1	2.97	2.55
NO	2.5	1.3	3.25	2.63
OP	2.0	1.2	2.40	1.78
PQ	1.8	1.2	2.16	1.58
QR	1.6	1.2	1.92	1.32
RS	0.5	1.2	0.60	–
				40.43

v   section total flow velocity   m/s

$$v = v_2 + v_8$$

$$Q = \sum q$$

$$= 40.43 \text{ m}^3/\text{s} \quad (40 \text{ m}^3/\text{s})$$

(D) This incorrect solution uses the total area and the total average velocity to calculate flow rate. Other definitions and equations are unchanged from the correct solution.

section	velocity at 0.2 depth (m/s)	velocity at 0.8 depth (m/s)	average velocity (m/s)
AB	–	–	–
BC	0.41	0.32	0.365
CD	0.44	0.32	0.38
DE	0.48	0.34	0.41
EF	0.48	0.33	0.405
FG	0.49	0.36	0.425
GH	0.49	0.35	0.42
III	0.51	0.37	0.44
IJ	0.50	0.36	0.43
JK	0.52	0.37	0.445
KL	0.50	0.38	0.44
LM	0.49	0.35	0.42
MN	0.50	0.36	0.43
NO	0.47	0.34	0.405
OP	0.43	0.31	0.37
PQ	0.41	0.32	0.365
QR	0.39	0.30	0.345
RS	–	–	–
			6.495

section	average depth (m)	section width (m)	flow area (m²)
AB	0.7	1.2	0.84
BC	1.9	1.1	2.09
CD	2.3	1.3	2.99
DE	2.7	1.3	3.51
EF	2.9	1.2	3.48
FG	3.0	1.1	3.30
GH	3.1	1.1	3.41
HI	2.9	1.2	3.48
IJ	2.9	1.2	3.48
JK	2.8	1.4	3.92
KL	3.1	1.2	3.72
LM	2.8	1.1	3.08
MN	2.7	1.1	2.97
NO	2.5	1.3	3.25
OP	2.0	1.2	2.40
PQ	1.8	1.2	2.16
QR	1.6	1.2	1.92
RS	0.5	1.2	0.60
			50.6

$$Q = \left(\sum v\right)\left(\sum A\right)$$

$$= \left(6.495 \frac{\text{m}}{\text{s}}\right)(50.6 \text{ m}^2)$$

$$= 329 \text{ m}^3/\text{s} \quad (330 \text{ m}^3/\text{s})$$

SOLUTION 13

d   rainfall depth   in
t   storm duration   hr

From the illustration, for mean annual precipitation of 27 in and storm duration of 2.5 hr, the rainfall depth is about 1.5 in.

i   rainfall intensity   in/hr

$$i = \frac{d}{t} = \frac{1.5 \text{ in}}{2.5 \text{ hr}}$$

$$= 0.60 \text{ in/hr}$$

The answer is (B).

Why Other Options Are Wrong

(A) This incorrect choice misreads the illustration by selecting a value for rainfall depth between the 20 in and 25 in mean annual precipitation. Definitions and equations are unchanged from the correct solution.

From the illustration, for the mean annual precipitation of 27 in and storm duration of 2.5 hr, the rainfall depth is about 1.0 in.

$$i = \frac{d}{t} = \frac{1.0 \text{ in}}{2.5 \text{ hr}}$$

$$= 0.40 \text{ in/hr}$$

(C) This incorrect choice assumes the rainfall intensity is equal to the rainfall depth.

From the illustration, for mean annual precipitation of 27 in and storm duration of 2.5 hr, the rainfall intensity is about 1.5 in/hr. Definitions are unchanged from the correct solution.

$$i = 1.5 \text{ in/hr}$$

(D) This incorrect choice takes the ratio of the mean annual precipitation and the storm duration as the rainfall intensity. Definitions are unchanged from the correct solution.

P_{ma} mean annual precipitation   in

$$i = \frac{P_{ma}}{d} = \frac{27 \text{ in}}{2.5 \text{ hr}}$$
$$= 10.8 \text{ in/hr} \quad (11 \text{ in/hr})$$

SOLUTION 14

Using the illustration, the 60 min duration and 5.8 in/hr intensity storm coordinates fall on the 2% frequency curve. A 2% frequency storm is one that occurs two times in every 100 years, or once every 50 years.

The answer is (D).

Why Other Options Are Wrong

(A) This choice would result if the 60 min duration 5.8 in/hr intensity storm was correctly identified with the 2% frequency curve, but incorrectly assumed 2% to be equivalent to 2 yr instead of 50 yr. This choice would also result if the 50% frequency storm were selected. A 50% frequency storm is one that occurs 50 times every 100 yr, or once every 2 yr.

(B) This choice would result from misreading the curve and selecting 10% for the storm frequency. A 10% storm is one that occurs 10 times in every 100 yr, or once every 10 yr.

(C) This incorrect choice would result if the 4% frequency curve were selected to represent the 60 min duration and 5.8 in/hr intensity storm. A 4% frequency storm is one that occurs four times in every 100 yr, or once every 25 yr.

SOLUTION 15

Q peak discharge m^3/s
Q_h hydrograph peak discharge m^3/s
Q_o measured peak discharge m^3/s
Q_{12} hydrograph discharge at 12 h m^3/s

From the illustration, at 12 h the discharge is 34 m^3/s and the gaging station measures 150 m^3/s. The unit hydrograph shows the peak occurring at 6 h with a discharge of 59 m^3/s.

$$\frac{Q}{Q_h} = \frac{Q_o}{Q_{12}}$$

$$Q = \frac{Q_h Q_o}{Q_{12}} = \frac{\left(59 \frac{m^3}{s}\right)\left(150 \frac{m^3}{s}\right)}{34 \frac{m^3}{s}}$$
$$= 260 \text{ m}^3/\text{s}$$

The answer is (D).

Why Other Options Are Wrong

(A) This incorrect choice takes the peak discharge from the unit hydrograph. Definitions are unchanged from the correct solution.

From the unit hydrograph,

$$Q = 59 \text{ m}^3/\text{s}$$

(B) This incorrect choice adds the measured discharge to the unit hydrograph discharge at 12 h. Definitions are unchanged from the correct solution.

$$Q = Q_h + Q_{12} = 150 \frac{m^3}{s} + 34 \frac{m^3}{s}$$
$$= 184 \text{ m}^3/\text{s} \quad (180 \text{ m}^3/\text{s})$$

(C) This incorrect choice adds the measured discharge to the peak discharge from the unit hydrograph. Definitions are unchanged from the correct solution.

$$Q = Q_o + Q_h = 150 \frac{m^3}{s} + 59 \frac{m^3}{s}$$
$$= 209 \text{ m}^3/\text{s} \quad (210 \text{ m}^3/\text{s})$$

SOLUTION 16

n period of interest   yr
P_F annual probability of a flood event   –
R acceptable risk of a flood event occurring   –

$$R = \frac{1\%}{100\%} = 0.01$$
$$P_F = 1 - (1)(1-R)^{1/n} = 1 - (1)(1-0.01)^{1/50}$$
$$= 0.00020 \quad (0.020\%)$$

The answer is (B).

Why Other Options Are Wrong

(A) This incorrect solution calculates the probability as a fraction instead of as percent, but expresses the result as percent. Definitions and equations are the same as used in the correct solution.

$$R = \frac{1\%}{100\%} = 0.01$$
$$P_F = 1 - (1)(1 - R)^{1/n}$$
$$= 1 - (1)(1 - 0.01)^{1/50}$$
$$= 0.00020\%$$

(C) This incorrect solution calculates the probability as 1% of the design life, expressing the result as a percent. Definitions and equations are the same as used in the correct solution.

$$R = \frac{1\%}{100\%} = 0.01$$
$$P_F = nR = (50)(0.01)$$
$$= 0.50\%$$

(D) This incorrect solution calculates the probability using the value for risk as a percent instead of as a fraction. Definitions and equations are the same as used in the correct solution.

$$P_F = 1 - (1)(1 - R)^{1/n}$$
$$= 1 - (1)(1 - 1)^{1/50}$$
$$= 1.0\%$$

SOLUTION 17

Assume runoff coefficients are the minimum values in the range of typical values listed in standard references.

land use	area (%)	runoff coefficient	land area (ac)
apartments	30	0.50	39.3
landscaped open space (park)	25	0.10	32.75
light industrial	45	0.50	58.95
			131

A land area for each use (subscript j designates each land use)   ac
C runoff coefficient for each land use (subscript ave designates average)   –

$$C_{\text{ave}} = \frac{\sum C_j A_j}{\sum A_j}$$
$$= \frac{\left(\begin{array}{c}(0.50)(39.3 \text{ ac}) + (0.10)(32.75 \text{ ac}) \\ + (0.50)(58.95 \text{ ac})\end{array}\right)}{131 \text{ ac}}$$
$$= 0.40$$

i storm intensity   in/hr

From the illustration, the storm intensity is about 6 in/hr.

Q runoff   ac-ft/hr

$$Q = C_{\text{ave}} i A = (0.40)\left(6 \, \frac{\text{in}}{\text{hr}}\right)(131 \text{ ac})\left(\frac{1 \text{ ft}}{12 \text{ in}}\right)$$
$$= 26 \text{ ac-ft/hr}$$

The answer is (B).

Why Other Options Are Wrong

(A) This incorrect solution calculates the arithmetic average instead of the weighted average of the runoff coefficients. Assumptions, definitions, and equations are unchanged from the correct solution.

n number of different land uses   –

$$C_{\text{ave}} = \frac{\sum C_j}{n} = \frac{0.50 + 0.10 + 0.50}{3}$$
$$= 0.37$$

From the illustration, the storm intensity is about 6 in/hr.

$$Q = CiA = (0.37)\left(6 \, \frac{\text{in}}{\text{hr}}\right)(131 \text{ ac})\left(\frac{1 \text{ ft}}{12 \text{ in}}\right)$$
$$= 24 \text{ ac-ft/hr}$$

(C) This incorrect solution calculates the runoff for the total area using the summed runoff coefficients. Other assumptions, definitions, and equations are unchanged from the correct solution.

C total runoff coefficient   –

$$C = \sum C_j = 0.50 + 0.10 + 0.50$$
$$= 1.1$$

From the illustration, the storm intensity is about 6 in/hr.

$$Q = CiA = (1.1)\left(6 \, \frac{\text{in}}{\text{hr}}\right)(131 \text{ ac})\left(\frac{1 \text{ ft}}{12 \text{ in}}\right)$$
$$= 72 \text{ ac-ft/hr}$$

(D) This incorrect solution misreads the units for intensity from the figure. Assumptions, definitions, and equations are unchanged from the correct solution.

land use	area (%)	runoff coefficient	land area (ac)
apartments	30	0.50	39.3
landscaped open space (park)	25	0.10	32.75
light industrial	45	0.50	58.95
			131

$$C_{\text{ave}} = \frac{\sum C_j A_j}{\sum A_j}$$

$$= \frac{\left(\begin{array}{c} (0.50)(39.3 \text{ ac}) + (0.10)(32.75 \text{ ac}) \\ + (0.50)(58.95 \text{ ac}) \end{array} \right)}{131 \text{ ac}}$$

$$= 0.40$$

i storm intensity   ft/hr

From the illustration, the storm intensity is about 6 ft/hr.

$$Q = C_{\text{ave}} i A$$

$$= (0.40) \left(6 \ \frac{\text{ft}}{\text{hr}} \right) (131 \text{ ac})$$

$$= 314 \text{ ac-ft/hr} \quad (310 \text{ ac-ft/hr})$$

WATER TREATMENT

SOLUTION 18

The typical average annual daily flow for planning purposes is 165 gal/person-day.

Q total annual flow   gal/yr

$$Q = (65{,}000 \text{ people}) \left(165 \ \frac{\text{gal}}{\text{person-day}} \right) \left(365 \ \frac{\text{day}}{\text{yr}} \right)$$

$$= 3.9 \times 10^9 \text{ gal/yr}$$

e annual electricity cost $

$$e = \left(3.9 \times 10^9 \ \frac{\text{gal}}{\text{yr}} \right) \left(\frac{\$0.12}{(10)^3 \text{ gal}} \right) (0.20)$$

$$= \$93{,}600/\text{yr} \quad (\$94{,}000/\text{yr})$$

The answer is (C).

Why Other Options Are Wrong

(A) The incorrect solution assumes an incorrect average annual daily flow of 100 gal/person-day. Definitions are unchanged from the correct solution.

$$Q = (65{,}000 \text{ people}) \left(100 \ \frac{\text{gal}}{\text{person-day}} \right) \left(365 \ \frac{\text{day}}{\text{yr}} \right)$$

$$= 2.4 \times 10^9 \text{ gal/yr}$$

$$e = \left(2.4 \times 10^9 \ \frac{\text{gal}}{\text{yr}} \right) \left(\frac{\$0.12}{(10)^3 \text{ gal}} \right) (0.20)$$

$$= \$57{,}600/\text{yr} \quad (\$58{,}000/\text{yr})$$

(B) This incorrect solution calculates the annual flow on the basis of a 5 day week instead of a 7 day week.

Definitions are unchanged from the correct solution.

$$Q = (65{,}000 \text{ people}) \left(165 \ \frac{\text{gal}}{\text{person-day}} \right)$$

$$\times \left(5 \ \frac{\text{day}}{\text{wk}} \right) \left(52 \ \frac{\text{wk}}{\text{yr}} \right)$$

$$= 2.8 \times 10^9 \text{ gal/yr}$$

$$e = \left(2.8 \times 10^9 \ \frac{\text{gal}}{\text{yr}} \right) \left(\frac{\$0.12}{(10)^3 \text{ gal}} \right) (0.20)$$

$$= \$67{,}200/\text{yr} \quad (\$67{,}000/\text{yr})$$

(D) This incorrect solution divides instead of multiplies by the percent attributable to electrical power. Definitions are unchanged from the correct solution.

The typical average annual daily flow for planning purposes is 165 gal/person-day.

$$Q = (65{,}000 \text{ people}) \left(165 \ \frac{\text{gal}}{\text{person-day}} \right) \left(365 \ \frac{\text{day}}{\text{yr}} \right)$$

$$= 3.9 \times 10^9 \text{ gal/yr}$$

$$e = \frac{\left(3.9 \times 10^9 \ \frac{\text{gal}}{\text{yr}} \right) \left(\frac{\$0.12}{(10)^3 \text{ gal}} \right)}{0.20}$$

$$= \$2{,}340{,}000/\text{yr} \quad (\$2{,}300{,}000/\text{yr})$$

SOLUTION 19

n	number of filters	—
Q	total flow rate to filters	m³/d
q_n	flow rate per filter for all filters operating in filtration mode	m³/d

For gross production per filter,

$$q_n = \frac{Q}{n} = \frac{28\,500 \ \frac{\text{m}^3}{\text{d}}}{5 \text{ filters}} = 5700 \text{ m}^3/\text{d}$$

m	number of daily backwash events/filter	1/d
t_b	time for backwash/filter	h
t_c	time for conditioning/filter	h
t_n	daily time when all filters are operating in filtration mode	h/d

$$t_n = 24 \ \frac{\text{h}}{\text{d}} - m(t_b + t_c)n$$

$$= 24 \ \frac{\text{h}}{\text{d}} - \left(\frac{2}{\text{d}} \right) \left(25 \ \frac{\text{min}}{\text{filter}} + 8 \ \frac{\text{min}}{\text{filter}} \right)$$

$$\times \left(\frac{1 \text{ h}}{60 \text{ min}} \right) (5 \text{ filters})$$

$$= 18.5 \text{ h/d}$$

q_{n-1} flow rate per filter when one filter
is backwashing m^3/d

$$q_{n-1} = \frac{Q}{n-1} = \frac{28\,500\,\dfrac{m^3}{d}}{4\text{ filters}}$$
$$= 7125\text{ m}^3/\text{d}$$

t_{n-1} daily time when all filters are
backwashing and conditioning   h/d

$$t_{n-1} = m(t_b + t_c)(n)$$
$$= \left(\frac{2}{d}\right)\left(25\,\frac{\min}{\text{filter}} + 8\,\frac{\min}{\text{filter}}\right)$$
$$\times \left(\frac{1\text{ h}}{60\text{ min}}\right)(5\text{ filters})$$
$$= 5.5\text{ h/d}$$

Q_g gross production per filter m^3/d

$$Q_g = \frac{q_n t_n + q_{n-1} t_{n-1}}{24\,\dfrac{h}{d}}$$
$$= \frac{\left(5700\,\dfrac{m^3}{d}\right)\left(18.5\,\dfrac{h}{d}\right) + \left(7125\,\dfrac{m^3}{d}\right)\left(5.5\,\dfrac{h}{d}\right)}{24\,\dfrac{h}{d}}$$
$$= 6027\text{ m}^3/\text{filter·d}$$

For water lost to backwash,

A filter bed area m^2
v_s filtration rate m^3/m^2·d

$$A = \frac{Q}{nv_s} = \frac{28\,500\,\dfrac{m^3}{d}}{(5\text{ filters})\left(225\,\dfrac{m^3}{m^2\cdot d}\right)}$$
$$= 25\text{ m}^2/\text{filter}$$

Q_b backwash wastewater flow rate $m^3/filter$·d
v_b backwash rate m^3/m^2·d

$$Q_b = v_b t_b Am$$
$$= \left(36\,\frac{m^3}{m^2\cdot h}\right)(25\text{ min})\left(25\,\frac{m^2}{\text{filter}}\right)$$
$$\times \left(\frac{2}{d}\right)\left(\frac{h}{60\text{ min}}\right)$$
$$= 750\text{ m}^3/\text{filter·d}$$

For water lost to conditioning,

$$Q_c = \frac{Qmt_c}{n} = \frac{\left(28\,500\,\dfrac{m^3}{d}\right)\left(\dfrac{2}{d}\right)(8\text{ min})}{(5\text{ filters})\left(\dfrac{1440\text{ min}}{d}\right)}$$
$$= 63\text{ m}^3/\text{filter·d}$$

Q_{net} net production per filter m^3/d

$$Q_{net} = Q_g - Q_b - Q_c$$
$$= 6027\,\frac{m^3}{d} - 750\,\frac{m^3}{d} - 63\,\frac{m^3}{d}$$
$$= 5214\text{ m}^3/\text{d}\quad(5200\text{ m}^3/\text{d})$$

The answer is (B).

Why Other Options Are Wrong

(A) This incorrect solution does not consider the increased filtration rate to the remaining filters when one filter is being backwashed. Other definitions and equations are unchanged from the correct solution.

For gross production per filter,

$$q_n = \frac{Q}{n} = \frac{28\,500\,\dfrac{m^3}{d}}{5\text{ filters}}$$
$$= 5700\text{ m}^3/\text{filter·d}$$

For water lost to backwash,

$$A = \frac{Q}{nv_s} = \frac{28\,500\,\dfrac{m^3}{d}}{(5\text{ filters})\left(225\,\dfrac{m^3}{m^2\cdot d}\right)}$$
$$= 25\text{ m}^2/\text{filter}$$
$$Q_b = v_b t_b Am$$
$$= \frac{\left(36\,\dfrac{m^3}{m^2\cdot h}\right)(25\text{ min})\left(25\,\dfrac{m^2}{\text{filter}}\right)\left(\dfrac{2}{d}\right)}{60\,\dfrac{\min}{h}}$$
$$= 750\text{ m}^3/\text{filter·d}$$

For water lost to conditioning,

$$Q_c = \frac{Qmt_c}{n} = \frac{\left(28\,500\,\dfrac{m^3}{d}\right)\left(\dfrac{2}{d}\right)(8\text{ min})}{(5\text{ filters})\left(\dfrac{1440\text{ min}}{d}\right)}$$
$$= 63\text{ m}^3/\text{filter·d}$$
$$Q_{net} = q_n - Q_b - Q_c$$
$$= 5700\,\frac{m^3}{d} - 750\,\frac{m^3}{d} - 63\,\frac{m^3}{d}$$
$$= 4887\text{ m}^3/\text{d}\quad(4900\text{ m}^3/\text{d})$$

(C) This incorrect solution does not consider the increased filtration rate to the remaining filters when one filter is being backwashed and accounts for only one backwash event per filter per day. Other definitions and equations are unchanged from the correct solution.

For gross production per filter,

$$q_n = \frac{Q}{n} = \frac{28\,500 \, \frac{\text{m}^3}{\text{d}}}{5 \text{ filters}}$$
$$= 5700 \text{ m}^3/\text{filter·d}$$

For water lost to backwash,

$$A = \frac{Q}{nv_s} = \frac{28\,500 \, \frac{\text{m}^3}{\text{d}}}{(5 \text{ filters})\left(225 \, \frac{\text{m}^3}{\text{m}^2 \cdot \text{d}}\right)}$$
$$= 25 \text{ m}^2/\text{filter}$$
$$Q_b = v_b t_b A m$$
$$= \frac{\left(36 \, \frac{\text{m}^3}{\text{m}^2 \cdot \text{h}}\right)(25 \text{ min})\left(25 \, \frac{\text{m}^2}{\text{filter}}\right)\left(\frac{1}{\text{d}}\right)}{60 \, \frac{\text{min}}{\text{h}}}$$
$$= 375 \text{ m}^3/\text{filter·d}$$

For water lost to conditioning,

$$Q_c = \frac{Q t_c}{n} = \frac{\left(28\,500 \, \frac{\text{m}^3}{\text{d}}\right)\left(8 \, \frac{\text{min}}{\text{d}}\right)}{(5 \text{ filters})\left(1440 \, \frac{\text{min}}{\text{d}}\right)}$$
$$= 32 \text{ m}^3/\text{filter·d}$$
$$Q_{\text{net}} = q_n - Q_b - Q_c$$
$$= 5700 \, \frac{\text{m}^3}{\text{d}} - 375 \, \frac{\text{m}^3}{\text{d}} - 32 \, \frac{\text{m}^3}{\text{d}}$$
$$= 5293 \text{ m}^3/\text{d} \quad (5300 \text{ m}^3/\text{d})$$

(D) This incorrect solution fails to account for two backwash events per day. Other definitions and equations are unchanged from the correct solution.

For gross production per filter,

$$q_n = \frac{Q}{n} = \frac{28\,500 \, \frac{\text{m}^3}{\text{d}}}{5 \text{ filters}}$$
$$= 5700 \, \frac{\text{m}^3}{\text{d}}$$
$$t_n = 24 \, \frac{\text{h}}{\text{d}} - (t_b + t_c)n$$
$$= 24 \, \frac{\text{h}}{\text{d}} - \left(25 \, \frac{\text{min}}{\text{filter·d}} + 8 \, \frac{\text{min}}{\text{filter·d}}\right)$$
$$\times \left(\frac{1 \text{ h}}{60 \text{ min}}\right)(5 \text{ filters})$$
$$= 21.25 \text{ h/d}$$

$$q_{n-1} = \frac{Q}{n-1} = \frac{28\,500 \, \frac{\text{m}^3}{\text{d}}}{4 \text{ filters}}$$
$$= 7125 \text{ m}^3/\text{d}$$
$$t_{n-1} = (t_b + t_c)(n)$$
$$= \left(25 \, \frac{\text{min}}{\text{filter}} + 8 \, \frac{\text{min}}{\text{filter}}\right)\left(\frac{1 \text{ h}}{60 \text{ min}}\right)(5 \text{ filters})$$
$$= 2.75 \text{ h/d}$$
$$Q_g = \frac{q_n t_n + q_{n-1} t_{n-1}}{24 \, \frac{\text{h}}{\text{d}}}$$
$$= \frac{\left(\left(5700 \, \frac{\text{m}^3}{\text{d}}\right)\left(21.25 \, \frac{\text{h}}{\text{d}}\right) + \left(7125 \, \frac{\text{m}^3}{\text{d}}\right)\left(2.75 \, \frac{\text{h}}{\text{d}}\right)\right)}{24 \, \frac{\text{h}}{\text{d}}}$$
$$= 5863 \text{ m}^3/\text{filter·d}$$

For water lost to backwash,

$$A = \frac{Q}{nv_s} = \frac{28\,500 \, \frac{\text{m}^3}{\text{d}}}{(5 \text{ filters})\left(225 \, \frac{\text{m}^3}{\text{m}^2 \cdot \text{d}}\right)}$$
$$= 25 \text{ m}^2/\text{filter}$$
$$Q_b = v_b t_b A$$
$$= \frac{\left(36 \, \frac{\text{m}^3}{\text{m}^2 \cdot \text{h}}\right)\left(25 \, \frac{\text{min}}{\text{d}}\right)\left(25 \, \frac{\text{m}^2}{\text{filter}}\right)}{60 \, \frac{\text{min}}{\text{h}}}$$
$$= 375 \text{ m}^3/\text{filter·d}$$

For water lost to conditioning,

$$Q_c = \frac{Q t_c}{n} = \frac{\left(28\,500 \, \frac{\text{m}^3}{\text{d}}\right)\left(8 \, \frac{\text{min}}{\text{d}}\right)}{(5 \text{ filters})\left(1440 \, \frac{\text{min}}{\text{d}}\right)}$$
$$= 32 \text{ m}^3/\text{filter·d}$$
$$Q_{\text{net}} = Q_g - Q_b - Q_c$$
$$= 5863 \, \frac{\text{m}^3}{\text{d}} - 375 \, \frac{\text{m}^3}{\text{d}} - 32 \, \frac{\text{m}^3}{\text{d}}$$
$$= 5456 \text{ m}^3/\text{d} \quad (5500 \text{ m}^3/\text{d})$$

SOLUTION 20

The minimum required tank capacity is determined using the slope of the average flow. This is determined by drawing the average flow line on the illustration and then drawing lines parallel to the average flow line that are also tangent to the maximum deviations from the average flow line. The vertical separation of the parallel lines represents the minimum required storage volume. Passing a line through the ordinate to the cumulative volume at 24 h draws the average flow line.

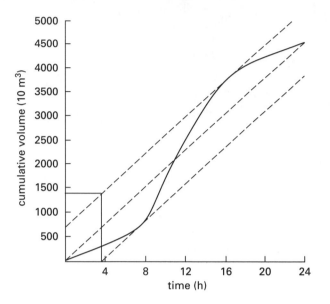

V storage volume   m^3

From the illustration, the vertical separation of the tangent parallel lines is

$$V = 13\,000 \text{ m}^3$$

The answer is (C).

Why Other Options Are Wrong

(A) This incorrect solution fails to multiply the required volume by 10 m^3 from the y-axis scale. The illustration and definitions are unchanged from the correct solution.

From the illustration, the vertical separation of the tangent parallel lines is

$$V = 1300 \text{ m}^3$$

(B) This incorrect solution takes the required storage volume as the maximum deviation from the average flow line. Definitions are unchanged from the correct solution.

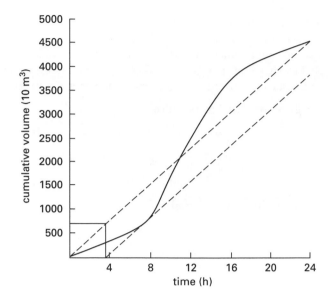

From the illustration, the vertical separation of the tangent parallel line and the average slope is

$$V = 6500 \text{ m}^3$$

(D) This incorrect solution assumes that the required tank capacity is equal to the total cumulative flow. Definitions are unchanged from the correct solution.

From the illustration, the total daily cumulative flow is

$$V = 45\,000 \text{ m}^3$$

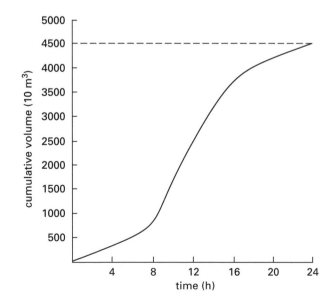

Depth Solutions

HYDRAULICS

SOLUTION 21

d_1 upstream water depth   m
Q water flow m^3/s
v_1 upstream water velocity   m/s
w channel width   m

$$v_1 = \frac{Q}{d_1 w} = \frac{2.93 \; \frac{m^3}{s}}{(0.15 \text{ m})(2.5 \text{ m})} = 7.8 \text{ m/s}$$

d_2 downstream water depth   m
g gravitational acceleration 9.81 m/s^2
v_2 downstream water velocity   m/s

$$d_2 = -(0.5)d_1 + \sqrt{\frac{2v_1^2 d_1}{g} + 0.25 d_1^2}$$

$$= -(0.5)(0.15 \text{ m}) + \sqrt{\left(\frac{(2)\left(7.8\frac{m}{s}\right)^2 (0.15 \text{ m})}{9.81 \frac{m}{s^2}}\right) + (0.25)(0.15\text{m})^2}$$

$$= 1.29 \text{ m}$$

$$v_2 = \frac{Q}{d_2 w} = \frac{2.93 \; \frac{m^3}{s}}{(1.29 \text{ m})(2.5 \text{ m})}$$
$$= 0.91 \text{ m/s}$$

ΔH head dissipated   m

$$\Delta H = d_1 + \frac{v_1^2}{g} - d_2 - \frac{v_2^2}{2g}$$
$$= 0.15 \text{ m}$$
$$+ \frac{\left(7.8\frac{m}{s}\right)^2}{(2)\left(9.81\frac{m}{s^2}\right)} - 1.29 \text{ m} - \frac{\left(0.91\frac{m}{s}\right)^2}{(2)\left(9.81\frac{m}{s^2}\right)}$$
$$= 1.92 \text{ m} \quad (1.9 \text{ m})$$

The answer is (C).

Why Other Options Are Wrong

(A) This incorrect solution fails to take the square root in the downstream depth calculation. Other definitions and equations are unchanged from the correct solution.

$$v_1 = \frac{Q}{d_1 w} = \frac{2.93 \; \frac{m^3}{s}}{(0.15 \text{ m})(2.5 \text{ m})}$$
$$= 7.8 \text{ m/s}$$

$$d_2 = -(0.5)d_1 + \frac{2v_1^2 d_1}{g} + 0.25 d_1^2$$

$$= -(0.5)(0.15 \text{ m}) + \frac{(2)\left(7.8\frac{m}{s}\right)^2 (0.15 \text{ m})}{9.81 \frac{m}{s^2}}$$

$$+ (0.25)(0.15 \text{ m})^2$$
$$= 1.79 \text{ m}$$

The units do not work.

$$v_2 = \frac{Q}{d_2 w} = \frac{2.93 \; \frac{m^3}{s}}{(1.79 \text{ m})(2.5 \text{ m})}$$
$$= 0.65 \text{ m/s}$$

$$\Delta H = d_1 + \frac{v_1^2}{2g} - d_2 - \frac{v_2^2}{2g}$$
$$= 0.15 \text{ m} + \frac{\left(7.8\frac{m}{s}\right)^2}{(2)\left(9.81\frac{m}{s^2}\right)}$$
$$- 1.79 \text{ m} - \frac{\left(0.65\frac{m}{s}\right)^2}{(2)\left(9.81\frac{m}{s^2}\right)}$$
$$= 1.44 \text{ m} \quad (1.4 \text{ m})$$

(B) This incorrect solution adds instead of subtracts the upstream depth term in the downstream depth calculation. Other definitions and equations are unchanged from the correct solution.

$$v_1 = \frac{Q}{d_1 w} = \frac{2.93 \; \frac{m^3}{s}}{(0.15 \text{ m})(2.5 \text{ m})} = 7.8 \text{ m/s}$$

$$d_2 = (0.5)d_1 + \sqrt{\frac{2v_1^2 d_1}{g} + 0.25d_1^2}$$

$$= (0.5)(0.15 \text{ m}) + \sqrt{\left(\frac{(2)\left(7.8 \frac{\text{m}}{\text{s}}\right)^2 (0.15 \text{ m})}{9.81 \frac{\text{m}}{\text{s}^2}} \right) + (0.25)(0.15 \text{ m})^2}$$

$$= 1.44 \text{ m}$$

$$v_2 = \frac{Q}{d_2 w} = \frac{2.93 \frac{\text{m}^3}{\text{s}}}{(1.44 \text{ m})(2.5 \text{ m})}$$

$$= 0.81 \text{ m/s}$$

$$\Delta H = d_1 + \frac{v_1^2}{2g} - d_2 - \frac{v_2^2}{2g}$$

$$= 0.15 \text{ m} + \frac{\left(7.8 \frac{\text{m}}{\text{s}}\right)^2}{(2)\left(9.81 \frac{\text{m}}{\text{s}^2}\right)} - 1.44 \text{ m}$$

$$- \frac{\left(0.81 \frac{\text{m}}{\text{s}}\right)^2}{(2)\left(9.81 \frac{\text{m}}{\text{s}^2}\right)}$$

$$= 1.78 \text{ m} \quad (1.8 \text{ m})$$

(D) This incorrect solution reverses the values for upstream and downstream depths in the energy dissipation equation. Other definitions and equations are unchanged from the correct solution.

$$v_1 = \frac{Q}{d_1 w} = \frac{2.93 \frac{\text{m}^3}{\text{s}}}{(0.15 \text{ m})(2.5 \text{ m})} = 7.8 \text{ m/s}$$

$$d_2 = -(0.5)d_1 + \sqrt{\frac{2v_1^2 d_1}{g} + 0.25d_1^2}$$

$$= -(0.5)(0.15 \text{ m}) + \sqrt{\left(\frac{(2)\left(7.8 \frac{\text{m}}{\text{s}}\right)^2 (0.15 \text{ m})}{9.81 \frac{\text{m}}{\text{s}^2}} \right) + (0.25)(0.15 \text{ m})^2}$$

$$= 1.29 \text{ m}$$

$$v_2 = \frac{Q}{d_2 w} = \frac{2.93 \frac{\text{m}^3}{\text{s}}}{(1.29 \text{ m})(2.5 \text{ m})} = 0.91 \text{ m/s}$$

$$\Delta H = d_2 + \frac{v_1^2}{2g} - d_1 - \frac{v_2^2}{2g}$$

$$= 1.29 \text{ m} + \frac{\left(7.8 \frac{\text{m}}{\text{s}}\right)^2}{(2)\left(9.81 \frac{\text{m}}{\text{s}^2}\right)} - 0.15 \text{ m}$$

$$- \frac{\left(0.91 \frac{\text{m}}{\text{s}}\right)^2}{(2)\left(9.81 \frac{\text{m}}{\text{s}^2}\right)}$$

$$= 4.2 \text{ m}$$

SOLUTION 22

Find the theoretical flow velocity at the channel discharge point. Note that the total head at the discharge point is kinetic head.

g gravitational acceleration   32.2 ft/sec^2
h upstream head   ft
v flow velocity   ft/sec

$$v = \sqrt{2gh} = \sqrt{(2)\left(32.2 \frac{\text{ft}}{\text{sec}^2}\right)(32 \text{ ft})}$$

$$= 45.4 \text{ ft/sec}$$

The criterion for baffled outlets usually sets the maximum flow velocity at 50 ft/sec. (See *Design of Small Canal Structures.* Denver, Colo.: U.S. Department of the Interior, 1978.) Because 45.4 ft/sec is less than 50 ft/sec, a baffled outlet can be used.

Calculate the channel cross-sectional area.

A_{ch} discharge channel cross-sectional area   ft^2
Q flow rate   ft^3/sec

$$A_{ch} = \frac{Q}{v} = \frac{120 \frac{\text{ft}^3}{\text{sec}}}{45.4 \frac{\text{ft}}{\text{sec}}} = 2.643 \text{ ft}^2$$

D_{ch} flow depth in channel at outlet   ft

For a square channel,

$$A_{ch} = D_{ch}^2$$
$$D_{ch} = \sqrt{A_{ch}} = \sqrt{2.643 \text{ ft}^2} = 1.626 \text{ ft}$$

Use the Froude number to relate the channel depth to the basic depth.

Fr   Froude number   –

$$\text{Fr} = \frac{v}{\sqrt{gD_{ch}}} = \frac{45.4 \frac{\text{ft}}{\text{sec}}}{\sqrt{\left(32.2 \frac{\text{ft}}{\text{sec}^2}\right)(1.626 \text{ ft})}}$$

$$= 6.274$$

A_b baffled outlet basin cross-sectional area   ft^2
D_b baffled outlet basin depth   ft
w baffled outlet basin width   ft

$$\text{Fr}wD_b\sqrt{gD_b} = Q$$

$$D_b^{3/2} = \frac{Q}{\text{Fr}W\sqrt{g}}$$

$$D_b = \left(\frac{Q}{\text{Fr}W\sqrt{g}}\right)^{2/3}$$

$$= \left(\frac{\left(120\ \dfrac{\text{ft}^3}{\text{sec}}\right)}{(6.274)(6\ \text{ft})\sqrt{32.2\ \dfrac{\text{ft}}{\text{sec}^2}}}\right)^{2/3}$$

$$= 0.68\ \text{ft}$$

The answer is (B).

Why Other Options Are Wrong

(A) This incorrect solution divides the area of the channel by the basin width to get basin depth. Other assumptions, definitions, and equations are unchanged from the correct solution.

$$v = \sqrt{2gh} = \sqrt{(2)\left(32.2\ \frac{\text{ft}}{\text{sec}^2}\right)(32\ \text{ft})}$$

$$= 45.4\ \text{ft/sec}$$

The criterion for baffled outlets usually sets the maximum flow velocity at 50 ft/sec. Because 45.5 ft/sec is less than 50 ft/sec, a baffled outlet can be used.

$$A_{ch} = \frac{Q}{v} = \frac{120\ \dfrac{\text{ft}^2}{\text{sec}}}{45.4\ \dfrac{\text{ft}}{\text{sec}}} = 2.643\ \text{ft}^2$$

$$D_b = \frac{A_{ch}}{w} = \frac{2.643\ \text{ft}^2}{6\ \text{ft}} = 0.44\ \text{ft}$$

(C) This incorrect solution does not take the square root of the denominator term of the Froude number equation. Other assumptions, definitions, and equations are unchanged from the correct solution.

$$v = \sqrt{2gh} = \sqrt{(2)\left(32.2\ \frac{\text{ft}}{\text{sec}^2}\right)(32\ \text{ft})}$$

$$= 45.4\ \text{ft/sec}$$

The criterion for baffled outlets usually sets the maximum flow velocity at 50 ft/sec. Because 45.5 ft/sec is less than 50 ft/sec, a baffled outlet can be used.

$$A_{ch} = \frac{Q}{v} = \frac{120\ \dfrac{\text{ft}^2}{\text{sec}}}{45.5\ \dfrac{\text{ft}}{\text{sec}}} = 2.643\ \text{ft}^2$$

For a square channel,

$$A_{ch} = D_{ch}^2$$

$$D_{ch} = \sqrt{A_{ch}} = \sqrt{2.643\ \text{ft}^2}$$

$$= 1.626\ \text{ft}$$

$$\text{Fr} = \frac{v}{gD_{ch}} = \frac{45.4\ \dfrac{\text{ft}}{\text{sec}}}{\left(32.2\ \dfrac{\text{ft}}{\text{sec}^2}\right)(1.626\ \text{ft})}$$

$$= 0.867\ \text{sec/ft}$$

The units do not cancel, but because the Froude number should be unitless, the units are ignored.

$$A_b = wD_b$$

$$\text{Fr} = \frac{v}{gD} = \frac{Q}{A_bgD_b} = \frac{Q}{wD_bgD_b}$$

$$0.867 = \frac{120\ \dfrac{\text{ft}^2}{\text{sec}}}{(6\ \text{ft})(D_b)\left(32.2\ \dfrac{\text{ft}}{\text{sec}^2}\right)(D_b)}$$

$$(6\ \text{ft})(D_b)\left(32.2\ \frac{\text{ft}}{\text{sec}^2}\right)(D_b) = \frac{120\ \dfrac{\text{ft}^3}{\text{sec}}}{0.867}$$

$$\left(193\ \frac{\text{ft}}{\text{sec}^2}\right)(D_b^2) = 138\ \text{ft}^3/\text{sec}$$

$$D_b^2 = 0.72\ \text{ft/sec}$$

Ignore units.

$$D_b = 0.85\ \text{ft}$$

Assume units are feet.

(D) This incorrect solution does not multiply by two in the velocity equation. Definitions, assumptions, and other equations are unchanged from the correct solution.

$$v = \sqrt{gh} = \sqrt{\left(32.2\ \frac{\text{ft}}{\text{sec}^2}\right)(32\ \text{ft})}$$

$$= 32.1\ \text{ft/sec}$$

The criterion for baffled outlets usually sets the maximum flow velocity at 50 ft/sec. Because 32.1 ft/sec is less than 50 ft/sec, a baffled outlet can be used.

$$A_{ch} = \frac{Q}{v} = \frac{120\ \dfrac{\text{ft}^2}{\text{sec}}}{32.1\ \dfrac{\text{ft}}{\text{sec}}} = 3.738\ \text{ft}^2$$

For a square channel,

$$A_{ch} = D_{ch}^2$$

$$D_{ch} = \sqrt{A_{ch}} = \sqrt{3.738 \text{ ft}^2} = 1.933 \text{ ft}$$

$$\text{Fr} = \frac{v}{\sqrt{gD_{ch}}} = \frac{32.1 \dfrac{\text{ft}}{\text{sec}}}{\sqrt{\left(32.2 \dfrac{\text{ft}}{\text{sec}^2}\right)(1.933 \text{ ft})}}$$

$$= 4.069$$

$$A_b = wD_b$$

$$\text{Fr} = \frac{v}{\sqrt{gD}} = \frac{Q}{A_b\sqrt{gD_b}} = \frac{Q}{wD_b\sqrt{gD_b}}$$

$$4.069 = \frac{120 \dfrac{\text{ft}^3}{\text{sec}}}{(6 \text{ ft})(D_b)\sqrt{\left(32.2 \dfrac{\text{ft}}{\text{sec}^2}\right)(D_b)}}$$

$$(6 \text{ ft})(D_b)\sqrt{\left(32.2 \dfrac{\text{ft}}{\text{sec}^2}\right)(D_b)} = \frac{120 \dfrac{\text{ft}^3}{\text{sec}}}{4.069}$$

$$(6 \text{ ft})^2(D_b)^2 \left(32.2 \dfrac{\text{ft}}{\text{sec}^2}\right)(D_b) = \left(\frac{29.5 \text{ ft}^3}{\text{sec}}\right)^2$$

$$\left(1159 \dfrac{\text{ft}^3}{\text{sec}^2}\right)(D_b)^3 = 870 \text{ ft}^6/\text{sec}^2$$

$$D_b^3 = 0.75 \text{ ft}^3$$

$$D_b = 0.91 \text{ ft}$$

SOLUTION 23

For a low-head siphon spillway, the orifice equation can be applied to determine the approximate required throat diameter. Because the length of low-head siphons is relatively short compared to the throat diameter, any head loss through the siphon can be assumed to be negligible.

A_o throat cross-sectional area   ft^2
D siphon throat diameter   ft

$$A_o = \pi\frac{D^2}{4}$$

C_d siphon entrance coefficient   –
g gravitational acceleration   32.2 ft/sec^2
h operating head   ft
Q discharge flow   ft^3/sec

$$Q = C_d A_o \sqrt{2gh}$$
$$= \frac{C_d \pi D^2 \sqrt{2gh}}{4}$$

$$D = \sqrt{\frac{4Q}{C_d \pi \sqrt{2gh}}}$$

$$= \sqrt{\frac{(4)\left(90 \dfrac{\text{ft}^3}{\text{sec}}\right)}{0.6\pi\sqrt{(2)\left(32.2 \dfrac{\text{ft}}{\text{sec}^2}\right)(4 \text{ ft})}}}$$

$$= 3.45 \text{ ft} \quad (3.5 \text{ ft})$$

Because low-head siphons operate at negative pressure, the limitation of sub-atmospheric pressure should be checked. The vortex flow equation can be used for this purpose.

R_c radius to throat centerline   ft
R_s radius to throat diameter   ft

$$R_s = R_c + \frac{D}{2}$$

$$R_s = 2 \text{ ft} + \frac{3.45 \text{ ft}}{2} = 3.725 \text{ ft}$$

p_a atmospheric pressure head   ft

$$Q \le R_c \sqrt{(0.7)(2gp_a)} \ln \frac{R_s}{R_c}$$

$$= 2 \text{ ft}\sqrt{(0.7)(2)\left(32.2 \dfrac{\text{ft}}{\text{sec}^2}\right)(34 \text{ ft})} \ln\left(\frac{3.725 \text{ ft}}{2 \text{ ft}}\right)$$

$$= 48.7 \frac{\text{ft}^3}{\text{sec}}$$

$$48.7 \text{ ft}^3/\text{sec} < 90 \text{ ft}^3/\text{sec} \quad [\text{OK}]$$

The answer is (D).

Why Other Options Are Wrong

(A) This incorrect solution fails to take the square root of the head and gravity constant term. Other assumptions, definitions, and equations are the same as used in the correct solution.

$$A_o = \pi\frac{D^2}{4}$$

$$Q = C_d A_o \sqrt{2gh} = \frac{C_d \pi D^2 \sqrt{2gh}}{4}$$

$$D = \sqrt{\frac{4Q}{C_d \pi 2gh}}$$

$$= \sqrt{\frac{(4)\left(90 \dfrac{\text{ft}^3}{\text{sec}}\right)}{0.6\pi(2)\left(32.2 \dfrac{\text{ft}}{\text{sec}^2}\right)(4 \text{ ft})}}$$

$$= 0.861 \text{ ft}^{1/2} \text{ sec}^{1/2}$$

The units do not work. Ignore incorrect units and use units of feet in following calculations.

$$R_s = R_c + \frac{D}{2}$$

$$R_s = 2 \text{ ft} + \frac{0.861 \text{ ft}}{2}$$

$$= 2.431 \text{ ft}$$

$$Q \leq R_c \sqrt{(0.7)(2gp_a)} \ln \frac{R_s}{R_c}$$

$$= 2 \text{ ft} \sqrt{\begin{array}{c}(0.7)(2)\left(32.2 \dfrac{\text{ft}}{\text{sec}^2}\right) \\ \times (34 \text{ ft})\end{array}}$$

$$\times \ln \left(\frac{2.431 \text{ ft}}{2 \text{ ft}}\right)$$

$$= 15.3 \frac{\text{ft}^3}{\text{sec}}$$

$$15.3 \text{ ft}^3/\text{sec} < 90 \text{ ft}^3/\text{sec} \quad [\text{OK}]$$

(B) This incorrect solution takes the square root of the operating head only instead of the head and gravity constant term. Other assumptions, definitions, and equations are the same as used in the correct solution.

$$A_o = \pi \frac{D^2}{4}$$

$$Q = C_d A_o \sqrt{2gh} = \frac{C_d \pi D^2 \sqrt{2gh}}{4}$$

$$D = \sqrt{\frac{4Q}{C_d \pi 2g\sqrt{h}}}$$

$$= \sqrt{\frac{(4)\left(90 \dfrac{\text{ft}^3}{\text{sec}}\right)}{0.6\pi(2)\left(32.2 \dfrac{\text{ft}}{\text{sec}^2}\right)\sqrt{(4 \text{ ft})}}}$$

$$= 1.22 \text{ ft}^{3/4} \text{ sec}^{1/2} \quad \left(1.2 \text{ ft}^{3/4} \text{ sec}^{1/2}\right)$$

The units do not work. Ignore incorrect units and use units of feet in following calculations.

$$R_s = R_c + \frac{D}{2}$$

$$R_s = 2 \text{ ft} + \frac{1.22 \text{ ft}}{2} = 2.61 \text{ ft}$$

$$Q \leq R_c \sqrt{(0.7)(2gp_a)} \ln \left(\frac{R_s}{R_c}\right)$$

$$= (2 \text{ ft}) \sqrt{(0.7)(2)\left(32.2 \dfrac{\text{ft}}{\text{sec}^2}\right)(34 \text{ ft})}$$

$$\times \ln \left(\frac{2.61 \text{ ft}}{2 \text{ ft}}\right)$$

$$= 20.8 \text{ ft}^3/\text{sec}$$

$$20.8 \text{ ft}^3/\text{scc} < 90 \text{ ft}^3/\text{sec} \quad [\text{OK}]$$

(C) This incorrect solution reverses the values for pressure head and operating head. Other definitions and equations are the same as used in the correct solution.

$$A_o = \pi \frac{D^2}{4}$$

h atmospheric pressure head   ft

$$Q = C_d A_o \sqrt{2gh} = \frac{C_d \pi D^2 \sqrt{2gh}}{4}$$

$$D = \sqrt{\frac{4Q}{C_d \pi \sqrt{2gh}}}$$

$$= \sqrt{\frac{(4)\left(90 \dfrac{\text{ft}^3}{\text{sec}}\right)}{0.6\pi\sqrt{(2)\left(32.2 \dfrac{\text{ft}}{\text{sec}^2}\right)(34 \text{ ft})}}}$$

$$= 2.020 \text{ ft} \quad (2.0 \text{ ft})$$

$$R_s = R_c + \frac{D}{2}$$

$$R_s = 2 \text{ ft} + \frac{2.021 \text{ ft}}{2}$$

$$= 3.01 \text{ ft}$$

p_a operating head   ft

$$Q \leq R_c \sqrt{(0.7)(2gp_a)} \ln \left(\frac{R_s}{R_c}\right)$$

$$= (2 \text{ ft}) \sqrt{(0.7)(2)\left(32.2 \dfrac{\text{ft}}{\text{sec}^2}\right)(4 \text{ ft})}$$

$$\times \ln \left(\frac{3.01 \text{ ft}}{2 \text{ ft}}\right)$$

$$= 11 \frac{\text{ft}^3}{\text{sec}}$$

$$11 \text{ ft}^3/\text{sec} < 90 \text{ ft}^3/\text{sec} \quad [\text{OK}]$$

SOLUTION 24

g gravitational acceleration   32.2 ft/sec^2
h head   ft
v velocity   ft/sec

$$v = \sqrt{2gh} = \sqrt{(2)\left(32.2 \dfrac{\text{ft}}{\text{sec}^2}\right)(42 \text{ ft})}$$

$$= 52 \text{ ft/sec}$$

A_c chute cross-sectional area   ft^2
Q flow   ft^3/sec

$$A_c = \frac{Q}{v} = \frac{200 \ \frac{ft^3}{sec}}{52 \ \frac{ft}{sec}}$$

$$= 3.8 \ ft^2$$

D_c chute flow depth   ft
w_c chute width   ft

$$D_c = \frac{A_c}{w_c} = \frac{3.8 \ ft^2}{3 \ ft}$$

$$= 1.267 \ ft$$

Fr Froude number –

$$Fr = \frac{v}{\sqrt{gD_c}} = \frac{52 \ \frac{ft}{sec}}{\sqrt{\left(32.2 \ \frac{ft}{sec^2}\right)(1.267 \ ft)}}$$

$$= 8.14$$

w_b basin width   ft

$$\frac{w_b}{D_c} = 0.875 \ Fr + 2.7$$

$$w_b = D_c(0.875 \ Fr + 2.7)$$

$$= (1.267 \ ft)\big((0.875)(8.14) + 2.7\big)$$

$$= 12.44 \ ft \quad (12 \ ft)$$

The answer is (C).

Why Other Options Are Wrong

(A) This incorrect solution assumes the depth of flow and minimum required basin width to be equal. Other definitions and equations are unchanged from the correct solution.

$$v = \sqrt{2gh} = \sqrt{(2)\left(32.2 \ \frac{ft}{sec^2}\right)(42 \ ft)}$$

$$= 52 \ ft/sec$$

$$A_c = \frac{Q}{v} = \frac{200 \ \frac{ft^3}{sec}}{52 \ \frac{ft}{sec}}$$

$$= 3.8 \ ft^2$$

$$w_b = D_c$$

$$D_c = \frac{A_c}{w_c} = \frac{3.8 \ ft^2}{3 \ ft}$$

$$= 1.267 \ ft$$

$$w_b = 1.267 \ ft \quad (1.3 \ ft)$$

(B) This incorrect solution inverts the values for area and flow in the area equation. Other definitions and equations are unchanged from the correct solution.

$$v = \sqrt{2gh} = \sqrt{(2)\left(32.2 \ \frac{ft}{sec^2}\right)(42 \ ft)}$$

$$= 52 \ ft/sec$$

$$A_c = \frac{Q}{v} = \frac{52 \ \frac{ft^3}{sec}}{200 \ \frac{ft}{sec}}$$

$$= 0.26 \ ft^2$$

$$D_c = \frac{A_c}{w_c} = \frac{0.26 \ ft^2}{3 \ ft}$$

$$= 0.0867 \ ft$$

$$Fr = \frac{v}{\sqrt{gD_c}}$$

$$= \frac{52 \ \frac{ft}{sec}}{\sqrt{\left(32.2 \ \frac{ft}{sec^2}\right)(0.0867 \ ft)}}$$

$$= 31.12$$

$$\frac{w_b}{D_c} = 0.875 \ Fr + 2.7$$

$$w_b = D_c(0.875 \ Fr + 2.7)$$

$$= (0.0867 \ ft)\big((0.875)(31.12) + 2.7\big)$$

$$= 2.59 \ ft \quad (2.6 \ ft)$$

(D) This incorrect solution takes the square root of the head instead of taking the square root of the head and gravity constant term. Other definitions and equations are unchanged from the correct solution.

$$v = 2g\sqrt{h}$$

$$= (2)\left(32.2 \ \frac{ft}{sec^2}\right)\sqrt{(42 \ ft)}$$

$$= 417 \ ft^{3/2}/sec^2$$

The units do not work. Assume units are ft/sec, consistent with velocity.

$$A_c = \frac{Q}{v} = \frac{200 \ \frac{ft^3}{sec}}{417 \ \frac{ft}{sec}}$$

$$= 0.480 \ ft^2$$

$$D_c = \frac{A_c}{w_c} = \frac{0.480 \ ft^2}{3 \ ft}$$

$$= 0.16 \ ft$$

$$Fr = \frac{v}{\sqrt{gD_c}}$$

$$= \frac{417 \ \frac{ft}{sec}}{\sqrt{\left(32.2 \ \frac{ft}{sec^2}\right)(0.16 \ ft)}}$$

$$= 183.7$$

$$\frac{w_b}{D_c} = 0.875 \text{ Fr} + 2.7$$

$$w_b = D_c(0.875 \text{ Fr} + 2.7)$$

$$= (0.16 \text{ ft})\big((0.875)(183.7) + 2.7\big)$$

$$= 26 \text{ ft}$$

SOLUTION 25

Inside diameters for schedule-40 steel pipe are found in standard references.

	inside	
pipe	diameter (in)	area (ft^2)
A	2.469	0.0332
B	1.610	0.0141
C	1.049	0.00600

A pipe cross-sectional area ft^2
Q water flow rate ft^3/sec
v water velocity ft/sec

$$v = \frac{Q}{A}$$

$$v_A = \frac{0.76 \frac{\text{ft}^3}{\text{sec}}}{0.0332 \text{ ft}^2} = 23 \text{ ft/sec}$$

g gravitational acceleration 32.2 ft/sec^2
p pressure lbf/ft^2
γ water specific weight assume 62.4 lbf/ft^3

Because no elevation data are given, assume that the elevation head is equal at all points for all three pipes. For such a case, the energy equation between pipes A and B is

$$\frac{p_A}{\gamma} + \frac{v_A^2}{2g} = \frac{p_B}{\gamma} + \frac{v_B^2}{2g}$$

$$\frac{\left(100 \frac{\text{lbf}}{\text{in}^2}\right)\left(144 \frac{\text{in}^2}{\text{ft}^2}\right)}{62.4 \text{ lbf/ft}^3} + \frac{\left(23 \frac{\text{ft}}{\text{sec}}\right)^2}{(2)\left(32.2 \frac{\text{ft}}{\text{sec}^2}\right)}$$

$$= \frac{\left(90 \frac{\text{lbf}}{\text{in}^2}\right)\left(144 \frac{\text{in}^2}{\text{ft}^2}\right)}{62.4 \frac{\text{lbf}}{\text{ft}^3}} + \frac{v_B^2}{(2)\left(32.2 \frac{\text{ft}}{\text{sec}^2}\right)}$$

$$v_B = 45 \text{ ft/sec}$$

The continuity equation is

$$A_A v_A = A_B v_B + A_C v_C$$

$$(0.0332 \text{ ft}^2)\left(23 \frac{\text{ft}}{\text{sec}}\right) = (0.0141 \text{ ft}^2)\left(45 \frac{\text{ft}}{\text{sec}}\right)$$

$$+ (0.00600 \text{ ft}^2)v_C$$

$$v_C = 22 \text{ ft/sec}$$

$$Q = vA$$

$$Q_C = \left(22 \frac{\text{ft}}{\text{sec}}\right)(0.00600 \text{ ft}^2)$$

$$= 0.13 \text{ ft}^3/\text{sec}$$

The answer is (A).

Why Other Options Are Wrong

(B) This incorrect solution miscalculates the energy equation. Other assumptions, definitions, and equations are the same as used in the correct solution.

	inside	
pipe	diameter (in)	area (ft^2)
A	2.469	0.0332
B	1.610	0.0141
C	1.049	0.00600

$$v = \frac{Q}{A}$$

$$v_A = \frac{\left(0.76 \frac{\text{ft}^3}{\text{sec}}\right)}{0.0332 \text{ ft}^2}$$

$$= 23 \text{ ft/sec}$$

$$\frac{p_A}{\gamma} + \frac{v_A^2}{2g} = \frac{p_B}{\gamma} + \frac{v_B^2}{2g}$$

$$\frac{\left(100 \frac{\text{lbf}}{\text{in}^2}\right)\left(144 \frac{\text{in}^2}{\text{ft}^2}\right)}{62.4 \frac{\text{lbf}}{\text{ft}^3}} + \frac{23 \frac{\text{ft}}{\text{sec}}}{(2)\left(32.2 \frac{\text{ft}}{\text{sec}^2}\right)}$$

$$= \frac{\left(90 \frac{\text{lbf}}{\text{in}^2}\right)\left(144 \frac{\text{in}^2}{\text{ft}^2}\right)}{62.4 \frac{\text{lbf}}{\text{ft}^3}} + \frac{v_B^2}{(2)\left(32.2 \frac{\text{ft}}{\text{sec}^2}\right)}$$

$$v_B = 39 \text{ ft/sec}$$

$$A_A v_A = A_B v_B + A_C v_C$$

$$(0.0332 \text{ ft}^2)\left(23 \frac{\text{ft}}{\text{sec}}\right) = (0.0141 \text{ ft}^2)\left(39 \frac{\text{ft}}{\text{sec}}\right)$$

$$+ (0.00600 \text{ ft}^2)v_C$$

$$v_C = 36 \text{ ft/sec}$$

$$Q = vA$$

$$Q_C = \left(36 \frac{\text{ft}}{\text{sec}}\right)(0.00600 \text{ ft}^2)$$

$$= 0.22 \text{ ft}^3/\text{sec}$$

(C) This incorrect solution ignores the continuity equation. The solution requires an assumption regarding the pressure in pipe C. Other assumptions, definitions, and equations are the same as used in the correct solution.

pipe	inside diameter (in)	area (ft^2)
A	2.469	0.0332
B	1.610	0.0141
C	1.049	0.00600

$$v = \frac{Q}{A}$$

$$v_A = \frac{\left(0.76 \ \frac{\text{ft}^3}{\text{sec}}\right)}{0.0332 \ \text{ft}^2}$$
$$= 23 \ \text{ft/sec}$$

Assume that the pressure decreases approximately the same from pipe B to C as it did from pipe A to B. Let the pressure in pipe C be 80 psi.

$$\frac{p_A}{\gamma} + \frac{v_A^2}{2g} = \frac{p_C}{\gamma} + \frac{v_C^2}{2g}$$

$$\frac{\left(100 \ \frac{\text{lbf}}{\text{in}^2}\right)\left(144 \ \frac{\text{in}^2}{\text{ft}^2}\right)}{62.4 \ \frac{\text{lbf}}{\text{ft}^3}} + \frac{\left(23 \ \frac{\text{ft}}{\text{sec}}\right)^2}{(2)\left(32.2 \ \frac{\text{ft}}{\text{sec}^2}\right)}$$

$$= \frac{\left(80 \ \frac{\text{lbf}}{\text{in}^2}\right)\left(144 \ \frac{\text{in}^2}{\text{ft}^2}\right)}{62.4 \ \frac{\text{lbf}}{\text{ft}^3}} + \frac{v_C^2}{(2)\left(32.2 \ \frac{\text{ft}}{\text{sec}^2}\right)}$$

$$v_C = 59 \ \text{ft/sec}$$
$$Q = vA$$
$$Q_C = \left(59 \ \frac{\text{ft}}{\text{sec}}\right)(0.00600 \ \text{ft}^2)$$
$$= 0.35 \ \text{ft}^3/\text{sec}$$

(D) This incorrect solution misuses the continuity equation and ignores the energy equation. Other assumptions, definitions, and equations are the same as used in the correct solution.

pipe	inside diameter (in)	area (ft^2)
A	2.469	0.0332
B	1.610	0.0141
C	1.049	0.00600

$$v = \frac{Q}{A}$$

$$v_A = \frac{\left(0.76 \ \frac{\text{ft}^3}{\text{sec}}\right)}{0.0332 \ \text{ft}^2}$$
$$= 23 \ \text{ft/sec}$$

$$A_A v_A = A_B v_B = A_C v_C$$
$$0.76 \ \frac{\text{ft}^3}{\text{sec}} = (0.00600 \ \text{ft}^2)v_C$$
$$v_C = 127 \ \text{ft/sec}$$
$$Q = vA$$
$$Q_C = \left(127 \ \frac{\text{ft}}{\text{sec}}\right)(0.00600 \ \text{ft}^2)$$
$$= 0.76 \ \text{ft}^3/\text{sec}$$

SOLUTION 26

Assume uniform flow since flow velocity is relatively slow and slope is constant and shallow.

D pipe diameter m
R hydraulic radius m
θ angle of flow cross section radians

For flow half full in a circular culvert, θ is π radians.

$$R = 0.25\left(1 - \frac{\sin\theta}{\theta}\right)D$$
$$= 0.25\left(1 - \frac{\sin\pi}{\pi}\right)D$$
$$= 0.25D$$

d_2 flow depth in culvert m

$$d_2 = \frac{D}{2}$$

Standard concrete pipe sizes go up to 9 ft (2.74 m). When the culvert flows half full,

$$d_2 = \frac{2.74 \ \text{m}}{2} = 1.37 \ \text{m}$$

n Manning roughness coefficient –
S culvert slope m/m
v_2 flow velocity in culvert m/s

For concrete pipe, the Manning roughness coefficient is 0.013.

$$v_2 = \frac{R^{2/3}\sqrt{S}}{n}$$

$$= \frac{((0.25)(2)(1.37))^{2/3}\sqrt{0.002 \ \frac{\text{m}}{\text{m}}}}{0.013}$$

$$= 2.67 \ \text{m/s}$$

A_1 channel cross-sectional area   m²
A_2 culvert cross-sectional area   m²
v_1 flow velocity in channel   m/s

Apply the continuity equation.

$$A_1 v_1 = A_2 v_2$$
$$A_1 v_1 = (1.5 \text{ m})(8 \text{ m})\left(2.5 \, \frac{\text{m}}{\text{s}}\right)$$
$$= 30 \text{ m}^3/\text{s}$$
$$A_2 = \left(\frac{1}{2}\right)\left(\frac{\pi D^2}{4}\right)$$
$$= \left(\frac{1}{2}\right)(2.74 \text{ m}^2)$$
$$= 2.95 \text{ m}^2$$

Try one pipe.

$$A_2 v_2 = \left(2.95 \, \frac{\text{m}^2}{\text{pipe}}\right)\left(2.67 \, \frac{\text{m}}{\text{s}}\right)$$
$$= 7.88 \text{ m}^3/\text{s per pipe}$$

$$\text{number of 9 ft diameter pipes} = \frac{30 \, \dfrac{\text{m}^3}{\text{s}}}{7.88 \, \dfrac{\dfrac{\text{m}^3}{\text{s}}}{\text{pipe}}}$$
$$= 3.8 \text{ pipes}$$

Use four pipes, each with a diameter of 9 ft.

The answer is (C).

Why Other Options Are Wrong

(A) This incorrect solution uses the total pipe cross-sectional area instead of the area of flow when applying the continuity equation. Other assumptions, definitions, and equations are unchanged from the correct solution.

$$R = 0.25 \left(1 - \frac{\sin\theta}{\theta}\right) D$$
$$= 0.25 \left(1 - \frac{\sin\pi}{\pi}\right) D$$
$$= 0.25D$$
$$d_2 = \frac{D}{2} = \frac{2.74 \text{ m}}{2}$$
$$= 1.37 \text{ m}$$

For a concrete pipe, the Manning roughness coefficient is 0.013.

$$v_2 = \frac{R^{2/3}\sqrt{S}}{n}$$
$$= \frac{\left((0.25)(2)1.37\right)^{2/3}\sqrt{0.002 \, \dfrac{\text{m}}{\text{m}}}}{0.013}$$
$$= 2.67 \text{ m/s}$$
$$A_1 v_1 = A_2 v_2$$
$$= (1.5 \text{ m})(8 \text{ m})\left(2.5 \, \frac{\text{m}}{\text{s}}\right)$$
$$= 30 \text{ m}^3/\text{s}$$
$$A_2 = \frac{\pi D^2}{4}$$
$$= \frac{\pi(2.74 \text{ m}^2)}{4}$$
$$= 5.9 \text{ m}^2 \text{ per pipe}$$
$$A_2 v_2 = \left(5.9 \, \frac{\text{m}^2}{\text{pipe}}\right)\left(2.67 \, \frac{\text{m}}{\text{s}}\right)$$
$$= 15.75 \text{ m}^3/\text{s per pipe}$$

$$\text{number of 9 ft diameter pipes} = \frac{30 \, \dfrac{\text{m}^3}{\text{s}}}{15.75 \, \dfrac{\dfrac{\text{m}^3}{\text{s}}}{\text{pipe}}}$$
$$= 1.9 \text{ pipes}$$

Use two pipes, each with a diameter of 9 ft.

(B) This incorrect solution uses a different approach and makes an error in combining the exponents in the continuity equation. Other assumptions, definitions, and equations are unchanged from the correct solution.

$$R = 0.25 \left(1 - \frac{\sin\theta}{\theta}\right) D$$
$$= 0.25 \left(1 - \frac{\sin\pi}{\pi}\right) D$$
$$= 0.25D$$
$$d_2 = D/2$$

For a concrete pipe, the Manning roughness coefficient is 0.013.

$$v_2 = \frac{R^{2/3}\sqrt{S}}{n}$$
$$= \frac{\left((0.25)(2)d_2\right)^{2/3}\sqrt{0.002 \, \dfrac{\text{m}}{\text{m}}}}{0.013}$$
$$= 3.44(0.5 d_2)^{2/3}\text{m/s}$$
$$= 2.167 d_2^{2/3} \text{ m/s}$$

$$A_1 v_1 = A_2 v_2$$
$$= (1.5 \text{ m})(8 \text{ m}) \left(2.5 \ \frac{\text{m}}{\text{s}}\right)$$
$$= 30 \text{ m}^3/\text{s}$$
$$A_2 = \left(\frac{1}{2}\right) \frac{\pi D^2}{4}$$
$$= \left(\frac{1}{2}\right) \pi d_2^2 \text{ m}^2$$
$$A_2 v_2 = 30 \ \frac{\text{m}^3}{\text{s}}$$
$$= \left(\frac{1}{2}\right) \pi d_2^2 (2.167) d_2^{2/3} \ \frac{\text{m}^3}{\text{s}}$$
$$= 3.4 d_2^4 \text{ m}^3/\text{s}$$

Standard concrete pipe sizes go up to 9 ft (2.74 m).

Try three pipes.

$$\frac{A_2 v_2}{\text{number of pipes}} = \frac{\left(30 \ \dfrac{\text{m}^3}{\text{s}}\right)}{3}$$
$$= 10 \text{ m}^3/\text{s}$$
$$d_2 = \left(\frac{\left(10 \ \dfrac{\text{m}^3}{\text{s}}\right)}{3.4}\right)^{1/4}$$
$$= 1.31 \text{ m}$$
$$D = (2)(1.31 \text{ m})$$
$$= (2.6 \text{ m}) \left(3.28 \ \frac{\text{m}}{\text{ft}}\right)$$
$$= 8.5 \text{ ft} > 9 \text{ ft, OK.}$$

(D) This incorrect solution uses a different approach and uses $\pi/2$ instead of π for the angle of cross-sectional flow. Other assumptions, definitions, and equations are unchanged from the correct solution.

For flow half full in a circular culvert, θ is $\pi/2$ radians.

$$R = 0.25 \left(1 - \frac{\sin \theta}{\theta}\right) D$$
$$= 0.25 \left(1 - \frac{\sin \dfrac{\pi}{2}}{\dfrac{\pi}{2}}\right) D$$
$$= 0.091 D$$

For a concrete pipe, the Manning roughness coefficient is 0.013.

$$v_2 = \frac{R^{2/3} \sqrt{S}}{n}$$
$$= \frac{\left((0.091)(2) d_2\right)^{2/3} \sqrt{0.002 \ \dfrac{\text{m}}{\text{m}}}}{0.013}$$
$$= 1.1 d_2^{2/3} \text{ m/s}$$
$$A_1 v_1 = A_2 v_2$$
$$= (1.5 \text{ m})(8 \text{ m}) \left(2.5 \ \frac{\text{m}}{\text{s}}\right)$$
$$= 30 \text{ m}^3/\text{s}$$
$$A_2 = \left(\frac{1}{2}\right) \left(\frac{\pi D^2}{4}\right)$$
$$= \left(\frac{1}{2}\right) \pi d_2^2 \text{ m}^2$$
$$A_2 v_2 = 30 \ \frac{\text{m}^3}{\text{s}}$$
$$= \left(\frac{1}{2}\right) \pi d_2^2 \text{ m}^2 (1.1) d_2^{2/3} \ \frac{\text{m}}{\text{s}}$$
$$= 1.73 d_2^{8/3} \text{ m}^3/\text{s}$$

Standard concrete pipe sizes go up to 9 ft (2.74 m).

Try eight pipes.

$$\frac{A_2 v_2}{\text{number of pipes}} = \frac{\left(30 \ \dfrac{\text{m}^3}{\text{s}}\right)}{8}$$
$$= 3.75 \text{ m}^3/\text{s}$$
$$d_2 = \left(\frac{\left(3.75 \ \dfrac{\text{m}^3}{\text{s}}\right)}{1.73}\right)^{3/8}$$
$$= 1.34 \text{ m}$$
$$D = (2)(1.34 \text{ m})$$
$$= (2.7 \text{ m}) \left(3.28 \ \frac{\text{m}}{\text{ft}}\right)$$
$$= 8.9 \text{ ft} \approx 9 \text{ ft, OK.}$$

Use eight pipes, each with a diameter of 9 ft.

SOLUTION 27

A_1	pipe cross-sectional area	m^2
A_2	nozzle cross-sectional area	m^2
D_1	pipe diameter	m
D_2	nozzle diameter	m

$$A_1 = \frac{\pi D_1^2}{4}$$

$$= \frac{\pi\left((2.54 \text{ cm})\left(\frac{1 \text{ m}}{100 \text{ cm}}\right)\right)^2}{4}$$

$$= 5.1 \times 10^{-4} \text{ m}^2$$

$$A_2 = \frac{\pi D_2^2}{4}$$

$$= \frac{\pi\left((0.5 \text{ cm})\left(\frac{1 \text{ m}}{100 \text{ cm}}\right)\right)^2}{4}$$

$$= 2.0 \times 10^{-5} \text{ m}^2$$

g	gravitational acceleration	9.81 m/s^2
p_1	water pressure at lower elevation	kPa
p_2	water pressure at upper elevation	kPa
Q_1	flow rate	m^3/s
Q_2	flow rate	m^3/s
z_1	lower water level elevation	m
z_2	upper water level elevation	m
ρ	density of water	998.3 kg/m^3 at $20°\text{C}$

Choose $z_1 = 0$.

$$\frac{p_1}{\rho g} + z_1 + \frac{Q_1^2}{2gA_1^2} = \frac{p_2}{\rho g} + z_2 + \frac{Q_2^2}{2gA_2^2}$$

$$p_1 = p_2 + z_2\rho g + \frac{Q_2^2\rho}{2A_2^2} - \frac{Q_1^2\rho}{2A_1^2}$$

$$= 7000 \text{ kPa} + \frac{(340 \text{ m})\left(998.3 \frac{\text{kg}}{\text{m}^3}\right)\left(9.81 \frac{\text{m}}{\text{s}^2}\right)}{\left(\frac{\text{kg·m}}{\text{s}^2 \cdot \text{N}}\right)\left((10)^3 \frac{\text{N}}{\text{m}^2 \cdot \text{kPa}}\right)}$$

$$+ \frac{\left(0.002 \frac{\text{m}^3}{\text{s}}\right)^2\left(998.3 \frac{\text{kg}}{\text{m}^3}\right)}{\left(\frac{\text{kg·m}}{\text{s}^2 \cdot \text{N}}\right)\left((10)^3 \frac{\text{N}}{\text{m}^2 \cdot \text{kPa}}\right)(2)(2.0 \times 10^{-5} \text{ m}^2)^2}$$

$$- \frac{\left(0.002 \frac{\text{m}^3}{\text{s}}\right)^2\left(998.3 \frac{\text{kg}}{\text{m}^3}\right)}{\left(\frac{\text{kg·m}}{\text{s}^2 \cdot \text{N}}\right)\left((10)^3 \frac{\text{N}}{\text{m}^2 \cdot \text{kPa}}\right)(2)(5.1 \times 10^{-4} \text{ m}^2)^2}$$

$$= 15\,310 \text{ kPa} \quad (15\,000 \text{ kPa})$$

The answer is (D).

Why Other Options Are Wrong

(A) This incorrect solution fails to distribute the product of water density and gravitational acceleration to the other terms in the energy equation when solving for pressure. Other assumptions, definitions, and equations are unchanged from the correct solution.

$$A_1 = \frac{\pi D_1^2}{4} = \frac{\pi\left((2.54 \text{ cm})\left(\frac{1 \text{ m}}{100 \text{ cm}}\right)\right)^2}{4}$$

$$= 5.1 \times 10^{-4} \text{ m}^2$$

$$A_2 = \frac{\pi D_2^2}{4} = \frac{\pi\left((0.5 \text{ cm})\left(\frac{1 \text{ m}}{100 \text{ cm}}\right)\right)^2}{4}$$

$$= 2.0 \times 10^{-5} \text{ m}^2$$

Choose $z_1 = 0$.

$$p_1 = \frac{p_2}{\rho g} + z_2 + \frac{Q_2^2}{2gA_2^2} - \frac{Q_1^2}{2gA_1^2}$$

$$= \frac{(7000 \text{ kPa})\left(\frac{\text{kg·m}}{\text{s}^2 \cdot \text{N}}\right)\left((10)^3 \frac{\text{N}}{\text{m}^2 \cdot \text{kPa}}\right)}{\left(998.3 \frac{\text{kg}}{\text{m}^3}\right)\left(9.81 \frac{\text{m}}{\text{s}^2}\right)} + 340 \text{ m}$$

$$+ \frac{\left(0.002 \frac{\text{m}^3}{\text{s}}\right)^2}{(2)\left(9.81 \frac{\text{m}}{\text{s}^2}\right)(2.0 \times 10^{-5} \text{ m}^2)^2}$$

$$- \frac{\left(0.002 \frac{\text{m}^3}{\text{s}}\right)^2}{(2)\left(9.81 \frac{\text{m}}{\text{s}^2}\right)(5.1 \times 10^{-4} \text{ m}^2)^2}$$

$$= 1564 \text{ kPa} \quad (1600 \text{ kPa})$$

The units for all the terms are not the same and do not add. Assume units are kPa.

(B) This incorrect solution uses the wrong conversion from N/m^2 to kPa (10^3 N·kPa/m^2 instead of $10^3 \text{ N/m}^2 \cdot \text{kPa}$). Other assumptions, definitions, and equations are unchanged from the correct solution.

$$A_1 = \frac{\pi D_1^2}{4} = \frac{\pi\left((2.54 \text{ cm})\left(\frac{1 \text{ m}}{100 \text{ cm}}\right)\right)^2}{4}$$

$$= 5.1 \times 10^{-4} \text{ m}^2$$

$$A_2 = \frac{\pi D_2^2}{4}$$

$$= \frac{\pi\left((0.5 \text{ cm})\left(\frac{1 \text{ m}}{100 \text{ cm}}\right)\right)^2}{4}$$

$$= 2.0 \times 10^{-5} \text{ m}^2$$

Choose $z_1 = 0$.

$$p_1 = p_2 + \frac{z_2}{\rho g} + \frac{Q_2^2 \rho}{2g A_2^2} - \frac{Q_1^2 \rho}{2 A_1^2}$$

$$= 7000 \text{ kPa} + \frac{(340 \text{ m}) \left((10)^3 \frac{\text{N·kPa}}{\text{m}^2}\right) \left(\frac{\text{kg}}{\text{m·s}^2\text{·N}}\right)}{\left(998.3 \frac{\text{kg}}{\text{m}^3}\right) \left(9.81 \frac{\text{m}}{\text{s}^2}\right)}$$

$$+ \frac{\left(0.002 \frac{\text{m}^3}{\text{s}}\right)^2}{(2) \left(9.81 \frac{\text{m}}{\text{s}^2}\right) (2.0 \times 10^{-5} \text{ m}^2)^2}$$

$$- \frac{\left(0.002 \frac{\text{m}^3}{\text{s}}\right)^2 \left((10)^3 \frac{\text{N·kPa}}{\text{m}^2}\right) \left(\frac{\text{kg}}{\text{m·s}^2\text{·N}}\right)}{\left(\left(998.3 \frac{\text{kg}}{\text{m}^3}\right) (2) \left(9.81 \frac{\text{m}}{\text{s}^2}\right)^2 \times (5.1 \times 10^{-4} \text{ m}^2)^2\right)}$$

$$= 7544 \text{ kPa} \quad (7500 \text{ kPa})$$

(C) This incorrect solution fails to square the area in the energy equation. Other assumptions, definitions, and equations are unchanged from the correct solution.

$$A_1 = \frac{\pi D_1^2}{4} = \frac{\pi \left((2.54 \text{ cm}) \left(\frac{1 \text{ m}}{100 \text{ cm}}\right)\right)^2}{4}$$

$$= 5.1 \times 10^{-4} \text{ m}^2$$

$$A_2 = \frac{\pi D_2^2}{4} = \frac{\pi \left((0.5 \text{ cm}) \left(\frac{1 \text{ m}}{100 \text{ cm}}\right)\right)^2}{4}$$

$$= 2.0 \times 10^{-5} \text{ m}^2$$

Choose $z_1 = 0$.

$$p_1 = p_2 + z_2 \rho g + \frac{Q_2^2 \rho}{2 A_2} - \frac{Q_1^2 \rho}{2 A_1}$$

$$= 7000 \text{ kPa} + \frac{(340 \text{ m}) \left(998.3 \frac{\text{kg}}{\text{m}^3}\right) \left(9.81 \frac{\text{m}}{\text{s}^2}\right)}{\left((10)^3 \frac{\text{N}}{\text{m}^2\text{·kPa}}\right) \left(\frac{\text{kg·m}}{\text{s}^2\text{·N}}\right)}$$

$$+ \frac{\left(0.002 \frac{\text{m}^3}{\text{s}}\right)^2 \left(998.3 \frac{\text{kg}}{\text{m}^3}\right)}{\left((10)^3 \frac{\text{N}}{\text{m}^2\text{·kPa}}\right) \left(\frac{\text{kg·m}}{\text{s}^2\text{·N}}\right) (2)(2.0 \times 10^{-5} \text{ m}^2)}$$

$$- \frac{\left(0.002 \frac{\text{m}^3}{\text{s}}\right)^2 \left(998.3 \frac{\text{kg}}{\text{m}^3}\right)}{\left((10)^3 \frac{\text{N}}{\text{m}^2\text{·kPa}}\right) \left(\frac{\text{kg·m}}{\text{s}^2\text{·N}}\right) (2)(5.1 \times 10^{-4} \text{ m}^2)}$$

$$= 10\,331 \text{ kPa} \quad (10\,000 \text{ kPa})$$

The units for all terms are not the same and do not add. Assume units are kPa.

SOLUTION 28

For elevation head loss,

h_z head loss from elevation   ft

$$h_z = z_2 - z_1 = 3503 \text{ ft} - 3457 \text{ ft}$$
$$= 46 \text{ ft}$$

For equivalent length of threaded 2 in schedule-80 steel fittings, the following characteristics apply.

fitting	quantity	unit equivalent length (ft)	total equivalent length (ft)
couple	15	0.45	6.75
90° ell	8	8.5	68.0
45° ell	4	2.7	10.8
straight tee	6	7.7	46.2
globe valve	2	54	108
			239.75

For friction head loss in the pipe (including equivalent length of fittings),

D inside diameter for 2 in schedule-80 steel pipe
ε roughness coefficient for steel pipe

$$D = 1.939 \text{ in}$$
$$\varepsilon = 0.0002 \text{ ft}$$

$\dfrac{\varepsilon}{D}$ relative roughness   –

$$\frac{\varepsilon}{D} = \frac{(0.0002 \text{ ft}) \left(12 \frac{\text{in}}{\text{ft}}\right)}{1.939 \text{ in}}$$
$$= 0.00124$$

Re   Reynolds number   –
v   flow velocity   ft/sec
ν kinematic viscosity   ft^2/sec

$$\nu = 1.217 \times 10^{-5} \text{ ft/sec at } 60°\text{F}$$

$$\text{Re} = \frac{D\text{v}}{\nu} = \frac{(1.939 \text{ in}) \left(\frac{1 \text{ ft}}{12 \text{ in}}\right) \left(5 \frac{\text{ft}}{\text{sec}}\right)}{1.217 \times 10^{-5} \frac{\text{ft}^2}{\text{sec}}}$$

$$= 6.6 \times 10^4$$

f friction factor   –

$$f = \frac{0.25}{\left(\log\left(\dfrac{\frac{\varepsilon}{D}}{3.7} + \dfrac{5.74}{\text{Re}^{0.9}}\right)\right)^2}$$

$$= \frac{0.25}{\left(\log\left(\dfrac{0.00124}{3.7} + \dfrac{5.74}{(6.6 \times 10^4)^{0.9}}\right)\right)^2}$$

$$= 0.0419$$

Note that the friction factor could also be determined using a Moody diagram with the same input values for Reynolds number and relative roughness, Re and ε/D.

g	gravitational acceleration	ft/sec^2
h	total head loss	ft
h_f	head loss from friction	ft
L	pipe length	ft
L_e	equivalent pipe length of fittings	ft

$$g = 32.2 \text{ ft/sec}^2$$

$$h_f = \frac{f(L + L_e)\text{v}^2}{2Dg}$$

$$= \frac{(0.0419)(500 \text{ ft} + 239.75 \text{ ft})\left(5 \dfrac{\text{ft}}{\text{sec}}\right)^2}{(2)(1.939 \text{ in})\left(\dfrac{1 \text{ ft}}{12 \text{ in}}\right)\left(32.2 \dfrac{\text{ft}}{\text{sec}^2}\right)}$$

$$= 74 \text{ ft}$$

For total head loss,

$$h = h_z + h_f = 46 \text{ ft} + 74 \text{ ft}$$
$$= 120 \text{ ft}$$

The answer is (C).

Why Other Options Are Wrong

(A) This incorrect solution assumes that minor losses are negligible. Other assumptions, definitions, and equations are unchanged from the correct solution.

For elevation head loss,

$$h_z = z_2 - z_1 = 3503 \text{ ft} - 3457 \text{ ft}$$
$$= 46 \text{ ft}$$

Assume that minor losses are negligible.

For friction loss in the pipe,

$$D = 1.939 \text{ in}$$
$$\varepsilon = 0.000005 \text{ ft}$$

$$\frac{\varepsilon}{D} = \frac{(0.000005 \text{ ft})\left(12 \dfrac{\text{in}}{\text{ft}}\right)}{1.939 \text{ in}}$$
$$= 0.000031$$

$$\text{Re} = \frac{D\text{v}}{\nu} = \frac{(1.939 \text{ in})\left(\dfrac{1 \text{ ft}}{12 \text{ in}}\right)\left(5 \dfrac{\text{ft}}{\text{sec}}\right)}{1.217 \times 10^{-5} \dfrac{\text{ft}^2}{\text{sec}}}$$

$$= 6.6 \times 10^4$$

$$f = \frac{0.25}{\left(\log\left(\dfrac{\frac{\varepsilon}{D}}{3.7} + \dfrac{5.74}{\text{Re}^{0.9}}\right)\right)^2}$$

$$= \frac{0.25}{\left(\log\left(\dfrac{0.000031}{3.7} + \dfrac{5.74}{(6.6 \times 10^4)^{0.9}}\right)\right)^2}$$

$$= 0.0197$$

$$h_f = \frac{fL\text{v}^2}{2Dg} = \frac{(0.0197)(500 \text{ ft})\left(5 \dfrac{\text{ft}}{\text{sec}}\right)^2}{(2)(1.939 \text{ in})\left(\dfrac{1 \text{ ft}}{12 \text{ in}}\right)\left(32.2 \dfrac{\text{ft}}{\text{sec}^2}\right)}$$

$$= 24 \text{ ft}$$

For total head loss,

$$h = h_z + h_f$$
$$= 46 \text{ ft} + 24 \text{ ft}$$
$$= 70 \text{ ft}$$

(B) This incorrect solution uses pressure drop tables for schedule-40 steel pipe. Other assumptions, definitions, and equations are unchanged from the correct solution.

For elevation head loss,

$$h_z = z_2 - z_1$$
$$= 3503 \text{ ft} - 3457 \text{ ft}$$
$$= 46 \text{ ft}$$

For equivalent length of fittings,

fitting	quantity	unit equivalent length (ft)	total equivalent length (ft)
couple	15	0.45	6.75
90° ell	8	3.1	24.8
45° ell	4	2.7	10.8
straight tee	6	7.7	46.2
globe valve	2	54	108
			196.55

Assume pressure drop tables for schedule-40 steel pipe approximate pressure drop in schedule-80 PVC pipe. From pressure tables for schedule-40 steel pipe, the head loss per unit length is

$$h_{fo} = \frac{\text{head loss}}{\text{unit length}} = \frac{25.1 \dfrac{\text{lbf}}{\text{in}^2}}{1000 \text{ ft}}$$

For friction loss in the pipe (including equivalent length of fittings),

$$h_f = h_{fo}(L + L_e)$$

$$= \frac{\left(25.1 \; \frac{lbf}{in^2}\right)(500 \text{ ft} + 196.55 \text{ ft})(1 \text{ ft of water})}{(1000 \text{ ft})\left(0.43328 \; \frac{lbf}{in^2}\right)}$$

$$= 40 \text{ ft}$$

For total head loss,

$$h = h_z + h_f = 46 \text{ ft} + 40 \text{ ft}$$
$$= 86 \text{ ft}$$

(D) This incorrect solution assumes the equivalent length of the fittings represents the minor losses in feet of water. Other assumptions, definitions, and equations are unchanged from the correct solution.

For elevation head loss,

$$h_z = z_2 - z_1 = 3503 \text{ ft} - 3457 \text{ ft}$$
$$= 46 \text{ ft}$$

For minor losses,

h_m minor head loss (from fittings)   ft

fitting	quantity	unit equivalent length (ft)	total equivalent length (ft)
couple	15	0.45	6.75
90° ell	8	3.1	24.8
45° ell	4	2.7	10.8
straight tee	6	7.7	46.2
globe valve	2	54	108
			196.55

$$h_m = 197 \text{ ft}$$

For friction loss in the pipe,

$$D = 1.939 \text{ in}$$
$$\varepsilon = 0.000005 \text{ ft}$$
$$\frac{\varepsilon}{D} = \frac{(0.000005 \text{ ft})\left(12 \; \frac{in}{ft}\right)}{1.939 \text{ in}}$$
$$= 0.000031$$

$$\text{Re} = \frac{Dv}{\nu} = \frac{(1.939 \text{ in})\left(\frac{1 \text{ ft}}{12 \text{ in}}\right)\left(5 \; \frac{ft}{sec}\right)}{1.217 \times 10^{-5} \; \frac{ft^2}{sec}}$$
$$= 6.6 \times 10^4$$

$$f = \frac{0.25}{\left(\log\left(\dfrac{\frac{\varepsilon}{D}}{33.7} + \dfrac{5.74}{\text{Re}^{0.9}}\right)\right)^2}$$

$$= \frac{0.25}{\left(\log\left(\dfrac{0.000031}{33.7} + \dfrac{5.74}{(6.6 \times 10^4)^{0.9}}\right)\right)^2}$$

$$= 0.0197$$

$$h_f = \frac{fLv^2}{2Dg} = \frac{(0.0197)(500 \text{ ft})\left(5 \; \frac{ft}{sec}\right)^2}{(2)(1.939 \text{ in})\left(\frac{1 \text{ ft}}{12 \text{ in}}\right)\left(32.2 \; \frac{ft}{sec^2}\right)}$$

$$= 24 \text{ ft}$$

For total head loss,

$$h = h_z + h_f + h_m$$
$$= 46 \text{ ft} + 24 \text{ ft} + 197 \text{ ft}$$
$$= 267 \text{ ft} \quad (270 \text{ ft})$$

SOLUTION 29

Assume the inside pipe diameter is equal to the nominal diameter.

D inside pipe diameter   in
ε roughness coefficient for steel pipe   ft
$\dfrac{\varepsilon}{D}$ relative roughness   –

$$\varepsilon = 0.0002 \text{ ft}$$

$$\frac{\varepsilon}{D_6} = \frac{(0.0002 \text{ ft})\left(\dfrac{12 \text{ in}}{ft}\right)}{6 \text{ in}}$$
$$= 0.0004$$

$$\frac{\varepsilon}{D_4} = \frac{(0.0002 \text{ ft})\left(\dfrac{12 \text{ in}}{ft}\right)}{4 \text{ in}}$$
$$= 0.0006$$

A pipe cross-sectional area   ft^2

$$A = \pi\frac{D^2}{4}$$

$$A_6 = \frac{\pi(6 \text{ in})^2\left(\dfrac{1 \text{ ft}^2}{144 \text{ in}^2}\right)}{4}$$
$$= 0.196 \text{ ft}^2$$

$$A_4 = \frac{\pi(4 \text{ in})^2\left(\dfrac{1 \text{ ft}^2}{144 \text{ in}^2}\right)}{4}$$
$$= 0.087 \text{ ft}^2$$

Re   Reynolds number   –
f friction factor   –

Assume a Reynolds number of 5×10^5 and a temperature of $60°\text{F}$.

From a Moody diagram, estimate a friction factor using Re and ε/D,

$$f_6 = 0.017$$
$$f_4 = 0.0185$$

g gravitational acceleration 32.2 ft/sec^2
L pipe length   ft
v flow velocity   ft/sec

For parallel pipes, the friction losses must be equal.

$$\frac{f_6 L_6 v_6^2}{2 D_6 g} = \frac{f_4 L_4 v_4^2}{2 D_4 g}$$

$$\frac{(0.017)(1400 \text{ ft}) v_6^2}{\left((2)(6 \text{ in})\left(\dfrac{1 \text{ ft}}{12 \text{ in}}\right) \times \left(32.2 \dfrac{\text{ft}}{\text{sec}^2}\right)\right)} = \frac{(0.0185)(1400 \text{ ft}) v_4^2}{\left((2)(4 \text{ in})\left(\dfrac{1 \text{ ft}}{12 \text{ in}}\right) \times \left(32.2 \dfrac{\text{ft}}{\text{sec}^2}\right)\right)}$$

$$0.74 v_6^2 = 1.21 v_4^2$$
$$v_6 = 1.3 v_4$$

Q total flow rate ft^3/sec

$$Q = A_6 v_6 + A_4 v_4$$
$$2.4 \frac{\text{ft}^3}{\text{sec}} = (0.196 \text{ ft}^2)(1.3 v_4) + (0.087 \text{ ft}^2)(v_4)$$
$$v_4 = 7.0 \text{ ft/sec}$$
$$v_6 = (1.3)\left(7.0 \frac{\text{ft}}{\text{sec}}\right) = 9.1 \text{ ft/sec}$$

Check the Reynolds number assumption.

ν kinematic viscosity ft^2/sec

$$\text{Re} = \frac{Dv}{\nu}$$

$$\nu = 1.217 \times 10^{-5} \frac{\text{ft}^2}{\text{sec}} \text{ at } 60°\text{F}$$

$$\text{Re}_6 = \frac{(6 \text{ in})\left(\dfrac{1 \text{ ft}}{12 \text{ in}}\right)\left(9.1 \dfrac{\text{ft}}{\text{sec}}\right)}{1.217 \times 10^{-5} \dfrac{\text{ft}^2}{\text{sec}}} = 3.7 \times 10^5$$

$$\text{Re}_4 = \frac{(4 \text{ in})\left(\dfrac{1 \text{ ft}}{12 \text{ in}}\right)\left(7.0 \dfrac{\text{ft}}{\text{sec}}\right)}{1.217 \times 10^{-5} \dfrac{\text{ft}^2}{\text{sec}}} = 1.9 \times 10^5$$

$$f = \frac{0.25}{\left(\log\left(\dfrac{\frac{\varepsilon}{D}}{3.7} + \dfrac{5.74}{\text{Re}^{0.9}}\right)\right)^2}$$

$$f_6 = \frac{0.25}{\left(\log\left(\dfrac{0.0004}{3.7} + \dfrac{5.74}{(3.7 \times 10^5)^{0.9}}\right)\right)^2}$$
$$= 0.0174 \cong 0.017 \quad \text{[close enough]}$$

$$f = \frac{0.25}{\left(\log\left(\dfrac{0.0006}{3.7} + \dfrac{5.74}{(1.9 \times 10^5)^{0.9}}\right)\right)^2}$$
$$= 0.0195 \cong 0.0185 \quad \text{[close enough]}$$

$$Q = Av$$

$$Q_4 = (0.087 \text{ ft}^2)\left(7.0 \frac{\text{ft}}{\text{sec}}\right)$$
$$= 0.609 \text{ ft}^3/\text{sec} \quad (0.61 \text{ ft}^3/\text{sec})$$

The answer is (B).

Why Other Options Are Wrong

(A) This incorrect solution fails to square the velocity terms in the friction factor equation. Other assumptions, definitions, and equations are unchanged from the correct solution.

$$\frac{\varepsilon}{D_6} = \frac{(0.002 \text{ ft})\left(12 \dfrac{\text{in}}{\text{ft}}\right)}{6 \text{ in}} = 0.0004$$

$$\frac{\varepsilon}{D_4} = \frac{(0.0002 \text{ ft})\left(12 \dfrac{\text{in}}{\text{ft}}\right)}{4 \text{ in}} = 0.0006$$

$$A = \pi \frac{D^2}{4}$$

$$A_6 = \frac{\pi(6 \text{ in})^2\left(\dfrac{1 \text{ ft}^2}{144 \text{ in}^2}\right)}{4} = 0.196 \text{ ft}^2$$

$$A_4 = \frac{\pi(4 \text{ in})^2\left(\dfrac{1 \text{ ft}^2}{144 \text{ in}^2}\right)}{4} = 0.087 \text{ ft}^2$$

Assume a Reynolds number of 5×10^5 and a temperature of $60°\text{F}$.

From a Moody diagram, estimate a friction factor using Re and ε/D,

$$f_6 = 0.017$$
$$f_4 = 0.0185$$

For parallel pipes, the friction losses must be equal.

$$\frac{f_6 L_6 v_6}{2 D_6 g} = \frac{f_4 L_4 v_4}{2 D_4 g}$$

$$\frac{(0.017)(1400 \text{ ft}) v_6}{\left((2)(6 \text{ in})\left(\dfrac{1 \text{ ft}}{12 \text{ in}}\right) \times \left(32.2 \dfrac{\text{ft}}{\text{sec}^2}\right)\right)} = \frac{(0.0185)(1400 \text{ ft}) v_4}{\left((2)(4 \text{ in})\left(\dfrac{1 \text{ ft}}{12 \text{ in}}\right) \times \left(32.2 \dfrac{\text{ft}}{\text{sec}^2}\right)\right)}$$

$$0.74v_6 = 1.21v_4$$

$$v_6 = 1.64v_4$$

$$Q = A_6v_6 + A_4v_4$$

$$2.4 \ \frac{ft^3}{sec} = (0.196 \ ft^2)(1.64v_4)$$
$$+ (0.087 \ ft^2)(v_4)$$

$$v_4 = 5.9 \ ft/sec$$

$$v_6 = (1.64)\left(5.9 \ \frac{ft}{sec}\right)$$

$$= 9.7 \ ft/sec$$

Check the Reynolds number assumption.

$$Re = \frac{Dv}{\nu}$$

$$Re_6 = \frac{(6 \ in)\left(\frac{1 \ ft}{12 \ in}\right)\left(9.7 \ \frac{ft}{sec}\right)}{1.217 \times 10^{-5} \ \frac{ft^2}{sec}}$$

$$= 4.0 \times 10^5$$

$$Re_4 = \frac{(4 \ in)\left(\frac{1 \ ft}{12 \ in}\right)\left(5.9 \ \frac{ft}{sec}\right)}{1.217 \times 10^{-5} \ \frac{ft^2}{sec}}$$

$$= 1.6 \times 10^5$$

$$f = \frac{0.25}{\left(\log\left(\frac{\frac{\varepsilon}{D}}{3.7} + \frac{5.74}{Re^{0.9}}\right)\right)^2}$$

$$f_6 = \frac{0.25}{\left(\log\left(\frac{0.0004}{3.7} + \frac{5.74}{(4.0 \times 10^5)^{0.9}}\right)\right)^2}$$

$$= 0.0174 \cong 0.017 \quad \text{[close enough]}$$

$$f = \frac{0.25}{\left(\log\left(\frac{0.0006}{3.7} + \frac{5.74}{1.6 \times 10^5}\right)^{0.9}\right)^2}$$

$$= 0.0225 \cong 0.0185 \quad \text{[close enough]}$$

$$Q = Av$$

$$Q_4 = (0.087 \ ft^2)\left(5.9 \ \frac{ft}{sec}\right)$$

$$= 0.51 \ ft^3/sec$$

(C) This incorrect solution assumes a relative roughness of 0.002 as a mid-range value on the Moody diagram. Other assumptions, definitions, and equations are unchanged from the correct solution.

Assume $\varepsilon/D = 0.002$.

$$A = \pi\frac{D^2}{4}$$

$$A_6 = \frac{\pi(6 \ in)^2\left(\frac{1 \ ft^2}{144 \ in^2}\right)}{4} = 0.196 \ ft^2$$

$$A_4 = \frac{\pi(4 \ in)^2\left(\frac{1 \ ft^2}{144 \ in^2}\right)}{4} = 0.087 \ ft^2$$

Assume a Reynolds number of 5×10^5 and a temperature of $60°F$.

From a Moody diagram, estimate a friction factor using Re and ε/D,

$$f = 0.0235$$

For parallel pipes, the friction losses must be equal.

$$\frac{f_6 L_6 v_6^2}{2D_6 g} = \frac{f_4 L_4 v_4^2}{2D_4 g}$$

$$\frac{(0.0235)(1400 \ ft)v_6^2}{\left(\begin{array}{c}(2)(6 \ in)\left(\frac{1 \ ft}{12 \ in}\right)\\ \times \left(32.2 \ \frac{ft}{sec^2}\right)\end{array}\right)} = \frac{(0.0235)(1400 \ ft)v_4^2}{\left(\begin{array}{c}(2)(4 \ in)\left(\frac{1 \ ft}{12 \ in}\right)\\ \times \left(32.2 \ \frac{ft}{sec^2}\right)\end{array}\right)}$$

$$1.02v_6^2 = 1.53v_4^2$$

$$v_6 = 1.22v_4$$

$$Q = A_6v_6 + A_4v_4$$

$$2.4 \ \frac{ft^3}{sec} = (0.196 \ ft^2)(1.22v_4)$$
$$+ (0.087 \ ft^2)(v_4)$$

$$v_4 = 7.4 \ ft/sec$$

$$Q = Av$$

$$Q_4 = (0.087 \ ft^2)\left(7.4 \ \frac{ft}{sec}\right)$$

$$= 0.64 \ ft^3/sec$$

(D) This incorrect solution simply proportions the flow based on the area of each pipe. Other assumptions, definitions, and equations are unchanged from the correct solution.

Assume the inside pipe diameter is equal to the nominal diameter. Assume friction losses can be ignored since pipes are of equal length.

$$A = \pi\frac{D^2}{4}$$

$$A_6 = \frac{\pi(6 \ in)^2\left(\frac{1 \ ft^2}{144 \ in^2}\right)}{4} = 0.196 \ ft^2$$

$$A_4 = \frac{\pi(4 \ in)^2\left(\frac{1 \ ft^2}{144 \ in^2}\right)}{4} = 0.087 \ ft^2$$

A total combined cross-sectional
  area of both pipes   ft^2

$$A = A_6 + A_4 = 0.196 \text{ ft}^2 + 0.087 \text{ ft}^2$$
$$= 0.283 \text{ ft}^2$$
$$\frac{A}{Q} = \frac{A_4}{Q_4} = \frac{0.283 \text{ ft}^2}{2.4 \frac{\text{ft}^3}{\text{sec}}} = 0.118 \text{ sec/ft}$$
$$Q_4 = \frac{A_4}{\frac{A}{Q}} = \frac{0.087 \text{ ft}^2}{0.118 \frac{\text{sec}}{\text{ft}}} = 0.74 \text{ ft}^3/\text{sec}$$

SOLUTION 30

D inside pipe diameter   ft
$\dot{m}$ mass flow rate   lbm/sec
Q volumetric flow rate   ft^3/sec
ρ water density   62.4 lbm/ft^3

$$\dot{m} = Q\rho = \left(0.55 \frac{\text{ft}^3}{\text{sec}}\right)\left(62.4 \frac{\text{lbm}}{\text{ft}^3}\right)$$
$$= 34.3 \text{ lbm/sec}$$

A pipe cross-sectional area   ft^2
v   flow velocity   ft/sec

Assume the inside pipe diameter is equal to the nominal diameter.

$$A = \pi\frac{D^2}{4} = \pi\frac{(3 \text{ in})^2}{4}\left(\frac{1 \text{ ft}}{12 \text{ in}}\right)^2 = 0.049$$

$$v = \frac{Q}{A} = \frac{0.55 \frac{\text{ft}^3}{\text{sec}}}{0.049 \text{ ft}^2} = 11.2 \text{ ft/sec}$$

F_x force component in the x direction   lbf
g_c gravitational constant   32.2 lbm-ft/lbf-sec^2
p pressure   lbf/ft^2
θ pipe bend angle   degree

Because the elbows have a constant diameter, the pressure and area are constant.

$$F_x = pA(\cos\theta - 1) + \frac{\dot{m}v(\cos\theta - 1)}{g_c}$$
$$= \left(50 \frac{\text{lbf}}{\text{in}^2}\right)\left(144 \frac{\text{in}^2}{\text{ft}^2}\right)(0.049 \text{ ft}^2)(\cos 90° - 1)$$
$$+ \frac{\left(34.3 \frac{\text{lbm}}{\text{sec}}\right)\left(11.2 \frac{\text{ft}}{\text{sec}}\right)(\cos 90° - 1)}{32.2 \frac{\text{lbm-ft}}{\text{lbf-sec}^2}}$$
$$= -364.73 \text{ lbf}$$

F_y force component in the y direction   lbf

$$F_y = \left(pA + \frac{\dot{m}v}{g_c}\right)\sin\theta$$
$$= \left(50 \frac{\text{lbf}}{\text{in}^2}\right)\left(144 \frac{\text{in}^2}{\text{ft}^2}\right)(0.049 \text{ ft}^2)(\sin 90°)$$
$$+ \frac{\left(34.3 \frac{\text{lbm}}{\text{sec}}\right)\left(11.2 \frac{\text{ft}}{\text{sec}}\right)(\sin 90°)}{32.2 \frac{\text{lbm-ft}}{\text{lbf-sec}^2}}$$
$$= 364.73 \text{ lbf}$$

F_R resultant force   lbf

$$F_R = \sqrt{F_x^2 + F_y^2}$$
$$= \sqrt{(-364.73 \text{ lbf})^2 + (364.73 \text{ lbf})^2}$$
$$= 516 \text{ lbf}$$

The answer is (A).

Why Other Options Are Wrong

(B) This incorrect solution uses the volumetric flow rate instead of the mass flow rate. Other assumptions, definitions, and equations are the same as used in the correct solution.

$$v = \frac{Q}{A} = \frac{0.55 \frac{\text{ft}^3}{\text{sec}}}{0.049 \text{ ft}^2} = 11.2 \text{ ft/sec}$$
$$F_x = pA(\cos\theta - 1) + \frac{\dot{m}v(\cos\theta - 1)}{g_c}$$
$$= \left(50 \frac{\text{lbf}}{\text{in}^2}\right)\left(144 \frac{\text{in}^2}{\text{ft}^2}\right)(0.049 \text{ ft}^2)(\cos 90° - 1)$$
$$+ \frac{\left(0.55 \frac{\text{ft}^3}{\text{sec}}\right)\left(11.2 \frac{\text{ft}}{\text{sec}}\right)(\cos 90° - 1)}{32.2 \frac{\text{lbm-ft}}{\text{lbf-sec}^2}}$$
$$= -370 \text{ lbf}$$

The units for all terms are not the same and do not add. Assume units are lbf.

$$F_y = \left(pA + \frac{\dot{m}v}{g_c}\right)\sin\theta$$
$$= \left(50 \frac{\text{lbf}}{\text{in}^2}\right)\left(144 \frac{\text{in}^2}{\text{ft}^2}\right)(0.049 \text{ ft}^2)(\sin 90°)$$
$$+ \frac{\left(0.55 \frac{\text{ft}^3}{\text{sec}}\right)\left(11.2 \frac{\text{ft}}{\text{sec}}\right)(\sin 90°)}{32.2 \frac{\text{lbm-ft}}{\text{lbf-sec}^2}}$$
$$= 370 \text{ lbf}$$

The units for all terms are not the same and do not add. Assume units are lbf.

$$F_R = \sqrt{F_x^2 + F_y^2}$$
$$= \sqrt{(-370 \text{ lbf})^2 + (370 \text{ lbf})^2}$$
$$= 523 \text{ lbf}$$

(C) This incorrect solution enters degrees for the pipe bend angle but calculates the sin and cos as radians. (Uses sin 90 radians and cos 90 radians.) Other assumptions, definitions, and equations are the same as used in the correct solution.

$$\dot{m} = Q\rho = \left(0.55 \frac{\text{ft}^3}{\text{sec}}\right)\left(62.4 \frac{\text{lbm}}{\text{ft}^3}\right) = 34.3 \text{ lbm/sec}$$

$$v = \frac{Q}{A} = \frac{0.55 \frac{\text{ft}^3}{\text{sec}}}{0.049 \text{ ft}^2} = 11.2 \text{ ft/sec}$$

Because the elbows have a constant diameter, the pressure and area are constant.

$$F_x = pA(\cos\theta - 1) + \frac{\dot{m}v(\cos\theta - 1)}{g_c}$$
$$= \left(50 \frac{\text{lbf}}{\text{in}^2}\right)\left(144 \frac{\text{in}^2}{\text{ft}^2}\right)(0.049 \text{ ft}^2)(\cos 90 - 1)$$
$$+ \frac{\left(34.3 \frac{\text{lbm}}{\text{sec}}\right)\left(11.2 \frac{\text{ft}}{\text{sec}}\right)(\cos 90 - 1)}{32.2 \frac{\text{lbm-ft}}{\text{lbf-sec}^2}}$$
$$= -552 \text{ lbf}$$

$$F_y = \left(pA + \frac{\dot{m}v}{g_c}\right)\sin\theta$$
$$= \left(50 \frac{\text{lbf}}{\text{in}^2}\right)\left(144 \frac{\text{in}^2}{\text{ft}^2}\right)(0.049 \text{ ft}^2)(\sin 90)$$
$$+ \frac{\left(34.3 \frac{\text{lbm}}{\text{sec}}\right)\left(11.2 \frac{\text{ft}}{\text{sec}}\right)(\sin 90)}{32.2 \frac{\text{lbm-ft}}{\text{lbf-sec}^2}}$$
$$= 341 \text{ lbf}$$

$$F_R = \sqrt{F_x^2 + F_y^2}$$
$$= \sqrt{(-552 \text{ lbf})^2 + (341 \text{ lbf})^2}$$
$$= 649 \text{ lbf}$$

(D) This incorrect solution squares the velocity term in the force equations. Other assumptions, definitions, and equations are the same as used in the correct solution.

$$\dot{m} = Q\rho = \left(0.55 \frac{\text{ft}^3}{\text{sec}}\right)\left(62.4 \frac{\text{lbm}}{\text{ft}^3}\right)$$
$$= 34.3 \text{ lbm/sec}$$

$$v = \frac{Q}{A} = \frac{0.55 \frac{\text{ft}^3}{\text{sec}}}{0.049 \text{ ft}^2}$$
$$= 11.2 \text{ ft/sec}$$

$$F_x = pA(\cos\theta - 1) + \frac{\dot{m}v^2(\cos\theta - 1)}{g_c}$$
$$= \left(50 \frac{\text{lbf}}{\text{in}^2}\right)\left(144 \frac{\text{in}^2}{\text{ft}^2}\right)(0.049 \text{ ft}^2)(\cos 90° - 1)$$
$$+ \frac{\left(34.3 \frac{\text{lbm}}{\text{sec}}\right)\left(11.2 \frac{\text{ft}}{\text{sec}}\right)^2(\cos 90° - 1)}{32.2 \frac{\text{lbm-ft}}{\text{lbf-sec}}}$$
$$= -514 \text{ lbf}$$

The units for all terms are not the same and do not add. Assume units are lbf.

$$F_y = \left(pA + \frac{\dot{m}v^2}{g_c}\right)\sin\theta$$
$$= \left(50 \frac{\text{lbf}}{\text{in}^2}\right)\left(144 \frac{\text{in}^2}{\text{ft}^2}\right)(0.049 \text{ ft}^2)(\sin 90°)$$
$$+ \frac{\left(34.3 \frac{\text{lbm}}{\text{sec}}\right)\left(11.2 \frac{\text{ft}}{\text{sec}}\right)^2(\sin 90°)}{32.2 \frac{\text{lbm-ft}}{\text{lbf-sec}}}$$
$$= 514 \text{ lbf}$$

$$F_R = \sqrt{F_x^2 + F_y^2}$$
$$= \sqrt{(-514 \text{ lbf})^2 + (514 \text{ lbf})^2}$$
$$= 727 \text{ lbf}$$

SOLUTION 31

b channel base width   m
d water depth   m
R_h hydraulic radius   m
θ side-wall angle measured from
       the horizontal   degree

For a 1-to-1 side wall slope, θ is 45°.

$$R_h = \frac{bd\sin\theta + d^2\cos\theta}{b\sin\theta + 2d}$$
$$= \frac{b(1 \text{ m})(\sin 45°) + (1 \text{ m})^2(\cos 45°)}{b(\sin 45°) + (2)(1 \text{ m})}$$
$$= \frac{0.707b + 0.707}{0.707b + 2} \text{ m}$$

Because the side walls and base are constructed of different materials, use an average Manning roughness coefficient.

n Manning roughness coefficient   –

For concrete,
$$n = 0.013$$

For firm gravel,
$$n = 0.023$$
$$n_{ave} = \frac{0.013 + 0.023}{2} = 0.018$$

S channel slope   m/m
v flow velocity   m/s

$$v = \left(\frac{1}{n}\right) R_h^{2/3} \sqrt{S}$$

$$0.75 = \left(\frac{1}{0.018}\right)\left(\frac{0.707b + 0.707}{0.707b + 2}\ \text{m}\right)^{2/3} \sqrt{0.0005}$$

$$0.604^{3/2} = \frac{0.707b + 0.707}{0.707b + 2}$$

$$(0.469)(0.707b + 2) = 0.707b + 0.707$$
$$b = 0.62\ \text{m}$$

The answer is (A).

Why Other Options Are Wrong

(B) This incorrect solution solves for the water depth, using the given depth as the base width. Other assumptions, definitions, and equations are the same as used in the correct solution.

$$R_h = \frac{bd\sin\theta + d^2\cos\theta}{b\sin\theta + 2d}$$
$$= \frac{d(1\ \text{m})(\sin 45°) + d^2(\cos 45°)}{(1\ \text{m})(\sin 45°) + 2d}$$
$$= \frac{0.707d + 0.707d^2}{0.707 + 2d}$$

For concrete,
$$n = 0.013$$

For firm gravel,

$$n = 0.023$$
$$n_{ave} = \frac{0.013 + 0.023}{2} = 0.018$$
$$v = \left(\frac{1}{n}\right) R_h^{2/3}\sqrt{S}$$
$$0.75 = \left(\frac{1}{0.018}\right)\left(\frac{0.707d + 0.707d^2}{0.707 + 2d}\right)^{2/3}$$
$$\times \sqrt{0.0005}$$
$$0.604^{3/2} = \frac{0.707d + 0.707d^2}{0.707 + 2d}$$
$$(0.469)(0.707 + 2d) = 0.707d + 0.707d^2$$

Solve for d by trial and error.
$$d = 0.87\ \text{m}$$

(C) This incorrect solution uses the hydraulic radius equation for a rectangular instead of a trapezoidal channel. Other assumptions, definitions, and equations are the same as used in the correct solution.

$$R_h = \frac{bd}{b + 2d} = \frac{b(1\ \text{m})}{b + (2)(1\ \text{m})} = \frac{b}{b + 2}$$

For concrete,
$$n = 0.013$$

For firm gravel,
$$n = 0.023$$
$$n_{ave} = \frac{0.013 + 0.023}{2} = 0.018$$
$$v = \left(\frac{1}{n}\right) R_h^{2/3}\sqrt{S}$$
$$0.75 = \left(\frac{1}{0.018}\right)\left(\frac{b}{b+2}\right)^{2/3}\sqrt{0.0005}$$
$$0.604^{3/2} = \frac{b}{b+2}$$
$$(0.469)(b + 2) = b$$
$$b = 1.8\ \text{m}$$

(D) This incorrect solution makes a mathematical error with the exponent in the velocity equation. Other assumptions, definitions, and equations are the same as used in the correct solution.

$$R_h = \frac{b(1\ \text{m})(\sin 45°) + (1\ \text{m})^2(\cos 45°)}{b(\sin 45°) + (2)(1\ \text{m})}$$
$$= \frac{0.707b + 0.707}{0.707b + 2}$$

For concrete,
$$n = 0.013$$

For firm gravel,

$$n = 0.023$$
$$n_{ave} = \frac{0.013 + 0.023}{2}$$
$$= 0.018$$
$$v = \left(\frac{1}{n}\right) R_h^{2/3}\sqrt{S}$$
$$0.75 = \left(\frac{1}{0.018}\right)\left(\frac{0.707b + 0.707}{0.707b + 2}\right)^{2/3}$$
$$\times \sqrt{0.0005}$$
$$0.604^{2/3} = \frac{0.707b + 0.707}{0.707b + 2}$$
$$(0.715)(0.707b + 2) = 0.707b + 0.707$$
$$b = 3.6\ \text{m}$$

SOLUTION 32

d_1 upstream water depth   ft
Q flow rate ft^3/sec
v flow velocity   ft/sec
w channel width   ft

$$Q = vd_1w = \left(3\,\frac{\text{ft}}{\text{sec}}\right)(8\text{ ft})(15\text{ ft})$$

$$= 360\text{ ft}^3/\text{sec}$$

d_c critical water depth over
      the barrier   ft
g gravitational acceleration   32.2 ft/sec^2
y_c barrier height   ft

$$d_1 + \frac{Q^2}{2gw^2d_1^2} = d_c + \frac{Q^2}{2gw^2d_c^2} + y_c$$

$$= 8\text{ ft} + \frac{\left(360\,\dfrac{\text{ft}^3}{\text{sec}}\right)^2}{\left(\begin{array}{c}(2)\left(32.2\,\dfrac{\text{ft}}{\text{sec}^2}\right)(15\text{ ft})^2 \\ \times (8\text{ ft})^2\end{array}\right)}$$

$$= 8.14\text{ ft}$$

$$d_c = \left(\frac{Q^2}{gw^2}\right)^{1/3}$$

$$= \left(\frac{\left(360\,\dfrac{\text{ft}^3}{\text{sec}}\right)^2}{\left(32.2\,\dfrac{\text{ft}}{\text{sec}^2}\right)(15\text{ ft})^2}\right)^{1/3}$$

$$= 2.62\text{ ft}$$

$$8.14\text{ ft} = d_c + \frac{Q^2}{2gw^2d_c^2} + y_c$$

$$= 2.62\text{ ft} + \frac{\left(360\,\dfrac{\text{ft}^3}{\text{sec}}\right)^2}{\left(\begin{array}{c}(2)\left(32.2\,\dfrac{\text{ft}}{\text{sec}^2}\right) \\ \times (15\text{ ft})^2(2.62\text{ ft})^2\end{array}\right)} + y_c$$

$$y_c = 8.14\text{ ft} - 2.62\text{ ft} - 1.3\text{ ft}$$

$$= 4.2\text{ ft}$$

The answer is (D).

Why Other Options Are Wrong

(A) This incorrect solution uses the flow per unit width and includes the width in the critical depth calculation. The width should not be included when the flow per unit width is used. Where the width is included, it is squared in the denominator, and does not occur in the numerator. Other definitions and equations are unchanged from the correct solution.

q flow rate per unit width ft^2/sec

$$q = vd_1 = \left(3\,\frac{\text{ft}}{\text{sec}}\right)(8\text{ ft})$$

$$= 24\text{ ft}^2/\text{sec}$$

$$d_1 + \frac{q^2}{2gd_1^2} = d_c + \frac{q^2}{2gd_c^2} + y_c$$

$$= 8\text{ ft} + \frac{\left(24\,\dfrac{\text{ft}^3}{\text{sec}}\right)^2}{(2)\left(32.2\,\dfrac{\text{ft}}{\text{sec}^2}\right)(8\text{ ft})^2}$$

$$= 8.14\text{ ft}$$

$$d_c = \left(\frac{q^2w}{g}\right)^{1/3}$$

$$= \left(\frac{\left(24\,\dfrac{\text{ft}^3}{\text{sec}}\right)^2(15\text{ ft})}{\left(32.2\,\dfrac{\text{ft}}{\text{sec}^2}\right)}\right)^{1/3}$$

$$= 6.44\text{ ft}$$

$$8.14\text{ ft} = d_c + \frac{Q^2}{2gd_c^2} + y_c$$

$$= 6.44\text{ ft} + \frac{\left(24\,\dfrac{\text{ft}^3}{\text{sec}}\right)^2}{(2)\left(32\,\dfrac{\text{ft}}{\text{sec}^2}\right)(6.44\text{ ft})^2} + y_c$$

$$y_c = 8.14\text{ ft} - 6.44\text{ ft} - 0.22\text{ ft}$$

$$= 1.48\text{ ft}\quad(1.5\text{ ft})$$

(B) This incorrect solution solves for the critical depth instead of the barrier height. Other definitions and equations are unchanged from the correct solution.

$$Q = vd_1w = \left(3\,\frac{\text{ft}}{\text{sec}}\right)(8\text{ ft})(15\text{ ft})$$

$$= 360\text{ ft}^3/\text{sec}$$

$$d_c = \left(\frac{Q^2}{gw^2}\right)^{1/3} = \left(\frac{\left(360\,\dfrac{\text{ft}^3}{\text{sec}}\right)^2}{\left(32.2\,\dfrac{\text{ft}}{\text{sec}^2}\right)(15\text{ ft})^2}\right)^{1/3}$$

$$= 2.6\text{ ft}$$

(C) This incorrect solution fails to multiply the gravitational acceleration by 2 in the denominator of the energy equation. Other definitions and equations are unchanged from the correct solution.

$$Q = vd_1w = \left(3\,\frac{\text{ft}}{\text{sec}}\right)(8\text{ ft})(15\text{ ft})$$

$$= 360\text{ ft}^3/\text{sec}$$

$$d_1 + \frac{Q^2}{gw^2d_1^2} = d_c + \frac{Q^2}{gw^2d_c^2} + y_c$$

$$= 8 \text{ ft} + \frac{\left(360 \ \frac{\text{ft}^3}{\text{sec}}\right)^2}{\left(32.2 \ \frac{\text{ft}}{\text{sec}^2}\right)(15 \text{ ft})^2(8 \text{ ft})^2}$$

$$= 8.28 \text{ ft}$$

$$d_c = \left(\frac{Q^2}{gw^2}\right)^{1/3}$$

$$= \left(\frac{\left(360 \ \frac{\text{ft}^3}{\text{sec}}\right)^2}{\left(32.2 \ \frac{\text{ft}}{\text{sec}}\right)^2(15 \text{ ft})^2}\right)^{1/3}$$

$$= 2.62 \text{ ft}$$

$$8.28 \text{ ft} = d_c + \frac{Q^2}{gw^2d_c^2} + y_c$$

$$= 2.62 \text{ ft} + \frac{\left(360 \ \frac{\text{ft}^3}{\text{sec}}\right)^2}{\left(\left(32.2 \ \frac{\text{ft}}{\text{sec}^2}\right) \times (15 \text{ ft})^2(2.62 \text{ ft})^2\right)} + y_c$$

$$y_c = 8.28 \text{ ft} - 2.62 \text{ ft} - 2.61 \text{ ft}$$

$$= 3.0 \text{ ft}$$

SOLUTION 33

Behind the dam,

d_1 water depth behind the dam   m
E_1 energy line behind the dam   m

$$E_1 = d_1 = 21 \text{ m} + 2 \text{ m}$$
$$= 23 \text{ m}$$

At the spillway crest, the critical energy is equal to the water depth above the crest.

d_c critical depth at the spillway crest   m
E_c energy line at the spillway crest   m

$$E_c = 2 \text{ m}$$

$$d_c = \left(\frac{2}{3}\right) E_c = \left(\frac{2}{3}\right)(2 \text{ m})$$

$$= 1.33 \text{ m}$$

g gravitational acceleration 9.81 m/s^2
v_c critical water velocity   m/s

$$E_c = d_c + \frac{v_c^2}{2g}$$

$$2 \text{ m} = 1.33 \text{ m} + \frac{v_c^2}{(2)\left(9.81 \ \frac{\text{m}}{\text{s}^2}\right)}$$

$$v_c = 3.62 \text{ m/s}$$

q flow rate per unit width m^2/s

$$q = v_c d_c = \left(3.62 \ \frac{\text{m}}{\text{s}}\right)(1.33 \text{ m})$$

$$= 4.81 \text{ m}^2/\text{s}$$

At the toe of the dam,

d_2 water depth at the toe of the dam   m
E_2 energy line at the toe of the dam   m

$$E_1 = E_2 = d_2 + \frac{q^2}{2gd_2^2}$$

$$23 \text{ m} = d_2 + \frac{\left(4.81 \ \frac{\text{m}^2}{\text{s}}\right)^2}{(2)\left(9.81 \ \frac{\text{m}}{\text{s}^2}\right)d_2^2} = d_2 + \frac{1.18}{d_2^2}$$

Solve for d_2 by trial and error.

$$d_2 = 0.23 \text{ m}$$

F_d force on the dam per unit width   kN/m
ρ density of water 1000 kg/m^3

$$F_d = \rho g \left(\frac{d_1^2}{2} + \frac{q^2}{gd_1} - \frac{d_2^2}{2} - \frac{q^2}{gd_2}\right)$$

$$\rho g = \left(1000 \ \frac{\text{kg}}{\text{m}^3}\right)\left(9.81 \ \frac{\text{m}}{\text{s}^2}\right)\left(\frac{\text{N·s}^2}{\text{kg·m}}\right)\left(\frac{1 \text{ kN}}{10^3 \text{ N}}\right)$$

$$= 9.81 \text{ kN/m}^3$$

$$F_d = \left(9.81 \ \frac{\text{kN}}{\text{m}^3}\right)$$

$$\times \left(\begin{array}{c} \dfrac{(23 \text{ m})^2}{2} + \dfrac{\left(4.81 \ \frac{\text{m}^2}{\text{s}}\right)^2}{\left(9.81 \ \frac{\text{m}}{\text{s}^2}\right)(23 \text{ m})} \\[3ex] - \dfrac{(0.23 \text{ m})^2}{2} - \dfrac{\left(4.81 \ \frac{\text{m}^2}{\text{s}}\right)^2}{\left(9.81 \ \frac{\text{m}}{\text{s}^2}\right)(0.23 \text{ m})} \end{array}\right)$$

$$= 2495 \text{ kN/m} \quad (2500 \text{ kN/m})$$

The answer is (D).

Why Other Options Are Wrong

(A) This incorrect solution assumes that at every point the energy line is equal to the depth of the water above the crest and that this is also equal to the critical depth. Other assumptions, definitions, and equations are unchanged from the correct solution.

At the spillway crest, the critical energy is equal to water depth above the crest.

$$E_1 = E_c = d_c = 2 \text{ m}$$

$$d_c = \frac{v_c^2}{2g}$$

$$2 \text{ m} = \frac{v_c^2}{(2)\left(9.81 \frac{\text{m}}{\text{s}^2}\right)}$$

$$v_c = 6.26 \text{ m/s}$$

$$q = v_c d_c = \left(6.26 \frac{\text{m}}{\text{s}}\right)(2 \text{ m})$$

$$= 12.52 \text{ m}^2/\text{s}$$

At the toe of the dam,

$$E_1 = E_2 = d_2 + \frac{q^2}{2gd_2^2}$$

$$2 \text{ m} = d_2 + \frac{\left(12.52 \frac{\text{m}^2}{\text{s}}\right)^2}{(2)\left(9.81 \frac{\text{m}}{\text{s}^2}\right)d_2^2} = d_2 + \frac{7.99 \text{ m}^3}{d_2^2}$$

Solve for d_2 by trial and error.

$$d_2 = 1.51 \text{ m}$$

The negative sign is ignored.

$$F_d = \rho g \left(\frac{d_1^2}{2} + \frac{q^2}{gd_1} - \frac{d_2^2}{2} - \frac{q^2}{gd_2}\right)$$

$$\rho g = \left(1000 \frac{\text{kg}}{\text{m}^3}\right)\left(9.81 \frac{\text{m}}{\text{s}^2}\right)\left(\frac{\text{N·s}^2}{\text{kg·m}}\right)\left(\frac{1 \text{ kN}}{10^3 \text{ N}}\right)$$

$$= 9.81 \text{ kN/m}^3$$

$$F_d = \left(9.81 \frac{\text{kN}}{\text{m}^3}\right)$$

$$\times \left(\begin{array}{c} \dfrac{(2 \text{ m})^2}{2} + \dfrac{\left(12.52 \frac{\text{m}^2}{\text{s}}\right)^2}{\left(9.81 \frac{\text{m}}{\text{s}^2}\right)(2 \text{ m})} \\[20pt] - \dfrac{(1.51 \text{ m})^2}{2} - \dfrac{\left(12.52 \frac{\text{m}^2}{\text{s}}\right)^2}{\left(9.81 \frac{\text{m}}{\text{s}^2}\right)(1.51 \text{ m})} \end{array}\right)$$

$$= 17 \text{ kN/m}$$

The negative sign in the result is ignored.

(B) This incorrect solution fails to square the depth terms in the energy equation at the toe of the dam and in the force equation. Other assumptions, definitions, and equations are unchanged from the correct solution.

Behind the dam,

$$E_1 = d_1 = 21 \text{ m} + 2 \text{ m} = 23 \text{ m}$$

At the spillway crest, the critical energy is equal to water depth above the crest.

$$E_c = 2 \text{ m}$$

$$d_c = \left(\frac{2}{3}\right) E_c = \left(\frac{2}{3}\right) 2 \text{ m} = 1.33 \text{ m}$$

$$2 \text{ m} = 1.33 \text{ m} + \frac{v_c^2}{(2)\left(9.81 \frac{\text{m}}{\text{s}^2}\right)}$$

$$v_c = 3.62 \text{ m/s}$$

$$q = v_c d_c = \left(3.62 \frac{\text{m}}{\text{s}}\right)(1.33 \text{ m})$$

$$= 4.81 \text{ m}^2/\text{s}$$

At the toe of the dam,

$$E_1 = E_2 = d_2 + \frac{q^2}{2gd_2}$$

$$23 \text{ m} = d_2 + \frac{\left(4.81 \frac{\text{m}^2}{\text{s}}\right)^2}{(2)\left(9.81 \frac{\text{m}}{\text{s}^2}\right)d_2}$$

$$= d_2 + \frac{1.18 \text{ m}^3}{d_2}$$

$$d_2^2 - 23d_2 + 1.18 = 0$$

Solve for d_2 using the quadratic formula.

$$d_2 = 0.051 \text{ m}$$

$$F_d = \rho g \left(\frac{d_1}{2} + \frac{q^2}{gd_1} - \frac{d_2}{2} - \frac{q^2}{gd_2}\right)$$

$$\rho g = \left(1000 \frac{\text{kg}}{\text{m}^3}\right)\left(9.81 \frac{\text{m}}{\text{s}^2}\right)\left(\frac{\text{N·s}^2}{\text{kg·m}}\right)\left(\frac{1 \text{ kN}}{10^3 \text{ N}}\right)$$

$$= 9.81 \text{ kN/m}^3$$

$$F_d = \left(9.81 \frac{\text{kN}}{\text{m}^3}\right)$$

$$\times \left(\begin{array}{c} \dfrac{23 \text{ m}}{2} + \dfrac{\left(4.81 \frac{\text{m}^2}{\text{s}}\right)^2}{\left(9.81 \frac{\text{m}}{\text{s}^2}\right)(23 \text{ m})} \\[20pt] - \dfrac{0.051 \text{ m}}{2} - \dfrac{\left(4.81 \frac{\text{m}^2}{\text{s}}\right)^2}{\left(9.81 \frac{\text{m}}{\text{s}^2}\right)(0.051 \text{ m})} \end{array}\right)$$

$$= 340 \text{ kN/m}$$

The negative sign in the result is ignored.

(C) This incorrect solution uses the critical depth as the depth of the water above the spillway. Other assumptions, definitions, and equations are unchanged from the correct solution.

Behind the dam,

$$E_1 = d_1 = 21 \text{ m} + 2 \text{ m}$$
$$= 23 \text{ m}$$

At the spillway crest, the critical depth is equal to water depth above the crest.

$$d_c = 2 \text{ m}$$

$$d_c = \frac{\text{v}_c^2}{2g}$$

$$2 \text{ m} = \frac{\text{v}_c^2}{(2)\left(9.81 \dfrac{\text{m}}{\text{s}^2}\right)}$$

$$\text{v}_c = 6.26 \text{ m/s}$$

$$q = \text{v}_c d_c$$
$$= \left(6.26 \dfrac{\text{m}}{\text{s}}\right)(2 \text{ m})$$
$$= 12.52 \text{ m}^2/\text{s}$$

At the toe of the dam,

$$E_1 = E_2 = d_2 + \frac{q^2}{2gd_2^2}$$

$$23 \text{ m} = d_2 + \frac{\left(12.52 \dfrac{\text{m}^2}{\text{s}}\right)^2}{(2)\left(9.81 \dfrac{\text{m}}{\text{s}^2}\right)d_2^2}$$

$$= d_2 + \frac{8.0 \text{ m}^3}{d_2^2}$$

Solve for d_2 by trial and error.

$$d_2 = 0.6 \text{ m}$$

$$F_d = \rho g\left(\frac{d_1^2}{2} + \frac{q^2}{gd_1} - \frac{d_2^2}{2} - \frac{q^2}{gd_2}\right)$$

$$\rho g = \left(1000 \dfrac{\text{kg}}{\text{m}^3}\right)\left(9.81 \dfrac{\text{m}}{\text{s}^2}\right)\left(\dfrac{\text{N}\cdot\text{s}^2}{\text{kg}\cdot\text{m}}\right)\left(\dfrac{1 \text{ kN}}{10^3 \text{ N}}\right)$$
$$= 9.81 \text{ kN/m}^3$$

$$F_d = \left(9.81 \dfrac{\text{kN}}{\text{m}^3}\right)$$

$$\times \left(\begin{array}{c} \dfrac{(23 \text{ m})^2}{2} + \dfrac{\left(12.52 \dfrac{\text{m}^2}{\text{s}}\right)^2}{\left(9.81 \dfrac{\text{m}}{\text{s}^2}\right)(23 \text{ m})} \\[4mm] - \dfrac{(0.6 \text{ m})^2}{2} - \dfrac{\left(12.52 \dfrac{\text{m}^2}{\text{s}}\right)^2}{\left(9.81 \dfrac{\text{m}}{\text{s}^2}\right)(0.6 \text{ m})} \end{array} \right)$$

$$= 2338 \text{ kN/m} \quad (2300 \text{ kN/m})$$

SOLUTION 34

θ side slope angle measured from the horizontal   degrees

For 3-to-1 horizontal-to-vertical slope,

$$\theta = \tan^{-1}\left(\frac{1}{3}\right) = 18.4°$$

A cross-sectional area of channel   m^2
b base width   m
d water depth   m

$$A = \left(b + \frac{d}{\tan\theta}\right)d$$
$$= \left(2 \text{ m} + \frac{1 \text{ m}}{\tan 18.4°}\right)(1 \text{ m})$$
$$= 5.0 \text{ m}^2$$

P wetted perimeter   m

$$P = b + 2\left(\frac{d}{\sin\theta}\right) = 2 \text{ m} + (2)\left(\frac{1 \text{ m}}{\sin 18.4°}\right)$$
$$= 8.3 \text{ m}$$

S channel slope   m/m
ΔL change in distance   m
Δz change in elevation   m

$$S = \frac{\Delta z}{\Delta L} = \frac{20 \text{ m}}{(9 \text{ km})\left(1000 \dfrac{\text{m}}{\text{km}}\right)}$$
$$= 0.0022 \text{ m/m}$$

n Manning roughness coefficient   –
Q flow rate   m^3/s

For an earth-lined channel, the Manning roughness coefficient is 0.018.

$$Q = \frac{\left(\dfrac{1}{n}\right)A^{5/3}\sqrt{S}}{P^{2/3}}$$
$$= \frac{\left(\dfrac{1}{0.018}\right)(5 \text{ m}^2)^{5/3}\sqrt{0.0022 \dfrac{\text{m}}{\text{m}}}}{(8.3 \text{ m})^{2/3}}$$
$$= 9.3 \text{ m}^3/\text{s}$$

The answer is (B).

Why Other Options Are Wrong

(A) This incorrect solution miscalculates the side slope angle. Other assumptions, definitions, and equations are the same as used in the correct solution.

θ side slope angle measured from the horizontal   degrees

For 3-to-1 horizontal-to-vertical slope,

$$\theta = \cos^{-1}\left(\frac{1}{3}\right) = 70.5°$$

$$A = \left(b + \frac{d}{\tan\theta}\right)d = \left(2 \text{ m} + \frac{1 \text{ m}}{\tan 70.5°}\right)(1 \text{ m})$$

$$= 2.35 \text{ m}^2$$

$$P = b + 2\left(\frac{d}{\sin\theta}\right) = 2 \text{ m} + (2)\left(\frac{1 \text{ m}}{\sin 70.5°}\right)$$

$$= 4.1 \text{ m}$$

$$s = \frac{\Delta z}{\Delta L} = \frac{20 \text{ m}}{(9 \text{ km})\left(1000 \frac{\text{m}}{\text{km}}\right)}$$

$$= 0.0022 \text{ m/m}$$

$$Q = \frac{\left(\frac{1}{n}\right)A^{5/3}\sqrt{S}}{P^{2/3}}$$

$$= \frac{\left(\frac{1}{0.018}\right)(2.35 \text{ m}^2)^{5/3}\sqrt{0.0022 \frac{\text{m}}{\text{m}}}}{(4.1 \text{ m})^{2/3}}$$

$$= 4.2 \text{ m}^3/\text{s}$$

(C) This incorrect solution reverses the values for channel base and water depth. Other assumptions, definitions and equations are the same as used in the correct solution.

For 3-to-1 horizontal-to-vertical slope,

$$\theta = \tan^{-1}\left(\frac{1}{3}\right) = 18.4°$$

$$A = \left(b + \frac{d}{\tan\theta}\right)d$$

$$= \left(1 \text{ m} + \frac{2 \text{ m}}{\tan 18.4°}\right)(2 \text{ m})$$

$$= 14 \text{ m}^2$$

$$P = b + 2\left(\frac{d}{\sin\theta}\right)$$

$$= 1 \text{ m} + (2)\left(\frac{2 \text{ m}}{\sin 18.4°}\right)$$

$$= 13.7 \text{ m}$$

$$S = \frac{\Delta z}{\Delta L} = \frac{20 \text{ m}}{(9 \text{ km})\left(1000 \frac{\text{m}}{\text{km}}\right)}$$

$$= 0.0022 \text{ m/m}$$

$$Q = \frac{\left(\frac{1}{n}\right)A^{5/3}\sqrt{S}}{P^{2/3}}$$

$$= \frac{\left(\frac{1}{0.018}\right)(14 \text{ m}^2)^{5/3}\sqrt{0.0022 \frac{\text{m}}{\text{m}}}}{(13.7 \text{ m})^{2/3}}$$

$$= 37 \text{ m}^3/\text{s}$$

(D) This incorrect solution uses the equation for the hydraulic radius to calculate the wetted perimeter. Other assumptions, definitions, and equations are the same as used in the correct solution.

For 3-to-1 horizontal-to-vertical slope,

$$\theta = \tan^{-1}\left(\frac{1}{3}\right) = 18.4°$$

$$A = \left(b + \frac{d}{\tan\theta}\right)d$$

$$= \left(2 \text{ m} + \frac{1 \text{ m}}{\tan 18.4°}\right)(1 \text{ m})$$

$$= 5.0 \text{ m}^2$$

$$P = \frac{bd\sin\theta + d^2\cos\theta}{b\sin\theta + 2d}$$

$$= \frac{(2 \text{ m})(1 \text{ m})(\sin 18.4°) + (1 \text{ m})^2(\cos 18.4°)}{(1 \text{ m})(\sin 18.4°) + (2)(1 \text{ m})}$$

$$= 0.68 \text{ m}$$

$$S = \frac{\Delta z}{\Delta L} = \frac{20 \text{ m}}{(9 \text{ km})\left(1000 \frac{\text{m}}{\text{km}}\right)}$$

$$= 0.0022 \text{ m/m}$$

$$Q = \frac{\left(\frac{1}{n}\right)A^{5/3}\sqrt{S}}{P^{2/3}}$$

$$= \frac{\left(\frac{1}{0.018}\right)(5.0 \text{ m}^2)^{5/3}\sqrt{0.0022 \frac{\text{m}}{\text{m}}}}{(0.68 \text{ m})^{2/3}}$$

$$= 49 \text{ m}^3/\text{s}$$

SOLUTION 35

Express infiltration to each section as $\text{m}^3/\text{d·mm·km}$.

section	total infiltration to section (m^3/d)	pipe diameter (mm)	pipe length (km)	total unit infiltration ($\text{m}^3/\text{d·mm·km}$)	infiltration to section ($\text{m}^3/\text{d·mm·km}$)
1	2315	100	13.4	1.73	
		200	6.8	1.70	
		300	6.2	1.24	4.67
2	958	100	9.2	1.04	
		200	7.1	0.67	
		300	4.4	0.73	2.44
3	3996	100	24.9	1.60	
		200	12.1	1.65	
		300	11.9	1.12	4.37
4	1867	100	21.3	0.88	
		200	11.0	0.85	
		300	4.7	1.32	3.05

$$\text{total unit infiltration} = \frac{\text{total infiltration to section}}{(\text{pipe diameter})(\text{pipe length})}$$

The section with the highest infiltration rate when pipe diameter and pipe length are included is section 1. Section 1 should receive first priority for rehabilitation.

The answer is (A).

Why Other Options Are Wrong

(B) This choice is incorrect because section 2 has the lowest infiltration rate either based on units of m^3/d or units of $m^3/d{\cdot}mm{\cdot}km$. Although this may be the least expensive to rehabilitate because it has the shortest total pipe length, the contribution to infiltration from this section does not justify its rehabilitation ahead of other sections.

(C) This choice is incorrect because it selects, by observation of the infiltration data in units of m^3/d, the section with the highest overall infiltration rate, section 3.

(D) This choice is incorrect because section 4 does not present any characteristics that would place it ahead of other sections. It would be appropriate to rehabilitate section 4 before section 2, but not before section 1. Some economic justification based on total pipe length may exist for rehabilitating section 4 before section 3, but not before section 2.

SOLUTION 36

h elevation head above the orifice   ft

$$h = 100 \text{ ft} - 92.6 \text{ ft}$$
$$= 7.4 \text{ ft}$$

For a sharp-edged orifice, the orifice coefficient is 0.62.

A	orifice area	ft^2
C_d	orifice coefficient	–
g	gravitational acceleration	32.2 ft/sec^2
Q	discharge rate through the orifice	ft^3/sec

$$A = \frac{Q}{C_d\sqrt{2gh}} = \frac{20 \, \frac{\text{ft}^3}{\text{sec}}}{(0.62)\sqrt{(2)\left(32.2 \, \frac{\text{ft}}{\text{sec}^2}\right)(7.4 \text{ ft})}}$$

$$= 1.48 \text{ ft}^2$$

D orifice diameter   ft

$$D = \sqrt{\frac{4A}{\pi}} = \sqrt{\frac{(4)(1.48 \text{ ft}^2)}{\pi}}$$
$$= 1.4 \text{ ft}$$

The answer is (D).

Why Other Options Are Wrong

(A) This incorrect solution applies the square root only to the numerator in the final equation to determine diameter. Other assumptions, definitions, and equations are the same as used in the correct solution.

$$h = 100 \text{ ft} - 92.6 \text{ ft}$$
$$= 7.4 \text{ ft}$$

$$A = \frac{Q}{C_d\sqrt{2gh}} = \frac{20 \, \frac{\text{ft}^3}{\text{sec}}}{(0.62)\sqrt{(2)\left(32.2 \, \frac{\text{ft}}{\text{sec}^2}\right)(7.4 \text{ ft})}}$$

$$= 1.48 \text{ ft}^2$$

$$D = \frac{\sqrt{4A}}{\pi} = \frac{\sqrt{(4)(1.48 \text{ ft}^2)}}{\pi}$$

$$= 0.77 \text{ ft}$$

(B) This incorrect solution multiplies instead of divides the flow rate by the orifice coefficient in the area equation. Other assumptions, definitions, and equations are the same as used in the correct solution.

$$h = 100 \text{ ft} - 92.6 \text{ ft}$$
$$= 7.4 \text{ ft}$$

$$A = \frac{C_d Q}{\sqrt{2gh}} = \frac{(0.62)\left(20 \, \frac{\text{ft}^3}{\text{sec}}\right)}{\sqrt{(2)\left(32.2 \, \frac{\text{ft}}{\text{sec}^2}\right)(7.4 \text{ ft})}}$$

$$= 0.57 \text{ ft}^2$$

$$D = \sqrt{\frac{4A}{\pi}} = \sqrt{\frac{(4)(0.57 \text{ ft}^2)}{\pi}}$$

$$= 0.85 \text{ ft}$$

(C) This incorrect solution ignores the contraction and velocity losses corrected by the orifice coefficient. Other assumptions, definitions, and equations are the same as used in the correct solution.

Assume the orifice to have negligible loss from jet velocity or fluid contraction.

$$h = 100 \text{ ft} - 92.6 \text{ ft}$$
$$= 7.4 \text{ ft}$$

$$A = \frac{Q}{\sqrt{2gh}} = \frac{20 \, \frac{\text{ft}^3}{\text{sec}}}{\sqrt{(2)\left(32.2 \, \frac{\text{ft}}{\text{sec}^2}\right)(7.4 \text{ ft})}}$$

$$= 0.92 \text{ ft}^2$$

$$D = \sqrt{\frac{4A}{\pi}} = \sqrt{\frac{(4)(0.92 \text{ ft}^2)}{\pi}}$$

$$= 1.1 \text{ ft}$$

SOLUTION 37

Assume that friction head loss is negligible.

g gravitational acceleration   32.2 ft/sec^2
h pressure head loss   ft
p pressure   psig or lbf/in^2
ρ water density at given
    temperature   lbm/ft^3

$$h = \frac{pg_c}{g\rho} = \frac{\left(100 \; \frac{\text{lbf}}{\text{in}^2}\right)\left(32.2 \; \frac{\text{ft·lbm}}{\text{lbf·sec}^2}\right)\left(144 \; \frac{\text{in}^2}{\text{ft}^2}\right)}{\left(32.2 \; \frac{\text{ft}}{\text{sec}^2}\right)\left(62.3 \; \frac{\text{lbm}}{\text{ft}^3}\right)}$$

$$= 231 \text{ ft}$$

Assume that the Torricelli equation applies.

C_v nozzle coefficient   –
v jet velocity   ft/sec

$$v = C_v\sqrt{2gh} = (0.98)\sqrt{(2)\left(32.2 \; \frac{\text{ft}}{\text{sec}^2}\right)(231 \text{ ft})}$$

$$= 120 \text{ ft/sec}$$

A area of nozzle opening   ft^2
D nozzle opening diameter   ft

$$A = \pi\frac{D^2}{4} = \frac{\pi(0.5 \text{ in})^2\left(\frac{1 \text{ ft}^2}{144 \text{ in}^2}\right)}{4}$$

$$= 0.0014 \text{ ft}^2$$

Q flow rate   gal/min

$$Q = Av$$

$$= (0.0014 \text{ ft}^2)\left(120 \; \frac{\text{ft}}{\text{sec}}\right)\left(\frac{1 \text{ gal}}{0.134 \text{ ft}^3}\right)\left(60 \; \frac{\text{sec}}{\text{min}}\right)$$

$$= 75 \text{ gal/min}$$

The answer is (B).

Why Other Options Are Wrong

(A) This incorrect solution does not account for the difference between lbf and lbm when converting pressure to head. Other assumptions, definitions, and equations are the same as used in the correct solution.

$$h = \frac{p}{g\rho} = \frac{\left(100 \; \frac{\text{lbf}}{\text{in}^2}\right)\left(144 \; \frac{\text{in}^2}{\text{ft}^2}\right)}{\left(32.2 \; \frac{\text{ft}}{\text{sec}^2}\right)\left(62.3 \; \frac{\text{lbm}}{\text{ft}^3}\right)}$$

$$= 7.2 \text{ lbf-sec}^2 \text{ /lbm}$$

The units do not work. Assume units of feet for the following calculations.

$$v = C_v\sqrt{2gh} = (0.98)\sqrt{(2)\left(32.2 \; \frac{\text{ft}}{\text{sec}^2}\right)(7.2 \text{ ft})}$$

$$= 21 \text{ ft/sec}$$

$$A = \pi\frac{D^2}{4} = \frac{\pi(0.5 \text{ in})^2\left(\frac{1 \text{ ft}^2}{144 \text{ in}^2}\right)}{4}$$

$$= 0.0014 \text{ ft}^2$$

$$Q = Av$$

$$= (0.0014 \text{ ft}^2)\left(21 \; \frac{\text{ft}}{\text{sec}}\right)\left(\frac{1 \text{ gal}}{0.134 \text{ ft}^3}\right)\left(60 \; \frac{\text{sec}}{\text{min}}\right)$$

$$= 13 \text{ gal/min}$$

(C) This incorrect solution fails to properly convert in^2 to ft^2 in the pressure head and nozzle area equations. Other assumptions, definitions, and equations are the same as used in the correct solution.

$$h = \frac{p}{g\rho} = \frac{\left(100 \; \frac{\text{lbf}}{\text{in}^2}\right)\left(32.2 \; \frac{\text{ft-lbm}}{\text{lbf-sec}^2}\right)\left(12 \; \frac{\text{in}^2}{\text{ft}^2}\right)}{\left(32.2 \; \frac{\text{ft}}{\text{sec}^2}\right)\left(62.3 \; \frac{\text{lbm}}{\text{ft}^3}\right)}$$

$$= 19 \text{ ft}$$

$$v = C_v\sqrt{2gh} = (0.98)\sqrt{(2)\left(32.2 \; \frac{\text{ft}}{\text{sec}^2}\right)(19 \text{ ft})}$$

$$= 34 \text{ ft/sec}$$

$$A = \pi\frac{D^2}{4} = \frac{\pi(0.5 \text{ in})^2\left(\frac{1 \text{ ft}^2}{12 \text{ in}^2}\right)}{4} = 0.016 \text{ ft}^2$$

$$Q = Av$$

$$= (0.016 \text{ ft}^2)\left(34 \; \frac{\text{ft}}{\text{sec}}\right)\left(\frac{1 \text{ gal}}{0.134 \text{ ft}^3}\right)\left(60 \; \frac{\text{sec}}{\text{min}}\right)$$

$$= 244 \text{ gal/min} \quad (240 \text{ gal/min})$$

(D) This incorrect solution takes the square root of the head loss only, instead of the head loss-gravity constant term. Other assumptions, definitions, and equations are the same as used in the correct solution.

$$h = \frac{p}{g\rho} = \frac{\left(100 \; \frac{\text{lbf}}{\text{in}^2}\right)\left(32.2 \; \frac{\text{ft-lbm}}{\text{lbf-sec}^2}\right)\left(144 \; \frac{\text{in}^2}{\text{ft}^2}\right)}{\left(32.2 \; \frac{\text{ft}}{\text{sec}^2}\right)\left(62.3 \; \frac{\text{lbm}}{\text{ft}^3}\right)}$$

$$= 231 \text{ ft}$$

$$v = C_v 2g\sqrt{h} = (0.98)(2)\left(32.2 \; \frac{\text{ft}}{\text{sec}^2}\right)\sqrt{(231 \text{ ft})}$$

$$= 959 \text{ ft/sec}$$

$$A = \pi\frac{D^2}{4} = \frac{\pi(0.5 \text{ in}^2)\left(\frac{1 \text{ ft}^2}{144 \text{ in}^2}\right)}{4}$$

$$= 0.0014 \text{ ft}^2$$

$$Q = Av$$

$$= (0.0014 \text{ ft}^2)\left(959 \frac{\text{ft}^2}{\text{sec}}\right)\left(\frac{1 \text{ gal}}{0.134 \text{ ft}^3}\right)\left(60 \frac{\text{sec}}{\text{min}}\right)$$

$$= 600 \text{ gal/min}$$

SOLUTION 38

For design flow based on population,

P population   people

$$P = (346 \text{ houses})\left(4 \frac{\text{people}}{\text{house}}\right) = 1384 \text{ people}$$

Q_a average domestic flow demand   gal/min
q per capita flow demand   gal/person-day

Assume an average annual per capita daily flow of 165 gal/person-day.

$$Q_a = Pq$$

$$= (1384 \text{ people})\left(165 \frac{\text{gal}}{\text{person-day}}\right)\left(\frac{1 \text{ day}}{1440 \text{ min}}\right)$$

$$= 159 \text{ gal/min}$$

For design flow for domestic uses during fire demand, assume that the maximum daily flow multiplier is 1.5.

M peak daily flow multiplier   –
Q_p peak daily domestic flow demand   gal/min

$$Q_p = Q_a M = (1.5)\left(159 \frac{\text{gal}}{\text{min}}\right)$$

$$= 238 \text{ gal/min}$$

Alternative 1 for fire flow—design fire demand based on house separation distance.

Assume that a distance between 30 ft and 100 ft separates the houses.

Q_{fd} fire demand based on house separation distance   gal/min

From standard reference tables,

$$Q_{fd} = 750 \text{ gal/min}$$

Alternative 2 for fire flow—design fire demand based on dwelling type.

Calculate the fire demand for two-story houses as the worse case.

A dwelling area   ft
F dwelling factor, 1.5 for wood frame   –
Q_{fA} fire demand based on dwelling type   gal/min

$$Q_{fA} = 18F\sqrt{A} = (18)(1.5)\sqrt{2300 \text{ ft}^2} = 1295 \text{ gal/min}$$

Round to nearest 250 gal/min.

$$Q_{fA} = 1250 \text{ gal/min}$$

Alternative 3 for fire flow—design fire demand based on population.

Q_{fP} fire demand based on population   gal/min

$$Q_{fP} = 1020\sqrt{\frac{P}{1000}}\left(1 - 0.01\sqrt{\frac{P}{1000}}\right)$$

$$= 1020\sqrt{\frac{1384}{1000}}\left(1 - 0.01\sqrt{\frac{1384}{1000}}\right)$$

$$= 1186 \text{ gal/min}$$

The highest fire demand occurs based on dwelling type. Use the fire demand from dwelling type combined with peak daily flow for design flow.

Q_d design flow   gal/min

$$Q_d = Q_{fA} + Q_P = 1250 \frac{\text{gal}}{\text{min}} + 238 \frac{\text{gal}}{\text{min}}$$

$$= 1488 \text{ gal/min} \quad (1500 \text{ gal/min})$$

The answer is (D).

Why Other Options Are Wrong

(A) This incorrect solution considers only the house separation criteria with maximum domestic daily flow. Other assumptions, definitions, and equations are the same as used in the correct solution.

For design flow based on population,

$$P = (346 \text{ houses})\left(4 \frac{\text{people}}{\text{house}}\right)$$

$$= 1384 \text{ people}$$

$$Q_a = Pq$$

$$= (1384 \text{ people})\left(165 \frac{\text{gal}}{\text{person-day}}\right)$$

$$\times \left(\frac{1 \text{ day}}{1440 \text{ min}}\right)$$

$$= 159 \text{ gal/min}$$

For design flow for domestic uses during fire demand,

$$Q_p = Q_a M = (1.5)\left(159 \ \frac{\text{gal}}{\text{min}}\right)$$
$$= 238 \ \text{gal/min}$$

For fire demand,

$$Q_{\text{fd}} = 750 \ \text{gal/min} \quad \text{[from reference]}$$
$$Q_d = Q_{\text{fd}} + Q_P$$
$$= 750 \ \frac{\text{gal}}{\text{min}} + 238 \ \frac{\text{gal}}{\text{min}}$$
$$= 988 \ \text{gal/min} \quad (990 \ \text{gal/min})$$

(B) This incorrect solution does not include maximum domestic daily flow. Other assumptions, definitions, and equations are the same as used in the correct solution.

Alternative 1 for fire flow—design fire demand based on house separation distance.

$$Q_{\text{fd}} = 750 \ \text{gal/min} \quad \text{[from reference]}$$

Alternative 2 for fire flow—design fire demand based on dwelling type.

Calculate the fire demand for two-story houses as the worse case.

$$Q_{\text{fA}} = 18F\sqrt{A}$$
$$= (18)(1.5)\sqrt{2300 \ \text{ft}^2}$$
$$= 1295 \ \text{gal/min}$$

Round to nearest 250 gal/min.

$$Q_{\text{fA}} = 1250 \ \text{gal/min}$$

Alternative 3 for fire flow—design fire demand based on population.

$$P = (346 \ \text{houses})\left(4 \ \frac{\text{people}}{\text{house}}\right) = 1384 \ \text{people}$$

$$Q_{\text{fP}} = 1020\sqrt{\frac{P}{1000}}\left(1 - 0.01\sqrt{\frac{P}{1000}}\right)$$
$$= 1020\sqrt{\frac{1384}{1000}}\left(1 - 0.01\sqrt{\frac{P}{1000}}\right)$$
$$= 1186 \ \text{gal/min}$$

The highest fire demand occurs based on dwelling type.

$$Q_d = Q_{\text{fP}} = 1250 \ \text{gal/min} \quad (1300 \ \text{gal/min})$$

(C) This incorrect solution considers only the population criteria with maximum domestic daily flow. Other assumptions, definitions, and equations are the same as used in the correct solution.

$$P = (346 \ \text{houses})\left(4 \ \frac{\text{people}}{\text{house}}\right) = 1384 \ \text{people}$$

$$Q_a = Pq$$
$$= (1384 \ \text{people})\left(165 \ \frac{\text{gal}}{\text{person-day}}\right)\left(\frac{1 \ \text{day}}{1440 \ \text{min}}\right)$$
$$= 159 \ \text{gal/min}$$

$$Q_p = Q_a M = (1.5)\left(159 \ \frac{\text{gal}}{\text{min}}\right)$$
$$= 238 \ \text{gal/min}$$

For design fire demand,

$$Q_{\text{fP}} = 1020\sqrt{\frac{P}{1000}}\left(1 - 0.01\sqrt{\frac{P}{1000}}\right)$$
$$= 1020\sqrt{\frac{1384}{1000}}\left(1 - (0.01)\sqrt{\frac{P}{1000}}\right)$$
$$= 1186 \ \text{gal/min}$$

$$Q_d = Q_{\text{fP}} + Q_p = 1186 \ \frac{\text{gal}}{\text{min}} + 238 \ \frac{\text{gal}}{\text{min}}$$
$$= 1424 \ \text{gal/min} \quad (1400 \ \text{gal/min})$$

SOLUTION 39

d_e depth of irrigation water
applied per plot   cm/h·plot
d_T total annual depth of irrigation
water applied   cm/yr
t_p total weekly irrigation
time per plot   h/wk·plot

$$t_p = \frac{d_T}{d_e} = \frac{260 \ \frac{\text{cm}}{\text{yr}}}{\left(2 \ \frac{\text{cm}}{\text{h·plot}}\right)\left(52 \ \frac{\text{wk}}{\text{yr}}\right)}$$
$$= 2.5 \ \text{h/wk·plot}$$

t_T total available weekly irrigation time h/wk

$$t_T = \left(12 \ \frac{\text{h}}{\text{d}}\right)\left(7 \ \frac{\text{d}}{\text{wk}}\right)$$
$$= 84 \ \text{h/wk}$$

n number of plots –

$$n = \frac{t_T}{t_p} = \frac{84 \ \frac{\text{h}}{\text{wk}}}{2.5 \ \frac{\text{h}}{\text{wk·plot}}}$$
$$= 34 \ \text{plots}$$

A_p plot irrigated area   ha
A_T total irrigated area   ha

$$A_p = \frac{A_T}{n} = \frac{500 \text{ ha}}{34 \text{ plots}}$$
$$= 14.7 \text{ ha} \quad (15 \text{ ha})$$

The answer is (B).

Why Other Options Are Wrong

(A) This incorrect solution includes the three irrigation periods in the denominator of the weekly irrigation time equation. Other assumptions, definitions, and equations are unchanged from the correct solution.

$$t_p = \frac{d_T}{d_e} = \frac{\left(260 \ \frac{\text{cm}}{\text{yr}}\right)\left(\frac{1 \text{ wk}}{3 \text{ periods}}\right)}{\left(2 \ \frac{\text{cm}}{\text{h·plot}}\right)\left(52 \ \frac{\text{wk}}{\text{yr}}\right)}$$
$$= 0.83 \text{ h/period·plot}$$

$$t_T = \left(12 \ \frac{\text{h}}{\text{d}}\right)\left(7 \ \frac{\text{d}}{\text{wk}}\right)$$
$$= 84 \text{ h/wk}$$

$$n = \frac{t_T}{t_p} = \frac{84 \ \frac{\text{h}}{\text{wk}}}{0.83 \ \frac{\text{h}}{\text{period·plot}}}$$
$$= 101 \text{ plots·period/wk}$$

$$A_p = \frac{A_T}{n} = \frac{500 \text{ ha}}{101 \text{ plots}}$$
$$= 5 \text{ ha}$$

(C) This incorrect solution uses a 5 day week instead of a 7 day week. Other assumptions, definitions, and equations are unchanged from the correct solution.

$$t_p = \frac{d_T}{d_e} = \frac{260 \ \frac{\text{cm}}{\text{yr}}}{\left(2 \ \frac{\text{cm}}{\text{h·plot}}\right)\left(52 \ \frac{\text{wk}}{\text{yr}}\right)}$$
$$= 2.5 \text{ h/wk·plot}$$

$$t_T = \left(12 \ \frac{\text{h}}{\text{d}}\right)\left(5 \ \frac{\text{d}}{\text{wk}}\right) = 60 \text{ h/wk}$$

$$n = \frac{t_T}{t_p} = \frac{60 \ \frac{\text{h}}{\text{wk}}}{2.5 \ \frac{\text{h}}{\text{period·plot}}}$$
$$= 24 \text{ plots·period/wk}$$

$$A_p = \frac{A_T}{n} = \frac{500 \text{ ha}}{24 \text{ plots}}$$
$$= 20.8 \text{ ha} \quad (21 \text{ ha})$$

(D) This incorrect solution includes the three irrigation periods in the numerator of the weekly irrigation time

equation. Other assumptions, definitions, and equations are unchanged from the correct solution.

$$t_p = \frac{d_T}{d_e} = \frac{\left(260 \ \frac{\text{cm}}{\text{yr}}\right)\left(3 \ \frac{\text{periods}}{\text{wk}}\right)}{\left(2 \ \frac{\text{cm}}{\text{h·plot}}\right)\left(52 \ \frac{\text{wk}}{\text{yr}}\right)}$$
$$= 7.5 \text{ h·plot·period/wk}$$

$$t_T = \left(12 \ \frac{\text{h}}{\text{d}}\right)\left(7 \ \frac{\text{d}}{\text{wk}}\right) = 84 \text{ h/wk}$$

$$n = \frac{84 \ \frac{\text{h}}{\text{wk}}}{7.5 \ \frac{\text{h·plot·period}}{\text{wk}}} = 11.2 \text{ plots}$$

$$A_p = \frac{A_T}{n} = \frac{500 \text{ ha}}{11.2 \text{ plots}}$$
$$= 45 \text{ ha}$$

SOLUTION 40

At the spillway crest, the critical energy is equal to the water depth above the crest.

E_c energy line at the spillway crest   ft

$$E_c = 6 \text{ ft}$$

d_c critical depth at the spillway crest   ft

$$d_c = \left(\frac{2}{3}\right) E_c = \left(\frac{2}{3}\right)(6 \text{ ft}) = 4 \text{ ft}$$

g gravitational acceleration   32.2 ft/sec^2
v_c critical water velocity   ft/s

$$E_c = d_c + \frac{v_c^2}{2g}$$
$$6 \text{ ft} = 4 \text{ ft} + \frac{v_c^2}{(2)\left(32.2 \ \frac{\text{ft}}{\text{sec}^2}\right)}$$
$$v_c = 11.3 \text{ ft/sec}$$

Δt the duration of the period of interest   sec

$$\Delta t = (8 \text{ hr})\left(3600 \ \frac{\text{sec}}{\text{hr}}\right) = 28{,}800 \text{ sec}$$

O_1 outflow at the beginning of the
        period of interest   ft^3/sec

Assume that during dry weather conditions, outflow is negligible.

O_2 outflow at the end of the period
        of interest   ft^3/sec
w spillway width   ft

$$O_2 = v_c d_c w = \left(11.3 \; \frac{\text{ft}}{\text{sec}}\right)(4 \text{ ft})(30 \text{ ft})$$

$$= 1356 \text{ ft}^3/\text{sec}$$

I_1 inflow at beginning of the
     period of interest ft^3/sec
I_2 inflow at the end of the period
     of interest ft^3/sec
Δs the increase in reservoir storage
     during the period of interest ft^3

$$\Delta s = \left(\frac{\Delta t}{2}\right)(I_1 + I_2 - O_1 - O_2)$$

$$= \left(\frac{28{,}800 \text{ sec}}{2}\right)\left(\begin{array}{c}1100 \; \dfrac{\text{ft}^3}{\text{sec}} + 1550 \; \dfrac{\text{ft}^3}{\text{sec}} \\ - \, 0 - 1356 \; \dfrac{\text{ft}^3}{\text{sec}}\end{array}\right)$$

$$\times \left(\frac{1 \text{ ac-ft}}{43{,}560 \text{ ft}^3}\right)$$

$$= 428 \text{ ac-ft} \quad (430 \text{ ac-ft})$$

The answer is (B).

Why Other Options Are Wrong

(A) This incorrect solution uses the rise in the water level above the spillway instead of the critical depth to calculate the outflow from the reservoir during the 10-year storm. Other assumptions, definitions, and equations are unchanged from the correct solution.

$$E_c = 6 \text{ ft}$$

$$d_c = \left(\frac{2}{3}\right)E_c = \left(\frac{2}{3}\right)(6 \text{ ft})$$

$$= 4 \text{ ft}$$

$$E_c = d_c + \frac{v_c^2}{2g}$$

$$6 \text{ ft} = 4 \text{ ft} + \frac{v_c^2}{(2)\left(32.2 \; \dfrac{\text{ft}}{\text{sec}^2}\right)}$$

$$v_c = 11.3 \text{ ft/sec}$$

$$\Delta t = (8 \text{ hr})\left(3600 \; \frac{\text{sec}}{\text{hr}}\right)$$

$$= 28{,}800 \text{ sec}$$

d rise in reservoir water level
     above the spillway   ft

$$O_2 = v_c dw$$

$$= \left(11.3 \; \frac{\text{ft}}{\text{sec}}\right)(6 \text{ ft})(30 \text{ ft})$$

$$= 2034 \text{ ft}^3/\text{sec}$$

$$\Delta s = \left(\frac{\Delta t}{2}\right)(I_1 + I_2 - O_1 - O_2)$$

$$= \left(\frac{28{,}800 \text{ sec}}{2}\right)\left(\begin{array}{c}1100 \; \dfrac{\text{ft}^3}{\text{sec}} + 1550 \; \dfrac{\text{ft}^3}{\text{sec}} \\ - \, 0 - 2034 \; \dfrac{\text{ft}^3}{\text{sec}}\end{array}\right)$$

$$\times \left(\frac{1 \text{ ac-ft}}{43{,}560 \text{ ft}^3}\right)$$

$$= 204 \text{ ac-ft} \quad (200 \text{ ac-ft})$$

(C) This incorrect solution fails to divide by two in the velocity equation. Other assumptions, definitions, and equations are unchanged from the correct solution.

$$E_c = 6 \text{ ft}$$

$$d_c = \left(\frac{2}{3}\right)E_c$$

$$= \left(\frac{2}{3}\right)(6 \text{ ft})$$

$$= 4 \text{ ft}$$

$$E_c = d_c + \frac{v_c^2}{g}$$

$$6 \text{ ft} = 4 \text{ ft} + \frac{v_c^2}{\left(32.2 \; \dfrac{\text{ft}}{\text{sec}^2}\right)}$$

$$v_c = 8.0 \text{ ft/sec}$$

$$\Delta t = (8 \text{ hr})\left(3600 \; \frac{\text{sec}}{\text{hr}}\right)$$

$$= 28{,}800 \text{ sec}$$

d rise in reservoir water level
     above the spillway   ft

$$O_2 = v_c d_c w$$

$$= \left(8.0 \; \frac{\text{ft}}{\text{sec}}\right)(4 \text{ ft})(30 \text{ ft})$$

$$= 960 \text{ ft}^3/\text{sec}$$

$$\Delta s = \left(\frac{\Delta t}{2}\right)(I_1 + I_2 - O_1 - O_2)$$

$$= \left(\frac{28{,}800 \text{ sec}}{2}\right)\left(\begin{array}{c}1100 \; \dfrac{\text{ft}^3}{\text{sec}} + 1550 \; \dfrac{\text{ft}^3}{\text{sec}} \\ - \, 0 - 960 \; \dfrac{\text{ft}^3}{\text{sec}}\end{array}\right)$$

$$\times \left(\frac{1 \text{ ac-ft}}{43{,}560 \text{ ft}^3}\right)$$

$$= 559 \text{ ac-ft} \quad (560 \text{ ac-ft})$$

(D) This incorrect solution calculates the total storage volume during the 10-year storm instead of the increase in volume only. Other assumptions, definitions, and equations are unchanged from the correct solution.

$$E_c = 6 \text{ ft}$$

$$d_c = \left(\frac{2}{3}\right) E_c = \left(\frac{2}{3}\right) (6 \text{ ft})$$

$$= 4 \text{ ft}$$

$$E_c = d_c + \frac{v_c^2}{2g}$$

$$6 \text{ ft} = 4 \text{ ft} + \frac{v_c^2}{(2)\left(32.2 \dfrac{\text{ft}}{\text{sec}^2}\right)}$$

$$v_c = 11.3 \text{ ft/sec}$$

$$\Delta t = (8 \text{ hr})\left(3600 \dfrac{\text{sec}}{\text{hr}}\right)$$

$$= 28{,}800 \text{ sec}$$

$$O_2 = v_c d_c w = \left(11.3 \dfrac{\text{ft}}{\text{sec}}\right)(4 \text{ ft})(30 \text{ ft})$$

$$= 1356 \text{ ft}^3/\text{sec}$$

s_1 storage at the beginning of the period of interest ft^3

Convert ac-ft to ft^3.

$$s_1 = (160{,}000 \text{ ac-ft})\left(43{,}560 \dfrac{\text{ft}^2}{\text{ac}}\right)$$

$$= 6.97 \times 10^9 \text{ ft}^3$$

s_2 storage at the end of the period of interest ft^3

$$s_2 = \frac{\Delta t I_1}{2} + \frac{\Delta t I_2}{2} + s_1 - \frac{\Delta t O_1}{2} - \frac{\Delta t O_2}{2}$$

$$= \left(\frac{28{,}800 \text{ sec}}{2}\right)\left(1100 \dfrac{\text{ft}^3}{\text{sec}}\right) + \left(\frac{28{,}800 \text{ sec}}{2}\right)$$

$$\times \left(1550 \dfrac{\text{ft}^3}{\text{sec}}\right) + 6.97 \times 10^9 \text{ ft}^3$$

$$- \left(\frac{(28{,}800 \text{ sec})(0)}{2}\right) - (28{,}800 \text{ sec})$$

$$\times \left(\frac{1356 \dfrac{\text{ft}^3}{\text{sec}}}{2}\right)$$

$$= 7.0 \times 10^9 \text{ ft}^3$$

Convert ft^3 to ac-ft.

$$(7.0 \times 10^9 \text{ ft}^3)\left(\frac{1 \text{ ac-ft}}{43{,}560 \text{ ft}^3}\right) = 160{,}697 \text{ ac-ft}$$

$$(160{,}000 \text{ ac-ft})$$

SOLUTION 41

Because the Darcy friction factor and flow rate in units of ft^3/sec were given, use the applicable Darcy equation in the following form.

f Darcy friction factor   –
k loss coefficient   –
L pipe length   ft

$$k = \frac{0.025 f L}{D^5}$$

pipe	length (ft)	diameter (ft)	loss coefficient (unitless)
AB	3400	0.50	63
BC	2600	0.33	382
CA	1900	0.50	35

n exponent for Darcy equation   –
Q flow rate ft^3/sec
Δ incremental flow rate change ft^3/sec

For the Darcy equation, use an exponent value of 2. Assume initial flows of 1.25 ft^3/sec in pipe AB and 0.75 ft^3/sec in pipes BC and CA. Assume positive flow is clockwise.

$$\Delta = \frac{-\sum kQ^n}{n \sum |kQ^{n-1}|}$$

Trial 1,

$$\Delta_1 = \frac{-\left(\begin{array}{c}(63)\left(1.25 \dfrac{\text{ft}^3}{\text{sec}}\right)^2 - (382)\left(0.75 \dfrac{\text{ft}^3}{\text{sec}}\right)^2 \\ - (35)\left(0.75 \dfrac{\text{ft}^3}{\text{sec}}\right)^2\end{array}\right)}{(2)\left(\begin{array}{c}\left|(63)\left(1.25 \dfrac{\text{ft}^3}{\text{sec}}\right)\right| + \left|(382)\left(0.75 \dfrac{\text{ft}^3}{\text{sec}}\right)\right| \\ + \left|(35)\left(0.75 \dfrac{\text{ft}^3}{\text{sec}}\right)\right|\end{array}\right)}$$

$$= 0.17 \text{ ft}^3/\text{sec}$$

Trial 2, with corrected flows of 1.42 ft^3/sec in pipe AB and 0.58 ft^3/sec in pipes BC and CA,

$$\Delta_2 = \frac{-\left(\begin{array}{c}(63)\left(1.42 \dfrac{\text{ft}^3}{\text{sec}}\right)^2 - (382)\left(0.58 \dfrac{\text{ft}^3}{\text{sec}}\right)^2 \\ - (35)\left(0.58 \dfrac{\text{ft}^3}{\text{sec}}\right)^2\end{array}\right)}{(2)\left(\begin{array}{c}\left|(63)\left(1.42 \dfrac{\text{ft}^3}{\text{sec}}\right)\right| + \left|(382)\left(-0.58 \dfrac{\text{ft}^3}{\text{sec}}\right)\right| \\ + \left|(35)\left(-0.58 \dfrac{\text{ft}^3}{\text{sec}}\right)\right|\end{array}\right)}$$

$$= 0.020 \text{ ft}^3/\text{sec}$$

Trial 3, with corrected flows of 1.44 ft³/sec in pipe AB and 0.56 ft³/sec in pipes BC and CA,

$$\Delta_3 = \frac{-\left(\begin{array}{c}(63)\left(1.44\ \dfrac{\text{ft}^3}{\text{sec}}\right)^2 - (382)\left(0.56\ \dfrac{\text{ft}^3}{\text{sec}}\right)^2 \\ -\ (35)\left(0.56\ \dfrac{\text{ft}^3}{\text{sec}}\right)^2\end{array}\right)}{(2)\left(\begin{array}{c}\left|(63)\left(1.44\ \dfrac{\text{ft}^3}{\text{sec}}\right)\right| + \left|(382)\left(-0.56\ \dfrac{\text{ft}^3}{\text{sec}}\right)\right| \\ +\ \left|(35)\left(-0.56\ \dfrac{\text{ft}^3}{\text{sec}}\right)\right|\end{array}\right)}$$

$$= 0.00020\ \text{ft}^3/\text{sec}$$

Accept flows used in trial 3.

For pipe BC,

$$Q = 0.56\ \text{ft}^3/\text{sec}$$

A pipe cross-sectional area   ft²

$$A = \pi \frac{D^2}{4} = \frac{\pi (0.33)^2}{4} = 0.086\ \text{ft}^2$$

v   flow velocity   ft/sec

$$v = \frac{Q}{A} = \frac{0.56\ \dfrac{\text{ft}^3}{\text{sec}}}{0.086\ \text{ft}^2} = 6.5\ \text{ft/sec}$$

The answer is (C).

Why Other Options Are Wrong

(A) This incorrect solution presents the flow rate in velocity units and does not calculate the flow velocity. Other assumptions, definitions, and equations are unchanged from the correct solution.

pipe	length (ft)	diameter (ft)	loss coefficient (unitless)
AB	3400	0.50	63
BC	2600	0.33	382
CA	1900	0.50	35

Assume initial flows of 1.25 ft³/sec in pipe AB and 0.75 ft³/sec in pipes BC and CA.

$$\Delta = \frac{-\sum kQ^n}{n \sum |kQ^{n-1}|}$$

Trial 1,

$$\Delta_1 = \frac{-\left(\begin{array}{c}(63)\left(1.25\ \dfrac{\text{ft}^3}{\text{sec}}\right)^2 - (382)\left(0.75\ \dfrac{\text{ft}^3}{\text{sec}}\right)^2 \\ -\ (35)\left(0.75\ \dfrac{\text{ft}^3}{\text{sec}}\right)^2\end{array}\right)}{(2)\left(\begin{array}{c}\left|(63)\left(1.25\ \dfrac{\text{ft}^3}{\text{sec}}\right)\right| + \left|(382)\left(-0.75\ \dfrac{\text{ft}^3}{\text{sec}}\right)\right| \\ +\ \left|(35)\left(-0.75\ \dfrac{\text{ft}^3}{\text{sec}}\right)\right|\end{array}\right)}$$

$$= 0.17\ \text{ft}^3/\text{sec}$$

Trial 2, with corrected flows of 1.42 ft³/sec in pipe AB and 0.58 ft³/sec in pipes BC and CA,

$$\Delta_2 = \frac{-\left(\begin{array}{c}(63)\left(1.42\ \dfrac{\text{ft}^3}{\text{sec}}\right)^2 - (382)\left(0.58\ \dfrac{\text{ft}^3}{\text{sec}}\right)^2 \\ -\ (35)\left(0.58\ \dfrac{\text{ft}^3}{\text{sec}}\right)^2\end{array}\right)}{(2)\left(\begin{array}{c}\left|(63)\left(1.42\ \dfrac{\text{ft}^3}{\text{sec}}\right)\right| + \left|(382)\left(-0.58\ \dfrac{\text{ft}^3}{\text{sec}}\right)\right| \\ +\ \left|(35)\left(-0.58\ \dfrac{\text{ft}^3}{\text{sec}}\right)\right|\end{array}\right)}$$

$$= 0.020\ \text{ft}^3/\text{sec}$$

Trial 3, with corrected flows of 1.44 ft³/sec in pipe AB and 0.56 ft³/sec in pipes BC and CA,

$$\Delta_3 = \frac{-\left(\begin{array}{c}(63)\left(1.44\ \dfrac{\text{ft}^3}{\text{sec}}\right)^2 - (382)\left(0.56\ \dfrac{\text{ft}^3}{\text{sec}}\right)^2 \\ -\ (35)\left(0.56\ \dfrac{\text{ft}^3}{\text{sec}}\right)^2\end{array}\right)}{(2)\left(\begin{array}{c}\left|(63)\left(1.44\ \dfrac{\text{ft}^3}{\text{sec}}\right)\right| + \left|(382)\left(-0.56\ \dfrac{\text{ft}^3}{\text{sec}}\right)\right| \\ +\ \left|(35)\left(-0.56\ \dfrac{\text{ft}^3}{\text{sec}}\right)\right|\end{array}\right)}$$

$$= 0.00020\ \text{ft}^3/\text{sec}$$

Accept flows used in trial 3.

For pipe BC,

$$v = 0.56\ \text{ft}^3/\text{sec}$$

(B) This incorrect solution uses 0.50 ft diameter in place of 0.33 ft to calculate the flow velocity. Other assumptions, definitions, and equations are unchanged from the correct solution.

pipe	length (ft)	diameter (ft)	loss coefficient (unitless)
AB	3400	0.50	63
BC	2600	0.33	382
CA	1900	0.50	35

Assume initial flows of 1.25 ft³/sec in pipe AB and 0.75 ft³/sec in pipes BC and CA. Assume positive flow is clockwise.

$$\Delta = \frac{-\sum kQ^n}{n \sum |kQ^{n-1}|}$$

Trial 1,

$$\Delta_1 = \frac{-\left(\begin{array}{c}(63)\left(1.25 \dfrac{\text{ft}^3}{\text{sec}}\right)^2 - (382)\left(0.75 \dfrac{\text{ft}^3}{\text{sec}}\right)^2 \\ -(35)\left(0.75 \dfrac{\text{ft}^3}{\text{sec}}\right)^2\end{array}\right)}{(2)\left(\begin{array}{c}\left|(63)\left(1.25 \dfrac{\text{ft}^3}{\text{sec}}\right)\right| + \left|(382)\left(-0.75 \dfrac{\text{ft}^3}{\text{sec}}\right)\right| \\ + \left|(35)\left(-0.75 \dfrac{\text{ft}^3}{\text{sec}}\right)\right|\end{array}\right)}$$

$$= 0.17 \text{ ft}^3/\text{sec}$$

Trial 2, with corrected flows of 1.42 ft³/sec in pipe AB and 0.58 ft³/sec in pipes BC and CA,

$$\Delta_2 = \frac{-\left(\begin{array}{c}(63)\left(1.42 \dfrac{\text{ft}^3}{\text{sec}}\right)^2 - (382)\left(0.58 \dfrac{\text{ft}^3}{\text{sec}}\right)^2 \\ -(35)\left(0.58 \dfrac{\text{ft}^3}{\text{sec}}\right)^2\end{array}\right)}{(2)\left(\begin{array}{c}\left|(63)\left(1.42 \dfrac{\text{ft}^3}{\text{sec}}\right)\right| + \left|(382)\left(-0.58 \dfrac{\text{ft}^3}{\text{sec}}\right)\right| \\ + \left|(35)\left(-0.58 \dfrac{\text{ft}^3}{\text{sec}}\right)\right|\end{array}\right)}$$

$$= 0.020 \text{ ft}^3/\text{sec}$$

Trial 3, with corrected flows of 1.44 ft³/sec in pipe AB and 0.56 ft³/sec in pipes BC and CA,

$$\Delta_3 = \frac{-\left(\begin{array}{c}(63)\left(1.44 \dfrac{\text{ft}^3}{\text{sec}}\right)^2 - (382)\left(0.56 \dfrac{\text{ft}^3}{\text{sec}}\right)^2 \\ -(35)\left(0.56 \dfrac{\text{ft}^3}{\text{sec}}\right)^2\end{array}\right)}{(2)\left(\begin{array}{c}\left|(63)\left(1.44 \dfrac{\text{ft}^3}{\text{sec}}\right)\right| + \left|(382)\left(-0.56 \dfrac{\text{ft}^3}{\text{sec}}\right)\right| \\ + \left|(35)\left(-0.56 \dfrac{\text{ft}^3}{\text{sec}}\right)\right|\end{array}\right)}$$

$$= 0.00020 \text{ ft}^3/\text{sec}$$

Accept flows used in trial 3.

For pipe BC,

$$Q = 0.56 \text{ ft}^3/\text{sec}$$

$$A = \frac{\pi D^2}{4} = \frac{\pi (0.50 \text{ ft})^2}{4} = 0.196 \text{ ft}^2$$

$$v = \frac{Q}{A} = \frac{0.56 \dfrac{\text{ft}^3}{\text{sec}}}{0.196 \text{ ft}^2} = 2.9 \text{ ft/sec}$$

(D) This incorrect solution uses the flow rate for pipe AB instead of pipe BC to calculate flow velocity. Other assumptions, definitions, and equations are unchanged from the correct solution.

pipe	length (ft)	diameter (ft)	loss coefficient (unitless)
AB	3400	0.50	63
BC	2600	0.33	382
CA	1900	0.50	35

Assume initial flows of 1.25 ft³/sec in pipe AB and 0.75 ft³/sec in pipes BC and CA.

$$\Delta = \frac{-\sum kQ^n}{n \sum |kQ^{n-1}|}$$

Trial 1,

$$\Delta_1 = \frac{-\left(\begin{array}{c}(63)\left(1.25 \dfrac{\text{ft}^3}{\text{sec}}\right)^2 - (382)\left(0.75 \dfrac{\text{ft}^3}{\text{sec}}\right)^2 \\ -(35)\left(0.75 \dfrac{\text{ft}^3}{\text{sec}}\right)^2\end{array}\right)}{(2)\left(\begin{array}{c}\left|(63)\left(1.25 \dfrac{\text{ft}^3}{\text{sec}}\right)\right| + \left|(382)\left(-0.75 \dfrac{\text{ft}^3}{\text{sec}}\right)\right| \\ + \left|(35)\left(-0.75 \dfrac{\text{ft}^3}{\text{sec}}\right)\right|\end{array}\right)}$$

$$= 0.17 \text{ ft}^3/\text{sec}$$

Trial 2, with corrected flows of 1.42 ft³/sec in pipe AB and 0.58 ft³/sec in pipes BC and CA,

$$\Delta_2 = \frac{-\left(\begin{array}{c}(63)\left(1.42 \dfrac{\text{ft}^3}{\text{sec}}\right)^2 - (382)\left(0.58 \dfrac{\text{ft}^3}{\text{sec}}\right)^2 \\ -(35)\left(0.58 \dfrac{\text{ft}^3}{\text{sec}}\right)^2\end{array}\right)}{(2)\left(\begin{array}{c}\left|(63)\left(1.42 \dfrac{\text{ft}^3}{\text{sec}}\right)\right| + \left|(382)\left(-0.58 \dfrac{\text{ft}^3}{\text{sec}}\right)\right| \\ + \left|(35)\left(-0.58 \dfrac{\text{ft}^3}{\text{sec}}\right)\right|\end{array}\right)}$$

$$= 0.020 \text{ ft}^3/\text{sec}$$

Trial 3, with corrected flows of 1.44 ft³/sec in pipe AB and 0.56 ft³/sec in pipes BC and CA,

$$\Delta_3 = \frac{-\left(\begin{array}{c}(63)\left(1.44 \dfrac{\text{ft}^3}{\text{sec}}\right)^2 - (382)\left(0.56 \dfrac{\text{ft}^3}{\text{sec}}\right)^2 \\ -(35)\left(0.56 \dfrac{\text{ft}^3}{\text{sec}}\right)^2\end{array}\right)}{(2)\left(\begin{array}{c}\left|(63)\left(1.44 \dfrac{\text{ft}^3}{\text{sec}}\right)\right| + \left|(382)\left(-0.56 \dfrac{\text{ft}^3}{\text{sec}}\right)\right| \\ + \left|(35)\left(-0.56 \dfrac{\text{ft}^3}{\text{sec}}\right)\right|\end{array}\right)}$$

$$= 0.00020 \text{ ft}^3/\text{sec}$$

Accept flows used in trial 3.

For pipe BC,

$$Q = 1.44 \text{ ft}^3/\text{sec}$$

$$A = \pi\frac{D^2}{4} = \frac{\pi(0.33)^2}{4}$$

$$= 0.086 \text{ ft}^2$$

$$v = \frac{Q}{A} = \frac{1.44 \frac{\text{ft}^3}{\text{sec}}}{0.086 \text{ ft}^2}$$

$$= 16.7 \text{ ft/sec} \quad (17 \text{ ft/sec})$$

SOLUTION 42

A pipe cross-sectional ft^2
D pipe nominal diameter ft

$$A = \pi\frac{D^2}{4}$$

$$A_{DC} = A_{AC} = \left(\frac{\pi(8 \text{ in})^2}{4}\right)\left(\frac{1 \text{ ft}^2}{144 \text{ in}^2}\right)$$

$$= 0.35 \text{ ft}^2$$

$$A_{CB} = \left(\frac{\pi(10 \text{ in})^2}{4}\right)\left(\frac{1 \text{ ft}^2}{144 \text{ in}^2}\right)$$

$$= 0.55 \text{ ft}^2$$

Q flow rate ft^3/sec
v flow velocity ft/sec

$$v = \frac{Q}{A}$$

$$v_{DC} = \frac{4 \frac{\text{ft}^3}{\text{sec}}}{0.35 \text{ ft}^2}$$

$$= 11.4 \text{ ft/sec}$$

h_f friction head ft

From standard tables for head loss, the head loss corresponding to a velocity of 11.4 ft/sec in 1600 ft of 8 in diameter pipe is

$$h_{f,DC} = \left(\frac{21.1 \frac{\text{lbf}}{\text{in}^2}}{1000 \text{ ft}}\right)(1600 \text{ ft})\left(\frac{1 \text{ ft}}{0.433 \frac{\text{lbf}}{\text{in}^2}}\right)$$

$$= 78 \text{ ft}$$

p/γ pressure head ft
z elevation head ft

$$z_D = z_C + \frac{p_C}{\gamma} + h_{f,DC}$$

$$\frac{p_C}{\gamma} = z_D - z_C - h_{f,DC}$$

$$= 676 \text{ ft} - 523 \text{ ft} - 78 \text{ ft}$$

$$= 75 \text{ ft}$$

$$z_A = z_C + \frac{p_C}{\gamma} + h_{f,AC}$$

$$h_{f,AC} = z_A - z_C - \frac{p_C}{\gamma}$$

$$= 719 \text{ ft} - 523 \text{ ft} - 75 \text{ ft}$$

$$= 121 \text{ ft}$$

Convert head loss from ft to lbf/in^2 by

$$\frac{(121 \text{ ft})\left(0.433 \frac{\text{lbf}}{\text{in}^2}\right)}{1 \text{ ft}} = 52 \text{ lbf/in}^2$$

From standard tables for head loss, the velocity corresponding to 1200 ft of 8 in diameter pipe is

$$h_{f,AC} \text{ per 1000 ft} = \frac{\left(52 \frac{\text{lbf}}{\text{in}^2}\right)(1000 \text{ ft})}{1200 \text{ ft}}$$

$$= 43 \text{ lbf/in}^2$$

$$v_{AC} = 18 \text{ ft/sec}$$

$$z_c + \frac{p_C}{\gamma} = 523 \text{ ft} + 75 \text{ ft} = 598 \text{ ft}$$

$$z_A = 719 \text{ ft} > 598 \text{ ft}$$

$$z_D = 676 \text{ ft} > 598 \text{ ft}$$

Therefore, flow is from A to C and from D to C. Thus flow is from C to B.

$$Q_{AC} = 6.3 \text{ ft}^3/\text{sec}$$

$$Q_{CB} = 10.3 \text{ ft}^3/\text{sec}$$

$$v_{CB} = 19 \text{ ft/sec}$$

$$Q_{CB} = Q_{AC} + Q_{DC}$$

$$Q_{AC} = \left(18 \frac{\text{ft}}{\text{sec}}\right)(0.35 \text{ ft}^2) = 6.3 \text{ ft}^3/\text{sec}$$

$$Q_{CB} = Q_{AC} + Q_{DC} = 4 \frac{\text{ft}^3}{\text{sec}} + 6.3 \frac{\text{ft}^3}{\text{sec}}$$

$$= 10.3 \text{ ft}^3/\text{sec}$$

$$v_{CB} = \frac{10.3 \frac{\text{ft}^3}{\text{sec}}}{0.55 \text{ ft}^2} = 19 \text{ ft/sec}$$

From standard tables for head loss, the head loss corresponding to a velocity of 19 ft/sec in 1090 ft of 10 in diameter pipe is

$$h_{f,CB} = \left(\frac{40 \frac{\text{lbf}}{\text{in}^2}}{1000 \text{ ft}}\right)(1090 \text{ ft})\left(\frac{1 \text{ ft}}{0.433 \frac{\text{lbf}}{\text{in}^2}}\right)$$

$$= 100 \text{ ft}$$

$$z_B = z_C + \frac{p_C}{\gamma} - h_{f,CB} = 523 \text{ ft} + 75 \text{ ft} - 100 \text{ ft}$$

$$= 498 \text{ ft} \quad (500 \text{ ft})$$

The answer is (B).

Why Other Options Are Wrong

(A) This incorrect solution misreads the length of pipe CB as 1900 ft instead of 1090 ft. Other assumptions, definitions, and equations are unchanged from the correct solution.

$$A = \pi \frac{D^2}{4}$$

$$A_{DC} = A_{AC} = \left(\frac{\pi(8 \text{ in})^2}{4} \right) \left(\frac{1 \text{ ft}^2}{144 \text{ in}^2} \right)$$

$$= 0.35 \text{ ft}^2$$

$$A_{CB} = \left(\frac{\pi(10 \text{ in}^2)}{4} \right) \left(\frac{1 \text{ ft}^2}{144 \text{ in}^2} \right)$$

$$= 0.55 \text{ ft}^2$$

$$v = \frac{Q}{A}$$

$$v_{DC} = \frac{4 \frac{\text{ft}^3}{\text{sec}}}{0.35 \text{ ft}^2}$$

$$= 11.4 \text{ ft/sec}$$

From standard tables for head loss, the head loss corresponding to a velocity of 11.4 ft/sec in 1600 ft of 8 in diameter pipe is

$$h_{f,DC} = \left(\frac{21.1 \frac{\text{lbf}}{\text{in}^2}}{1000 \text{ ft}} \right) (1600 \text{ ft}) \left(\frac{1 \text{ ft}}{0.433 \frac{\text{lbf}}{\text{in}^2}} \right)$$

$$= 78 \text{ ft}$$

$$z_D = z_C + \frac{p_C}{\gamma} + h_{f,DC}$$

$$\frac{p_C}{\gamma} = z_D - z_C - h_{f,DC}$$

$$= 676 \text{ ft} - 523 \text{ ft} - 78 \text{ ft}$$

$$= 75 \text{ ft}$$

$$z_A = z_C + \frac{p_C}{\gamma} + h_{f,AC}$$

$$h_{f,AC} = z_A - z_C - \frac{p_C}{\gamma}$$

$$= 719 \text{ ft} - 523 \text{ ft} - 75 \text{ ft}$$

$$= 121 \text{ ft}$$

Convert head loss from ft to lbf/in² by

$$(121 \text{ ft}) \left(\frac{0.433 \frac{\text{lbf}}{\text{in}^2}}{1 \text{ ft}} \right) = 52 \text{ lbf/in}^2$$

From standard tables for head loss, the velocity corresponding to 1200 ft of 8 in diameter pipe is

$$h_{f,AC} \text{ per } 1000 \text{ ft} = \frac{\left(52 \frac{\text{lbf}}{\text{in}^2} \right) (1000 \text{ ft})}{1200 \text{ ft}}$$

$$= 43 \text{ lbf/in}^2$$

$$v_{AC} = 18 \text{ ft/sec}$$

$$Q_{CB} = Q_{AC} + Q_{DC}$$

$$Q_{AC} = \left(18 \frac{\text{ft}}{\text{sec}} \right) (0.35 \text{ ft}^2)$$

$$= 6.3 \text{ ft}^3/\text{sec}$$

$$Q_{CB} = Q_{AC} + Q_{DC}$$

$$= 4 \frac{\text{ft}^3}{\text{sec}} + 6.3 \frac{\text{ft}^3}{\text{sec}}$$

$$= 10.3 \text{ ft}^3/\text{sec}$$

$$v_{CB} = \frac{10.3 \frac{\text{ft}^3}{\text{sec}}}{0.55 \text{ ft}^2}$$

$$= 19 \text{ ft/sec}$$

From standard tables for head loss, the head loss corresponding to a velocity of 19 ft/sec in 1090 ft of 10 in diameter pipe is

$$h_{f,CB} = \left(\frac{40 \frac{\text{lbf}}{\text{in}^2}}{1000 \text{ ft}} \right) (1900 \text{ ft}) \left(\frac{1 \text{ ft}}{0.433 \frac{\text{lbf}}{\text{in}^2}} \right)$$

$$= 176 \text{ ft}$$

$$z_B = z_C + \frac{p_C}{\gamma} - h_{f,CB}$$

$$= 523 \text{ ft} + 75 \text{ ft} - 176 \text{ ft}$$

$$= 422 \text{ ft} \quad (420 \text{ ft})$$

(C) This incorrect solution fails to correct the values taken from head loss tables for actual pipe length. Other assumptions, definitions, and equations are unchanged from the correct solution.

$$A = \pi \frac{D^2}{4}$$

$$A_{DC} = A_{AC} = \left(\frac{\pi(8 \text{ in})^2}{4} \right) \left(\frac{1 \text{ ft}^2}{144 \text{ in}^2} \right)$$

$$= 0.35 \text{ ft}^2$$

$$A_{CB} = \left(\frac{\pi(10 \text{ in}^2)}{4} \right) \left(\frac{1 \text{ ft}^2}{144 \text{ in}^2} \right)$$

$$= 0.55 \text{ ft}^2$$

$$v = \frac{Q}{A}$$

$$v_{DC} = \frac{4 \frac{\text{ft}^3}{\text{sec}}}{0.35 \text{ ft}^2} = 11.4 \text{ ft/sec}$$

From standard tables for head loss, the head loss corresponding to a velocity of 11.4 ft/sec in 1600 ft of 8 in diameter pipe is

$$h_{f,DC} = \left(21.1 \frac{\text{lbf}}{\text{in}^2} \right) \left(\frac{1 \text{ ft}}{0.433 \frac{\text{lbf}}{\text{in}^2}} \right) = 49 \text{ ft}$$

$$z_D = z_C + \frac{p_C}{\gamma} + h_{f,DC}$$

$$\frac{p_C}{\gamma} = z_D - z_C - h_{f,DC}$$

$$= 676 \text{ ft} - 523 \text{ ft} - 49 \text{ ft}$$

$$= 104 \text{ ft}$$

$$z_A = z_C + \frac{p_C}{\gamma} + h_{f,AC}$$

$$h_{f,AC} = z_A - z_C - \frac{p_C}{\gamma}$$

$$= 719 \text{ ft} - 523 \text{ ft} - 104 \text{ ft}$$

$$= 92 \text{ ft}$$

Convert head loss from ft to lbf/in^2 by

$$(92 \text{ ft}) \left(\frac{0.433 \frac{\text{lbf}}{\text{in}^2}}{1 \text{ ft}} \right) = 40 \text{ lbf/in}^2$$

From standard tables for head loss, the velocity corresponding to 1200 ft of 8 in diameter pipe is

$$v_{AC} = 17 \text{ ft/sec}$$

$$Q_{CB} = Q_{AC} + Q_{DC}$$

$$Q_{AC} = \left(17 \frac{\text{ft}}{\text{sec}} \right) (0.35 \text{ ft}^2)$$

$$= 6.0 \text{ ft}^3/\text{sec}$$

$$Q_{CB} = Q_{AC} + Q_{DC} = 4 \frac{\text{ft}^3}{\text{sec}} + 6 \frac{\text{ft}^3}{\text{sec}}$$

$$= 10 \text{ ft}^3/\text{sec}$$

$$v_{CB} = \frac{10 \frac{\text{ft}^3}{\text{sec}}}{0.55 \text{ ft}^2}$$

$$= 18 \text{ ft/sec}$$

From standard tables for head loss, the head loss corresponding to a velocity of 18 ft/sec in 1090 ft of 10 in diameter pipe is

$$h_{f,CB} = \frac{\left(37 \frac{\text{lbf}}{\text{in}^2} \right) (1 \text{ ft})}{\left(0.433 \frac{\text{lbf}}{\text{in}^2} \right)} = 85 \text{ ft}$$

$$z_B = z_C + \frac{p_C}{\gamma} - h_{f,CB}$$

$$= 523 \text{ ft} + 104 \text{ ft} - 85 \text{ ft}$$

$$= 542 \text{ ft} \quad (540 \text{ ft})$$

(D) This incorrect solution misreads the diameter for pipe BC as 12 in instead of 10 in. Other assumptions, definitions, and equations are unchanged from the correct solution.

$$A = \frac{\pi D^2}{4}$$

$$A_{DC} = A_{AC} = \left(\frac{\pi (8 \text{ in})^2}{4} \right) \left(\frac{1 \text{ ft}^2}{144 \text{ in}^2} \right) = 0.35 \text{ ft}^2$$

$$A_{CB} = \left(\frac{\pi (12 \text{ in})^2}{4} \right) \left(\frac{1 \text{ ft}^2}{144 \text{ in}^2} \right) = 0.79 \text{ ft}^2$$

$$v = \frac{Q}{A}$$

$$v_{DC} = \frac{4 \frac{\text{ft}^3}{\text{sec}}}{0.35 \text{ ft}^2} = 11.4 \text{ ft/sec}$$

From standard tables for head loss, the head loss corresponding to a velocity of 11.4 ft/sec in 1600 ft of 8 in diameter pipe is

$$h_{f,DC} = \left(\frac{21.1 \frac{\text{lbf}}{\text{in}^2}}{1000 \text{ ft}} \right) (1600 \text{ ft}) \left(\frac{1 \text{ ft}}{0.433 \frac{\text{lbf}}{\text{in}^2}} \right)$$

$$= 78 \text{ ft}$$

$$z_D = z_C + \frac{p_C}{\gamma} + h_{f,DC}$$

$$\frac{p_C}{\gamma} = z_D - z_C - h_{f,DC}$$

$$= 676 \text{ ft} - 523 \text{ ft} - 78 \text{ ft}$$

$$= 75 \text{ ft}$$

$$z_A = z_C + \frac{p_C}{\gamma} + h_{f,AC}$$

$$h_{f,AC} = z_A - z_C - \frac{p_C}{\gamma}$$

$$= 719 \text{ ft} - 523 \text{ ft} - 75 \text{ ft}$$

$$= 121 \text{ ft}$$

Convert head loss from ft to lbf/in^2 by

$$\frac{(121 \text{ ft}) \left(0.433 \frac{\text{lbf}}{\text{in}^2} \right)}{1 \text{ ft}} = 52 \text{ lbf/in}^2$$

From standard tables for head loss, the velocity corresponding to 1200 ft of 8 in diameter pipe is

$$h_{f,AC} \text{ per 1000 ft} = \frac{\left(52 \frac{\text{lbf}}{\text{in}^2} \right) (1000 \text{ ft})}{1200 \text{ ft}}$$

$$= 43 \text{ lbf/in}^2$$

$$v_{AC} = 18 \text{ ft/sec}$$

$$Q_{CB} = Q_{AC} + Q_{DC}$$

$$Q_{AC} = \left(18 \frac{\text{ft}}{\text{sec}} \right) (0.35 \text{ ft}^2)$$

$$= 6.3 \text{ ft}^3/\text{sec}$$

$$Q_{CB} = Q_{AC} + Q_{DC} = 4\ \frac{\text{ft}^3}{\text{sec}} + 6.3\ \frac{\text{ft}^3}{\text{sec}}$$

$$= 10.3\ \text{ft}^3/\text{sec}$$

$$v_{CB} = \frac{10.3\ \dfrac{\text{ft}^3}{\text{sec}}}{0.79\ \text{ft}^2} = 13\ \text{ft/sec}$$

From standard tables for head loss, the head loss corresponding to a velocity of 18 ft/sec in 1090 ft of 13 in diameter pipe is

$$h_{f,CB} = \left(\frac{18.7\ \dfrac{\text{lbf}}{\text{in}^2}}{1000\ \text{ft}}\right)(1090\ \text{ft})\left(\frac{1\ \text{ft}}{0.433\ \dfrac{\text{lbf}}{\text{in}^2}}\right)$$

$$= 47\ \text{ft}$$

$$z_B = z_C + \frac{p_C}{\gamma} - h_{f,CB}$$

$$= 523\ \text{ft} + 75\ \text{ft} - 47\ \text{ft}$$

$$= 551\ \text{ft}\quad(550\ \text{ft})$$

SOLUTION 43

fitting	quantity	unit equivalent length (ft)	total equivalent length (ft)
standard 90° ell	4	8.5	34
couple	4	0.45	1.8
			36

L straight pipe length   ft
L_e total equivalent pipe length   ft
L_T total straight and equivalent pipe length   ft

$$L_T = L + L_e$$

$$= 140\ \text{ft} + 35\ \text{ft} + 3\ \text{ft} + 36\ \text{ft}$$

$$= 214\ \text{ft}$$

h_z head loss from elevation   ft
z_1 lower elevation   ft
z_2 upper elevation   ft

$$h_z = z_2 - z_1$$

$$= 1085\ \text{ft} - 1000\ \text{ft}$$

$$= 85\ \text{ft}$$

From head loss tables, determine pressure loss corresponding to flow to prepare the following table.

flow rate (gal/min)	pressure loss (lbf/in²)	head loss from friction (ft)	total head (ft)
50	20.6	10.18	95.18
60	29.6	14.63	99.63
70	38.6	19.08	104.08

h_f head loss from friction   ft
L_u unit pipe length   ft
p pressure loss   lbf/in²

$$h_f = \left(\frac{p}{L_u}\right)(L_T)$$

$$= \left(\frac{20.6\ \dfrac{\text{lbf}}{\text{in}^2}}{1000\ \text{ft}}\right)(214\ \text{ft})\left(\frac{1\ \text{ft}}{0.433\ \dfrac{\text{lbf}}{\text{in}^2}}\right)$$

$$= 10.18\ \text{ft}$$

Repeat calculation for other tabulated values.

h_T total head loss   ft

$$h_T = h_f + h_z = 10.18\ \text{ft} + 85\ \text{ft}$$

$$= 95.18\ \text{ft}$$

Repeat calculation for other tabulated values.

Combine the pump curves by adding the heads for series operation and plot the combined heads. Plot the total head calculated for each flow rate from the preceding table. The following illustration results. The intersection of the total head loss and the combined operating curve define the pumping capacity.

Q pumping capacity   gal/min

$$Q = 60\ \text{gal/min}$$

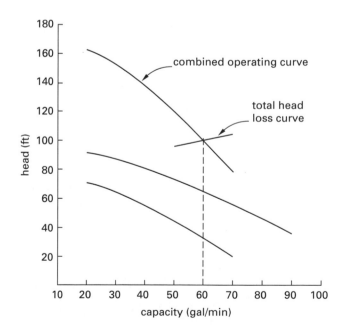

An alternative solution would involve trial-and-error that estimates a flow velocity from head loss tables. The alternative method follows.

For an initial head loss estimate, assume that the flow velocity is 5 ft³/sec. From head loss tables,

$$p = 24 \ \text{lbf/in}^2$$

$$h_f = \left(\frac{24 \ \frac{\text{lbf}}{\text{in}^2}}{1000 \ \text{ft}} \right) (214 \ \text{ft}) \left(\frac{1 \ \text{ft}}{0.433 \ \frac{\text{lbf}}{\text{in}^2}} \right) = 12 \ \text{ft}$$

$$h_T = h_f + h_z = 12 \ \text{ft} + 85 \ \text{ft} = 97 \ \text{ft}$$

Using the combined pump curves, find the flow rate corresponding to the total head of 97 ft.

$$Q = 60 \ \text{gal/min}$$

Check the flow-velocity assumption. From head loss tables, for 60 gal/min and 2 in schedule-40 steel pipe, the flow velocity is 5.74 ft/sec. This is close enough to the assumed 5 ft/sec. If not, then the process would be repeated with a revised flow-velocity estimate.

The answer is (B).

Why Other Options Are Wrong

(A) This incorrect solution ignores the lower pump, considering only head loss from the discharge of the upper pump to the upper reservoir. Other assumptions, definitions, and equations are the same as used in the correct solution.

fitting	quantity	unit equivalent length (ft)	total equivalent length (ft)
standard 90° ell	3	8.5	25.5
couple	4	0.45	1.8
			27.3

$$L_T = L + L_e$$
$$= 140 \ \text{ft} + 3 \ \text{ft} + 27.3 \ \text{ft}$$
$$= 170 \ \text{ft}$$

For an intial head loss estimate, assume that the flow velocity is 5 ft³/sec. From head loss tables,

$$p = 24 \ \text{lbf/in}^2$$

$$h_f = \left(\frac{p}{L_u} \right) (L_T)$$

$$= \left(\frac{24 \ \frac{\text{lbf}}{\text{in}^2}}{1000 \ \text{ft}} \right) (170 \ \text{ft}) \left(\frac{1 \ \text{ft}}{0.433 \ \frac{\text{lbf}}{\text{in}^2}} \right)$$

$$= 9.5 \ \text{ft}$$

$$h_z = z_2 - z_1 = 1085 \ \text{ft} - 1025 \ \text{ft}$$
$$= 60 \ \text{ft}$$

$$h_T = h_f + h_z = 9.5 \ \text{ft} + 60 \ \text{ft}$$
$$= 69.5 \ \text{ft}$$

$$Q = 56 \ \text{gal/min}$$

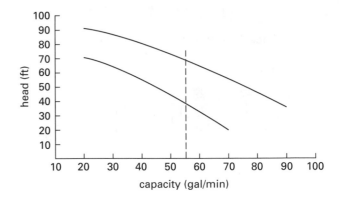

Check the flow-velocity assumption. From head loss tables, for 56 gal/min and 2 in schedule-40 steel pipe, the flow velocity is about 5 ft/sec. Assumption of 5 ft/sec was correct.

(C) This incorrect solution ignores friction and minor losses. Other assumptions, definitions, and equations are the same as used in the correct solution.

$$h_z = z_2 - z_1 = 1085 \ \text{ft} - 1000 \ \text{ft}$$
$$= 85 \ \text{ft}$$

$$Q = 65 \ \text{gal/min}$$

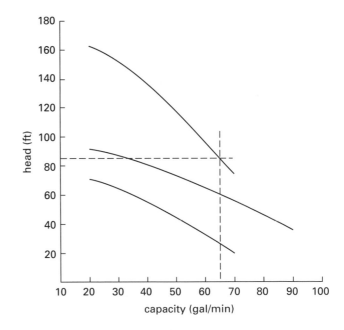

(D) This incorrect solution does not combine pump curves for series operation. Other assumptions, definitions, and equations are the same as used in the correct solution.

For the lower pump,

fitting	quantity	unit equivalent length (ft)	total equivalent length (ft)
standard 90° ell	1	8.5	8.5

$$L_T = L + L_e = 35 \text{ ft} + 3 \text{ ft} + 8.5 \text{ ft}$$
$$= 46.5 \text{ ft}$$

For an initial head loss estimate, assume that the flow velocity is 5 ft³/sec. From head loss tables,

$$p = 24 \text{ lbf/in}^2$$

$$h_f = \left(\frac{24 \frac{\text{lbf}}{\text{in}^2}}{1000 \text{ ft}} \right) (46.5 \text{ ft}) \left(\frac{1 \text{ ft}}{0.433 \frac{\text{lbf}}{\text{in}^2}} \right)$$

$$= 2.6 \text{ ft}$$
$$h_z = z_2 - z_1 = 1025 \text{ ft} - 1000 \text{ ft}$$
$$= 25 \text{ ft}$$
$$h_T = h_f + h_z = 2.6 \text{ ft} + 25 \text{ ft}$$
$$= 27.6 \text{ ft}$$

Use the total head loss and the operating curve for the lower pump to find the lower pump capacity.

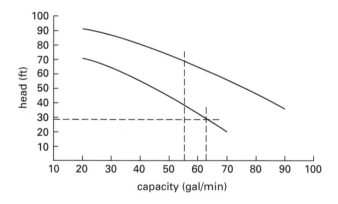

Q_l lower pump capacity   gal/min

$$Q_l = 63 \text{ gal/min}$$

For the upper pump,

fitting	quantity	unit equivalent length (ft)	total equivalent length (ft)
standard 90° ell	3	8.5	25.5
couple	4	0.45	1.8
			27.3

$$L_T = L + L_e = 140 \text{ ft} + 27.3 \text{ ft}$$
$$= 167 \text{ ft}$$

For the initial head loss estimate, assume that the flow velocity is 5 ft/sec. From head loss tables,

$$p = 24 \text{ lbf/in}^2$$

$$h_f = \left(\frac{24 \frac{\text{lbf}}{\text{in}^2}}{1000 \text{ ft}} \right) (167 \text{ ft}) \left(\frac{1 \text{ ft)}}{0.433 \frac{\text{lbf}}{\text{in}^2}} \right)$$

$$= 9.3 \text{ ft}$$
$$h_z = z_2 - z_1 = 1085 \text{ ft} - 1025 \text{ ft}$$
$$= 60 \text{ ft}$$
$$h_T = h_f + h_z = 60 \text{ ft} + 9.3 \text{ ft}$$
$$= 69.3 \text{ ft}$$

Use the total head loss and the operating curve for the upper pump to find the upper pump capacity.

Q_u upper pump capacity   gal/min

$$Q_u = 56 \text{ gal/min}$$
$$Q = Q_l + Q_u = 63 \frac{\text{gal}}{\text{min}} + 56 \frac{\text{gal}}{\text{min}}$$
$$= 119 \text{ gal/min} \quad (120 \text{ gal/min})$$

Check the flow-velocity assumption. From head loss tables, for 63 gal/min and 56 gal/min and 2 in schedule-40 steel pipe, the flow velocity is close enough to 5 ft/sec to accept assumption.

SOLUTION 44

Q flow rate   gal/min
ω operation (rotating) speed   rev/min

For the same pump operated at different speeds but with constant head and efficiency, the flow rates are proportional to the operating speed.

$$\frac{Q_1}{\omega_1} = \frac{Q_2}{\omega_2}$$

$$Q_2 = \frac{Q_1 \omega_2}{\omega_1} = \frac{\left(850 \frac{\text{gal}}{\text{min}} \right) \left(2200 \frac{\text{rev}}{\text{min}} \right)}{1750 \frac{\text{rev}}{\text{min}}}$$

$$= 1069 \text{ gal/min} \quad (1100 \text{ gal/min})$$

The answer is (D).

Why Other Options Are Wrong

(A) This incorrect solution reduces the pumping rate by the efficiency and confuses the proportionality equation. Other definitions and equations are unchanged from the correct solution.

E_f pump fractional efficiency   –

$$\frac{E_f Q_1}{\omega_1} = \frac{Q_2}{\omega_2}$$

$$Q_2 = \frac{E_f Q_1 \omega_1}{\omega_2} = \frac{(0.78)\left(850\ \dfrac{\text{gal}}{\text{min}}\right)\left(1750\ \dfrac{\text{rev}}{\text{min}}\right)}{2200\ \dfrac{\text{rev}}{\text{min}}}$$

$$= 527\ \text{gal/min}\quad (530\ \text{gal/min})$$

(B) This incorrect solution confuses the proportionality equation. Other definitions and equations are unchanged from the correct solution.

$$Q_2 = \frac{Q_1 \omega_1}{\omega_2} = \frac{\left(850\ \dfrac{\text{gal}}{\text{min}}\right)\left(1750\ \dfrac{\text{rev}}{\text{min}}\right)}{2200\ \dfrac{\text{rev}}{\text{min}}}$$

$$= 676\ \text{gal/min}\quad (680\ \text{gal/min})$$

(C) This incorrect solution reduces the pumping rate by the efficiency. Other definitions and equations are unchanged from the correct solution.

E_f pump fractional efficiency   –

$$\frac{E_f Q_1}{\omega_1} = \frac{Q_2}{\omega_2}$$

$$Q_2 = \frac{E_f Q_1 \omega_2}{\omega_1} = \frac{(0.78)\left(850\ \dfrac{\text{gal}}{\text{min}}\right)\left(2200\ \dfrac{\text{rev}}{\text{min}}\right)}{1750\ \dfrac{\text{rev}}{\text{min}}}$$

$$= 833\ \text{gal/min}\quad (830\ \text{gal/min})$$

SOLUTION 45

h	elevation head	ft
n	number of stages	–
N_s	specific speed	–
Q	pump discharge	gal/min
ω	rotating speed	rev/min

$$N_s = \frac{\omega\sqrt{Q}}{\left(\dfrac{h}{n}\right)^{3/4}}$$

$$\left(\frac{h}{n}\right)^{3/4} = \frac{\omega\sqrt{Q}}{N_s}$$

$$n = h\left(\frac{N_s}{\omega\sqrt{Q}}\right)^{4/3}$$

$$= (350\ \text{ft})\left(\frac{2300}{\left(1750\ \dfrac{\text{rev}}{\text{min}}\right)\sqrt{\left(800\ \dfrac{\text{gal}}{\text{min}}\right)}}\right)^{4/3}$$

$$= 5.8\quad (6\ \text{stages})$$

The answer is (C).

Why Other Options Are Wrong

(A) This choice incorrectly rearranges the specific speed equation. Other definitions and equations are unchanged from the correct solution.

$$N_s = \frac{\omega\sqrt{Q}}{\left(\dfrac{h}{n}\right)^{3/4}}$$

$$n^{3/4} = \frac{\omega\sqrt{Q}}{N_s h^{3/4}} = \frac{\left(1750\ \dfrac{\text{rev}}{\text{min}}\right)\sqrt{800\ \dfrac{\text{gal}}{\text{min}}}}{(2300)(350\ \text{ft})^{3/4}}$$

$$= 0.27$$

$$n = 0.27^{4/3}$$

$$= 0.17\quad (1\ \text{stage})$$

(B) This incorrect choice takes the square root of both the flow and rotational speed in the number of stages equation and fails to multiply by the head, although the equation is correct. Other definitions and equations are unchanged from the correct solution.

$$N_s = \frac{\omega\sqrt{Q}}{\left(\dfrac{h}{n}\right)^{3/4}}$$

$$\left(\frac{h}{n}\right)^{3/4} = \frac{\omega\sqrt{Q}}{N_s}$$

$$n = \left(\frac{N_s}{\sqrt{\omega Q}}\right)^{4/3}$$

$$= \left(\frac{2300}{\sqrt{\left(1750\ \dfrac{\text{rev}}{\text{min}}\right)\left(800\ \dfrac{\text{gal}}{\text{min}}\right)}}\right)^{4/3}$$

$$= 2.4\quad (3\ \text{stages})$$

(D) This incorrect choice includes a math error in the application of exponents in developing the equation for the number of stages. Other definitions and equations are unchanged from the correct solution.

$$N_s = \frac{\omega\sqrt{Q}}{\left(\dfrac{h}{n}\right)^{3/4}}$$

$$\left(\frac{h}{n}\right)^{3/4} = \frac{\omega\sqrt{Q}}{N_s}$$

$$n = h\left(\frac{N_s}{\omega\sqrt{Q}}\right)^{3/4}$$

$$= (350\text{ ft})\left(\frac{2300}{\left(1750\ \frac{\text{rev}}{\text{min}}\right)\sqrt{800\ \frac{\text{gal}}{\text{min}}}}\right)^{3/4}$$

$$= 35\text{ stages}$$

SOLUTION 46

d_z pipe centerline depth below grade ft
z pipe centerline elevation ft
z_g ground surface elevation ft

$$z_g = z + d_z$$

C conversion from lbf/in^2
   to ft of water (0.433 lbf/in^2)/ft of water
d_p water pressure depth ft
p water pressure lbf/in^2

$$d_p = \frac{p}{C}$$

HGL elevation of hydraulic grade line ft

$$\text{HGL} = z + d_p$$

When the hydraulic grade line is greater than the ground surface elevation, water overflows through the manhole onto the street. The calculation results are summarized in the table.

location	ground surface elevation (ft)	pipe centerline elevation (ft)	water pressure (lbf/in^2)
MH A	108.2	100	0
MH B	96.5	86	4.2
MH C	83.4	71	6.8
MH D	60.8	54	3.1
MH E	49.2	43	1.2

location	water pressure depth (ft)	hydraulic grade line elevation (ft)	hydraulic grade line above ground surface
MH A	0	100	no
MH B	9.7	95.7	no
MH C	16	87.0	yes
MH D	7.2	61.2	yes
MH E	2.8	45.8	no

Water overflows through manholes C and D.

The answer is (C).

Why Other Options Are Wrong

(A) This incorrect solution calculates the hydraulic grade line by adding the pressure depth to the water depth instead of to the centerline elevation. Other assumptions, definitions, and equations are unchanged from the correct solution.

$$\text{HGL} = d_z + d_p$$

location	ground surface elevation (ft)	pipe centerline elevation (ft)	water pressure (lbf/in^2)
MH A	108.2	8.2	0
MH B	96.5	10.5	4.2
MH C	83.4	12.4	6.8
MH D	60.8	6.8	3.1
MH E	49.2	6.2	1.2

location	water pressure depth (ft)	hydraulic grade line elevation (ft)	hydraulic grade line above ground surface
MH A	0	8.2	no
MH B	9.7	20.0	no
MH C	16	28.4	no
MH D	7.2	14.0	no
MH E	2.8	9.0	no

Water does not overflow through any manhole.

(B) This incorrect solution assumes that when the hydraulic grade line is less than the ground surface elevation, overflow through the manhole occurs. Other assumptions, definitions, and equations are unchanged from the correct solution.

location	ground surface elevation (ft)	pipe centerline elevation (ft)	water pressure (lbf/in^2)
MH A	108.2	100	0
MH B	96.5	86	4.2
MH C	83.4	71	6.8
MH D	60.8	54	3.1
MH E	49.2	43	1.2

location	water pressure depth (ft)	hydraulic grade line elevation (ft)	hydraulic grade line above ground surface
MH A	0	100	yes
MH B	9.7	95.7	yes
MH C	16	87.0	no
MH D	7.2	61.2	no
MH E	2.8	45.8	yes

Water overflows through manholes A, B, and E.

(D) This incorrect solution miscalculates the water pressure depth. Other assumptions, definitions, and equations are unchanged from the correct solution.

γ specific weight of water   32.2 lbf/ft^3

$$d_p = \frac{p}{\gamma}$$

location	ground surface elevation (ft)	pipe centerline elevation (ft)	water pressure (lbf/in^2)
MH A	108.2	100	0
MH B	96.5	86	4.2
MH C	83.4	71	6.8
MH D	60.8	54	3.1
MH E	49.2	43	1.2

location	water pressure depth (ft)	hydraulic grade line elevation (ft)	hydraulic grade line above ground surface
MH A	0	100	no
MH B	19	105	yes
MH C	30	101	yes
MH D	14	68.0	yes
MH E	5.4	48.4	no

Water overflows through manholes B, C, and D.

SOLUTION 47

b channel bottom width   m
d_c critical water depth   m
w channel width at water surface   m
θ side slope angle measured from
      the horizontal   degree

For a trapezoidal channel with 1-to-1 side slopes, the side slope angle is 45°.

$$w = b + 2d_c \tan\theta = 8 \text{ m} + 2d_c \tan 45°$$
$$= 8 \text{ m} + 2d_c$$

A channel cross-sectional area   m^2
Fr   Froude number   –

g gravitational acceleration   9.81 m/s^2
Q flow rate   m^3/s

At critical flow for a non-rectangular channel,

$$\text{Fr} = 1 = \sqrt{\frac{Q^2 w}{g A^3}}$$

Solve for A in terms of d_c using Froude equation.

$$A = \left(\frac{Q^2 w}{g}\right)^{1/3} = \left(\frac{\left(22 \frac{\text{m}^3}{\text{s}}\right)^2 w}{9.81 \frac{\text{m}}{\text{s}^2}}\right)^{1/3}$$
$$= 3.7 w^{1/3}$$
$$= (3.7)(8 \text{ m} + 2d_c)^{1/3}$$

Solve for A in terms of d_c using channel dimensions.

$$A = bd_c + \frac{d_c^2}{\tan\theta} = (8 \text{ m})d_c + \frac{d_c^2}{\tan 45°}$$
$$= (8 \text{ m})d_c + d_c^2$$

Solve for d_c using results from above calculations for A.

$$(8 \text{ m})d_c + d_c^2 = (3.7)(8 \text{ m} + 2d_c)^{1/3}$$
$$d_c = 0.89 \text{ m}$$
$$A = (8 \text{ m})(0.89 \text{ m}) + (0.89)^2$$
$$= 7.9 \text{ m}^2$$

R hydraulic radius   m

$$R = \frac{bd_c \sin\theta + d_c^2 \cos\theta}{b \sin\theta + 2d_c}$$
$$= \frac{(8 \text{ m})(0.89 \text{ m})(\sin 45°) + (0.89 \text{ m})^2(\cos 45°)}{(8 \text{ m})(\sin 45°) + (2)(0.89 \text{ m})}$$
$$= 0.75 \text{ m}$$

n Manning roughness coefficient   –
S channel slope   m/m

Use 0.013 for the Manning roughness coefficient for concrete lining.

$$\frac{Q}{A} = \left(\frac{1}{n}\right) R^{2/3} \sqrt{S}$$

$$S = \left(\frac{Qn}{AR^{2/3}}\right)^2 = \left(\frac{\left(22 \frac{\text{m}^3}{\text{s}}\right)(0.013)}{(7.9 \text{ m}^2)(0.75 \text{ m})^{2/3}}\right)^2$$
$$= 0.0019 \text{ m/m}$$

The answer is (C).

Why Other Options Are Wrong

(A) This incorrect solution uses the equation for the wetted perimeter as the equation for the hydraulic radius. Other assumptions, definitions, and equations are unchanged from the correct solution.

$$w = b + 2d_c \tan \theta = 8 \text{ m} + 2d_c \tan 45°$$
$$= 8 \text{ m} + 2d_c$$

Solve for A in terms of d_c using Froude equation.

$$A = \left(\frac{Q^2 w}{g} \right)^{1/3} = \left(\frac{\left(22 \frac{\text{m}^3}{\text{s}} \right)^2 w}{9.81 \frac{\text{m}}{\text{s}^2}} \right)^{1/3}$$

$$= 3.7 w^{1/3}$$
$$= 3.7 \left(8 \text{ m} + 2d_c \right)^{1/3}$$

Solve for A in terms of d_c using channel dimensions.

$$A = bd_c + \frac{d_c^2}{\tan \theta} = (8 \text{ m})d_c + \frac{d_c^2}{\tan 45°}$$
$$= (8 \text{ m})d_c + d_c^2$$

Solve for d_c using results from above calculations for A.

$$(8 \text{ m})d_c + d_c^2 = 3.7 \left(8 \text{ m} + 2d_c \right)^{1/3}$$

Solve for $d_c = 0.89$ m.

$$A = (8 \text{ m})(0.89 \text{ m}) + (0.89 \text{ m})^2$$
$$= 7.9 \text{ m}^2$$
$$R = b + \frac{2d_c}{\sin \theta} = 8 \text{ m} + \frac{(2)(0.89 \text{ m})}{\sin 45°}$$
$$= 10.1 \text{ m}$$
$$S = \left(\frac{Qn}{AR^{2/3}} \right)^2 = \left(\frac{\left(22 \frac{\text{m}^3}{\text{s}} \right)(0.013)}{(7.9 \text{ m}^2)(10.1 \text{ m})^{2/3}} \right)^2$$

$$= 0.000060 \text{ m/m}$$

(B) This incorrect solution makes an error when calculating the area using the Froude number equation. Other assumptions, definitions, and equations are unchanged from the correct solution.

$$w = b + 2d_c \tan \theta = 8 \text{ m} + 2d_c \tan 45°$$
$$= 8 \text{ m} + 2d_c$$
$$\text{Fr} = 1 = \sqrt{\frac{Q^2 w}{gA^3}}$$

Solve for A in terms of d_c using Froude equation.

$$A = \sqrt{\frac{Q^2 w}{g}} = \sqrt{\frac{\left(22 \frac{\text{m}^3}{\text{s}} \right)^2 w}{9.81 \frac{\text{m}}{\text{s}^2}}}$$

$$= 3.7 \sqrt{w}$$
$$= 3.7 \sqrt{8 \text{ m} + 2d_c}$$

Solve for A in terms of d_c using channel dimensions.

$$A = bd_c + \frac{d_c^2}{\tan \theta} = (8 \text{ m})d_c + \frac{d_c^2}{\tan 45°}$$
$$= (8 \text{ m})d_c + d_c^2$$

Solve for d_c using results from above calculations for A.

$$(8 \text{ m})d_c + d_c^2 = 3.7 \sqrt{8 \text{ m} + 2d_c}$$

Solve for $d_c = 1.3$ m.

$$A = (8 \text{ m})(1.3 \text{ m}) + (1.3 \text{ m})^2$$
$$= 12 \text{ m}^2$$
$$R = \frac{bd_c \sin \theta + d_c^2 \cos \theta}{b \sin \theta + 2d_c}$$
$$= \frac{(8 \text{ m})(1.3 \text{ m})(\sin 45°) + (1.3 \text{ m})^2 (\cos 45°)}{(8 \text{ m})(\sin 45°) + (2)(1.3 \text{ m})}$$
$$= 1.035 \text{ m}$$
$$S = \left(\frac{Qn}{AR^{2/3}} \right)^2 = \left(\frac{\left(22 \frac{\text{m}^3}{\text{s}} \right)(0.013)}{(12 \text{ m}^2)(1.035 \text{ m})^{2/3}} \right)^2$$

$$= 0.00054 \text{ m/m}$$

(D) This incorrect solution miscalculates the channel width at the water surface. Other assumptions, definitions, and equations are unchanged from the correct solution.

$$w = b + 2d_c \tan \theta = 8 \text{ m} + 2d_c \tan 45°$$
$$= 8 \text{ m} + 2d_c$$
$$= 4 \text{ m} + d_c$$

Solve for A in terms of d_c using Froude equation.

$$A = \left(\frac{Q^2 w}{g} \right)^{1/3} = \left(\frac{\left(22 \frac{\text{m}^3}{\text{s}} \right)^2 w}{9.81 \frac{\text{m}}{\text{s}^2}} \right)^{1/3}$$

$$= 3.7 w^{1/3}$$
$$= 3.7 \left(4 \text{ m} + d_c \right)^{1/3}$$

Solve for A in terms of d_c using channel dimensions.

$$A = bd_c + \frac{d_c^2}{\tan\theta} = (8\text{ m})d_c + \frac{d_c^2}{\tan 45°}$$
$$= (8\text{ m})d_c + d_c^2$$

Solve for d_c using results from above calculations for A.

$$(8\text{ m})d_c + d_c^2 = 3.7(4\text{ m} + d_c)^{1/3}$$

Solve for $d_c = 0.71$ m.

$$A = (8\text{ m})(0.71\text{ m}) + (0.71\text{ m})^2$$
$$= 6.2\text{ m}^2$$
$$R = \frac{bd_c\sin\theta + d_c^2\cos\theta}{b\sin\theta + 2d_c}$$
$$= \frac{(8\text{ m})(0.71\text{ m})(\sin 45°) + (0.71\text{ m})^2(\cos 45°)}{(8\text{ m})(\sin 45°) + (2)(0.71\text{ m})}$$
$$= 0.62\text{ m}$$

$$S = \left(\frac{Qn}{AR^{2/3}}\right)^2 = \left(\frac{\left(22\,\frac{\text{m}^3}{\text{s}}\right)(0.013)}{(6.2\text{ m}^2)(0.62\text{ m})^{2/3}}\right)^2$$
$$= 0.25\text{ m/m}$$

SOLUTION 48

From standard reference tables, the inside cross sectional area of a 12 in diameter welded steel pipe (standard) is about 0.785 ft^2.

A	pipe cross sectional	ft^2
g	gravitational acceleration	32.2 ft/sec^2
h_f	head loss	ft
k, n	loss coefficients	–
Q	flow rate	ft^3/sec
Re	Reynolds number	–
v	velocity	ft/sec
ϵ/D	relative roughness	0.0002 ft
ν	kinematic viscosity	ft^2/sec

$$A = 0.785\text{ ft}^2$$
$$v_1 = \frac{Q_1}{A} = \frac{2.2\,\frac{\text{ft}^3}{\text{sec}}}{0.785\text{ ft}^2}$$
$$= 2.8\text{ ft/sec}$$
$$v_2 = \frac{Q_2}{A} = \frac{4.4\,\frac{\text{ft}^3}{\text{sec}}}{0.785\text{ ft}^2}$$
$$= 5.6\text{ ft/sec}$$
$$\nu = 1.059 \times 10^{-5}\text{ ft}^2/\text{sec at }70°\text{F} \quad [\text{assume temperature}]$$

$$\text{Re}_1 = \frac{Dv_1}{\nu} = \frac{(1\text{ ft})\left(2.8\,\frac{\text{ft}}{\text{sec}}\right)}{1.059 \times 10^{-5}\,\frac{\text{ft}^2}{\text{sec}}}$$
$$= 2.6 \times 10^5$$

$$\text{Re}_2 = \frac{Dv_2}{\nu} = \frac{(1\text{ ft})\left(5.6\,\frac{\text{ft}}{\text{sec}}\right)}{1.059 \times 10^{-5}\,\frac{\text{ft}^2}{\text{sec}}}$$
$$= 5.3 \times 10^5$$

f Darcy friction factor   –
L length   5280 ft

From the Moody diagram,

$$f_1 = 0.0168$$
$$f_2 = 0.0152$$
$$h_f = \frac{fLv^2}{2Dg} = kQ^n$$

$$k = \frac{f_1 Lv_1^2}{2DgQ_1^n} = \frac{(0.0168)(5280\text{ ft})\left(2.8\,\frac{\text{ft}}{\text{sec}}\right)^2}{(2)(1\text{ ft})\left(32.2\,\frac{\text{ft}}{\text{sec}^2}\right)\left(2.2\,\frac{\text{ft}^3}{\text{sec}}\right)^n}$$
$$= \frac{10.8}{2.2^n}$$

$$k = \frac{f_2 Lv_2^2}{2DgQ_2^n} = \frac{(0.0152)(5280\text{ ft})\left(5.6\,\frac{\text{ft}}{\text{sec}}\right)^2}{(2)(1\text{ ft})\left(32.2\,\frac{\text{ft}}{\text{sec}^2}\right)\left(4.4\,\frac{\text{ft}^3}{\text{sec}}\right)^n}$$
$$= \frac{39.1}{4.4^n}$$

$$\frac{10.8}{2.2^n} = \frac{39.1}{4.4^n}$$
$$\left(\frac{4.4}{2.2}\right)^n = \frac{39.1}{10.8}$$
$$\log 2^n = \log 3.62$$
$$n = 1.86$$
$$k = \frac{10.8}{(2.2)^{1.86}}$$
$$= 2.49 \quad (2.5)$$

The answer is (B).

Why Other Options Are Wrong

(A) This incorrect solution improperly evaluates the log function. Other assumptions, definitions, and equations are unchanged from the correct solution.

$$A = 0.785\text{ ft}^2$$
$$v_1 = \frac{Q_1}{A} = \frac{2.2\,\frac{\text{ft}^3}{\text{sec}}}{0.785\text{ ft}^2}$$
$$= 2.8\text{ ft/sec}$$

$$v_2 = \frac{Q_2}{A} = \frac{4.4\ \frac{\text{ft}^3}{\text{sec}}}{0.785\ \text{ft}^2}$$
$$= 5.6\ \text{ft/sec}$$
$$\nu = 1.059 \times 10^{-5}\ \text{ft}^2/\text{sec at } 70°\text{F} \quad [\text{assume temperature}]$$

$$\text{Re}_1 = \frac{Dv_1}{\nu} = \frac{(1\ \text{ft})\left(2.8\ \frac{\text{ft}}{\text{sec}}\right)}{1.059 \times 10^{-5}\ \frac{\text{ft}^2}{\text{sec}}}$$
$$= 2.6 \times 10^5$$

$$\text{Re}_2 = \frac{Dv_2}{\nu} = \frac{(1\ \text{ft})\left(5.6\ \frac{\text{ft}}{\text{sec}}\right)}{1.059 \times 10^{-5}\ \frac{\text{ft}^2}{\text{sec}}}$$
$$= 5.3 \times 10^5$$

From the Moody diagram,

$$f_1 = 0.0168$$
$$f_2 = 0.0152$$

$$k = \frac{f_1 L v_1^2}{2DgQ_1^n} = \frac{(0.0168)(5280\ \text{ft})\left(2.8\ \frac{\text{ft}}{\text{sec}}\right)^2}{(2)(1\ \text{ft})\left(32.2\ \frac{\text{ft}}{\text{sec}^2}\right)\left(2.2\ \frac{\text{ft}^3}{\text{sec}}\right)^n}$$
$$= \frac{10.8}{2.2^n}$$

$$k = \frac{f_2 L v_2^2}{2DgQ_2^n} = \frac{(0.0152)(5280\ \text{ft})\left(5.6\ \frac{\text{ft}}{\text{sec}}\right)^2}{(2)(1\ \text{ft})\left(32.2\ \frac{\text{ft}}{\text{sec}^2}\right)\left(4.4\ \frac{\text{ft}^3}{\text{sec}}\right)^n}$$
$$= \frac{39.1}{4.4^n}$$
$$\frac{10.8}{2.2^n} = \frac{39.1}{4.4^n}$$
$$\left(\frac{4.4}{2.2}\right)^n = \frac{39.1}{10.8}$$
$$\log 2^n = \log 36.2$$
$$n = 5.18$$
$$k = \frac{10.8}{(2.2)^{5.18}} = 0.18$$

(C) This incorrect solution misapplies the loss coefficient n in the final calculation of the loss coefficient k. Other assumptions, definitions, and equations are unchanged from the correct solution.

$$A = 0.785\ \text{ft}^2$$

$$v_1 = \frac{Q_1}{A} = \frac{2.2\ \frac{\text{ft}^3}{\text{sec}}}{0.785\ \text{ft}^2}$$
$$= 2.8\ \text{ft/sec}$$

$$v_2 = \frac{Q_2}{A} = \frac{4.4\ \frac{\text{ft}^3}{\text{sec}}}{0.785\ \text{ft}^2}$$
$$= 5.6\ \text{ft/sec}$$
$$\nu = 1.059 \times 10^{-5}\ \text{ft}^2/\text{sec at } 70°\text{F} \quad [\text{assume temperature}]$$

$$\text{Re}_1 = \frac{Dv_1}{\nu} = \frac{(1\ \text{ft})\left(2.8\ \frac{\text{ft}}{\text{sec}}\right)}{1.059 \times 10^{-5}\ \frac{\text{ft}^2}{\text{sec}}}$$
$$= 2.6 \times 10^5$$

$$\text{Re}_2 = \frac{Dv_2}{\nu} = \frac{(1\ \text{ft})\left(5.6\ \frac{\text{ft}}{\text{sec}}\right)}{1.059 \times 10^{-5}\ \frac{\text{ft}^2}{\text{sec}}}$$
$$= 5.3 \times 10^5$$

From the Moody diagram,

$$f_1 = 0.0168$$
$$f_2 = 0.0152$$

$$k = \frac{f_1 L v_1^2}{2DgQ_1^n} = \frac{(0.0168)(5280\ \text{ft})\left(2.8\ \frac{\text{ft}}{\text{sec}}\right)^2}{(2)(1\ \text{ft})\left(32.2\ \frac{\text{ft}}{\text{sec}^2}\right)\left(2.2\ \frac{\text{ft}^3}{\text{sec}}\right)^n}$$
$$= \frac{10.8}{2.2^n}$$

$$k = \frac{f_2 L v_2^2}{2DgQ_2^n} = \frac{(0.0152)(5280\ \text{ft})\left(5.6\ \frac{\text{ft}}{\text{sec}}\right)^2}{(2)(1\ \text{ft})\left(32.2\ \frac{\text{ft}}{\text{sec}^2}\right)\left(4.4\ \frac{\text{ft}^3}{\text{sec}}\right)^n}$$
$$= \frac{39.1}{4.4^n}$$
$$\frac{10.8}{2.2^n} = \frac{39.1}{4.4^n}$$
$$\left(\frac{4.4}{2.2}\right)^n = \frac{39.1}{10.8}$$
$$\log 2^n = \log 3.62$$
$$n = 1.86$$
$$k = \left(\frac{10.8}{2.2}\right)^{1.86} = 19.3 \quad (19)$$

(D) This incorrect solution uses a Manning roughness coefficient from standard references for the value of n. Other assumptions, definitions, and equations are unchanged from the correct solution.

From standard reference tables, use a Manning roughness coefficient of 0.012 for welded steel pipe.

$$A = 0.785\ \text{ft}^2$$

$$v_1 = \frac{Q_1}{A} = \frac{2.2 \ \frac{\text{ft}^3}{\text{sec}}}{0.785 \ \text{ft}^2}$$
$$= 2.8 \ \text{ft/sec}$$

$$v_2 = \frac{Q_2}{A} = \frac{4.4 \ \frac{\text{ft}^3}{\text{sec}}}{0.785 \ \text{ft}^2}$$
$$= 5.6 \ \text{ft/sec}$$

$$\nu = 1.059 \times 10^{-5} \ \text{ft}^2/\text{sec at } 70°\text{F} \quad [\text{assume temperature}]$$

$$\text{Re}_1 = \frac{D v_1}{\nu}$$
$$= \frac{(1 \ \text{ft}) \left(2.8 \ \frac{\text{ft}}{\text{sec}} \right)}{1.059 \times 10^{-5} \ \frac{\text{ft}^2}{\text{sec}}}$$
$$= 2.6 \times 10^5$$

$$\text{Re}_2 = \frac{D v_2}{\nu}$$
$$= \frac{(1 \ \text{ft}) \left(5.6 \ \frac{\text{ft}}{\text{sec}} \right)}{1.059 \times 10^{-5} \ \frac{\text{ft}^2}{\text{sec}}}$$
$$= 5.3 \times 10^5$$

From the Moody diagram,

$$f_1 = 0.0168$$
$$f_2 = 0.0152$$

$$k = \frac{f_1 L v_1^2}{2 D g Q_1^n}$$
$$= \frac{(0.0168)(5280 \ \text{ft}) \left(2.8 \ \frac{\text{ft}}{\text{sec}} \right)^2}{(2)(1 \ \text{ft}) \left(32.2 \ \frac{\text{ft}}{\text{sec}^2} \right) \left(2.2 \ \frac{\text{ft}^3}{\text{sec}} \right)^{0.012}}$$
$$= 10.70$$

$$k = \frac{f_2 L v_2^2}{2 D g Q_2^n}$$
$$= \frac{(0.0152)(5280 \ \text{ft}) \left(5.6 \ \frac{\text{ft}}{\text{sec}} \right)^2}{(2)(1 \ \text{ft}) \left(32.2 \ \frac{\text{ft}}{\text{sec}^2} \right) \left(4.4 \ \frac{\text{ft}^3}{\text{sec}} \right)^{0.012}}$$
$$= 38.4$$

$$k = \frac{k_1 + k_2}{2} = \frac{10.7 + 38.4}{2}$$
$$= 24.6 \quad (25)$$

SOLUTION 49

From the illustration, for a capacity of 75 gal/min, the net positive suction head is 14 ft.

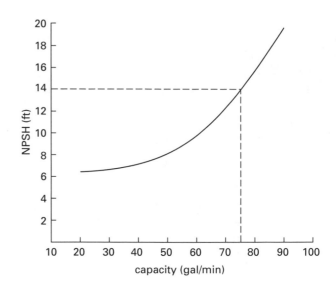

h_f	friction and minor head losses	ft
NPSH	net positive suction head	ft
p_o	atmospheric pressure at 4573 ft above MSL	12.2 lbf/in^2
p_v	vapor pressure of water at 48°F	0.178 lbf/in^2
z_s	pump height above the water surface	ft
γ	specific weight of water at 48°F	62.4 lbf/ft^3

$$z_s = \left(\left(\frac{12.2 \ \frac{\text{lbf}}{\text{in}^2}}{62.4 \ \frac{\text{lbf}}{\text{ft}^3}} \right) \left(144 \ \frac{\text{in}^2}{\text{ft}^2} \right) \right) - 14 \ \text{ft}$$
$$- 6.7 \ \text{ft} - \left(\left(\frac{0.178 \ \frac{\text{lbf}}{\text{in}^2}}{62.4 \ \frac{\text{lbf}}{\text{ft}^3}} \right) \left(144 \ \frac{\text{in}^2}{\text{ft}^2} \right) \right)$$
$$= 7.0 \ \text{ft}$$

z_p maximum pump elevation   ft

$$z_p = 4573 \ \text{ft} + 7.0 \ \text{ft} = 4580 \ \text{ft}$$

The answer is (B).

Why Other Options Are Wrong

(A) This incorrect solution subtracts the pump height from the water surface elevation instead of adding to it. The figure and other assumptions, definitions, and equations are unchanged from the correct solution.

$$z_s = \frac{p_o}{\gamma} - \mathrm{NPSH} - h_f - \frac{p_v}{\gamma}$$

$$= \left(\left(\frac{12.2\,\frac{\mathrm{lbf}}{\mathrm{in}^2}}{62.4\,\frac{\mathrm{lbf}}{\mathrm{ft}^3}}\right)\left(144\,\frac{\mathrm{in}^2}{\mathrm{ft}^2}\right)\right) - 14\text{ ft} - 6.7\text{ ft}$$

$$\quad - \left(\left(\frac{0.178\,\frac{\mathrm{lbf}}{\mathrm{in}^2}}{62.4\,\frac{\mathrm{lbf}}{\mathrm{ft}^2}}\right)\left(144\,\frac{\mathrm{in}^2}{\mathrm{ft}^2}\right)\right)$$

$$= 7.0\text{ ft}$$

$$z_p = 4573\text{ ft} - 7.0\text{ ft} = 4566\text{ ft}$$

(C) This incorrect solution uses the atmospheric pressure at sea level instead of at 4573 ft. The figure and other assumptions, definitions, and equations are unchanged from the correct solution.

p_o atmospheric pressure   14.7 lbf/in^2

$$z_s = \frac{p_o}{\gamma} - \mathrm{NPSH} - h_f - \frac{p_v}{\gamma}$$

$$= \left(\left(\frac{14.7\,\frac{\mathrm{lbf}}{\mathrm{in}^2}}{62.4\,\frac{\mathrm{lbf}}{\mathrm{ft}^3}}\right)\left(144\,\frac{\mathrm{in}^2}{\mathrm{ft}^2}\right)\right) - 14\text{ ft}$$

$$\quad - 6.7\text{ ft} - \left(\left(\frac{0.178\,\frac{\mathrm{lbf}}{\mathrm{in}^2}}{62.4\,\frac{\mathrm{lbf}}{\mathrm{ft}^3}}\right)\left(144\,\frac{\mathrm{in}^2}{\mathrm{ft}^2}\right)\right)$$

$$= 13\text{ ft}$$

$$z_p = 4573\text{ ft} + 13\text{ ft} = 4586\text{ ft}$$

(D) This incorrect solution adds instead of subtracts the net positive suction head. The figure and other assumptions, definitions, and equations are unchanged from the correct solution.

$$z_s = \frac{p_o}{\gamma} - \mathrm{NPSH} - h_f - \frac{p_v}{\gamma}$$

$$= \left(\left(\frac{12.2\,\frac{\mathrm{lbf}}{\mathrm{in}^2}}{62.4\,\frac{\mathrm{lbf}}{\mathrm{ft}^3}}\right)\left(144\,\frac{\mathrm{in}^2}{\mathrm{ft}^2}\right)\right) + 14\text{ ft}$$

$$\quad - 6.7\text{ ft} - \left(\left(\frac{0.178\,\frac{\mathrm{lbf}}{\mathrm{in}^2}}{62.4\,\frac{\mathrm{lbf}}{\mathrm{ft}^3}}\right)\left(144\,\frac{\mathrm{in}^2}{\mathrm{ft}^2}\right)\right)$$

$$= 35\text{ ft}$$

$$z_p = 4573\text{ ft} + 35\text{ ft} = 4608\text{ ft}$$

SOLUTION 50

C	Chezy coefficient	–
D	pipe diameter	m

f	friction factor	–
g	gravitational acceleration	9.81 m/s^2
R	hydraulic radius	m
Re	Reynolds number	–
S	slope	m/m
v	flow velocity	m/s
ε	specific roughness	1.5×10^{-6} m typical for PVC pipe
ε/D	relative roughness	–
ν	kinematic viscosity	1.371×10^{-6} m^2/s at 10°C

An equation is needed to relate the friction factor to other known values. The required equation can be developed as follows.

$$C = \sqrt{\frac{8g}{f}}$$

$$\mathrm{v} = C\sqrt{R}\sqrt{S} = \sqrt{\frac{8g}{f}}\sqrt{R}\sqrt{S}$$

$$\mathrm{Re} = \frac{4R\mathrm{v}}{\nu} = \frac{4R\left(C\sqrt{R}\sqrt{S}\right)}{\nu}$$

$$= \frac{4R\sqrt{\frac{8g}{f}}\sqrt{R}\sqrt{S}}{\nu}$$

$$\frac{1}{\sqrt{f}} = -2\log\left(\frac{\frac{\varepsilon}{D}}{3.7} + \frac{2.51}{\mathrm{Re}\sqrt{f}}\right)$$

$$= -2\log\left(\frac{\frac{\varepsilon}{D}}{3.7} + \frac{2.51\nu}{4R\sqrt{\frac{8g}{f}}\sqrt{R}\sqrt{S}\sqrt{f}}\right)$$

$$= -2\log\left(\frac{\frac{\varepsilon}{D}}{3.7} + \frac{0.22\nu}{\sqrt{g}R^{3/2}\sqrt{S}}\right)$$

$$\frac{\varepsilon}{D} = \frac{1.5 \times 10^{-6}\text{ m}}{(100\text{ mm})\left(\frac{1\text{ m}}{10^3\text{ mm}}\right)}$$

$$= 0.000015$$

d flow depth   m

The hydraulic radius can be determined from reference tables using the ratio of flow depth to pipe diameter.

$$\frac{d}{D} = \frac{43\text{ mm}}{100\text{ mm}} = 0.43$$

$$R = 0.023\text{ m}$$

$$\frac{1}{\sqrt{f}} = -2\log\left(\begin{array}{c} \dfrac{0.000015}{3.7} \\ \\ + \dfrac{(0.22)\left(1.371 \times 10^{-6}\ \frac{m^2}{s}\right)}{\sqrt{9.81\ \frac{m}{s^2}}\,(0.023\ m)^{3/2}} \\ \times \sqrt{0.03\ \frac{m}{m}} \end{array}\right)$$

$$= 7.57$$
$$f = 0.0174 \quad (0.017)$$

The answer is (B).

Why Other Options Are Wrong

(A) This incorrect option includes a mathematical error in the development of the friction loss equation. Other assumptions, definitions, and equations are the same as used in the correct solution.

$$\frac{1}{\sqrt{f}} = -2\log\left(\frac{\frac{\varepsilon}{D}}{3.7} + \frac{0.078\nu}{\sqrt{g}R^{3/2}\sqrt{S}}\right)$$

$$\frac{\varepsilon}{D} = \frac{1.5 \times 10^{-6}\ m}{(100\ mm)\left(\frac{1\ m}{10^3\ mm}\right)} = 0.000015$$

The hydraulic radius can be determined from reference tables using the ratio of flow depth to pipe diameter.

$$\frac{d}{D} = \frac{43\ mm}{100\ mm} = 0.43$$
$$R = 0.023\ m$$

$$\frac{1}{\sqrt{f}} = -2\log\left(\begin{array}{c} \dfrac{0.000015}{3.7} \\ \\ + \dfrac{(0.078)\left(1.371 \times 10^{-6}\ \frac{m^2}{s}\right)}{\sqrt{9.81\ \frac{m}{s^2}}\,(0.023\ m)^{3/2}} \\ \times \sqrt{0.03\ \frac{m}{m}} \end{array}\right)$$

$$= 8.44$$
$$f = 0.014$$

(C) This incorrect option fails to multiply by two after taking the log in the friction factor equation. Other assumptions, definitions, and equations are the same as used in the correct solution.

$$\frac{1}{\sqrt{f}} = -\log\left(\frac{\frac{\varepsilon}{D}}{3.7} + \frac{0.22\nu}{\sqrt{g}R^{3/2}\sqrt{S}}\right)$$

$$\frac{\varepsilon}{D} = \frac{1.5 \times 10^{-6}\ m}{(100\ mm)\left(\frac{1\ m}{10^3\ mm}\right)}$$
$$= 0.000015$$

The hydraulic radius can be determined from reference tables using the ratio of flow depth to pipe diameter.

$$\frac{d}{D} = \frac{43\ mm}{100\ mm} = 0.43$$
$$R = 0.023\ m$$

$$\frac{1}{\sqrt{f}} = -\log\left(\begin{array}{c} \dfrac{0.000015}{3.7} \\ \\ + \dfrac{(0.22)\left(1.371 \times 10^{-6}\ \frac{m^2}{s}\right)}{\sqrt{9.81\ \frac{m}{s^2}}\,(0.023\ m)^{3/2}} \\ \times \sqrt{0.03\ \frac{m}{m}} \end{array}\right)$$

$$= 3.79$$
$$f = 0.070$$

(D) This incorrect option uses the value for absolute viscosity with the kinematic viscosity units. Other assumptions, definitions, and equations are the same as used in the correct solution.

ν kinematic viscosity 1.31×10^{-3} m²/s at 10°C

$$\frac{1}{\sqrt{f}} = -2\log\left(\frac{\frac{\varepsilon}{D}}{3.7} + \frac{0.22\nu}{\sqrt{g}R^{3/2}\sqrt{S}}\right)$$

$$\frac{\varepsilon}{D} = \frac{1.5 \times 10^{-6}\ m}{(100\ mm)\left(\frac{1\ m}{10^3\ mm}\right)}$$
$$= 0.000015$$
$$\frac{d}{D} = \frac{43\ mm}{100\ mm} = 0.43$$
$$R = 0.023\ m$$

$$\frac{1}{\sqrt{f}} = -2\log\left(\begin{array}{c}\dfrac{0.000015}{3.7}\\[3mm]+\left(\begin{array}{c}\dfrac{(0.22)\left(1.31\times10^{-3}\ \dfrac{\text{m}^2}{\text{s}}\right)}{\sqrt{9.81\ \dfrac{\text{m}}{\text{s}^2}}(0.023\ \text{m})^{3/2}}\\[5mm]\times\sqrt{0.03\ \dfrac{\text{m}}{\text{m}}}\end{array}\right)\end{array}\right)$$

$$= 1.63$$
$$f = 0.37$$

SOLUTION 51

For steel pipe, assume the pipe is elastic and correct the bulk modulus of elasticity for use in the speed of sound in the pipe calculation.

D_p	pipe inside diameter	mm
E	corrected bulk modulus of elasticity	N/m^2
E_p	pipe material modulus of elasticity	$200\times10^9\ \text{N/m}^2$
E_w	water bulk modulus	$2.07\times10^9\ \text{N/m}^2$
t_p	pipe wall thickness	mm

From standard reference tables, 200 mm schedule-40 steel pipe has an inside diameter of 202.7 mm and a wall thickness of 8.18 mm.

$$E = \frac{E_w t_p E_p}{t_p E_p + D_p E_w}$$

$$= \frac{\left(\begin{array}{c}\left(2.07\times10^9\ \dfrac{\text{N}}{\text{m}^2}\right)(8.18\ \text{mm})\\[3mm]\times\left(200\times10^9\ \dfrac{\text{N}}{\text{m}^2}\right)\end{array}\right)}{\left(\begin{array}{c}(8.18\ \text{mm})\left(200\times10^9\ \dfrac{\text{N}}{\text{m}^2}\right)\\[3mm]+(202.7\ \text{mm})\left(2.07\times10^9\ \dfrac{\text{N}}{\text{m}^2}\right)\end{array}\right)}$$

$$= 1.6\times10^9\ \text{N/m}^2$$

a	speed of sound in the pipe	m/s
ρ	water density	$1000\ \text{kg/m}^3$

$$a = \sqrt{\frac{E}{\rho}}$$

$$= \sqrt{\frac{\left(1.6\times10^9\ \dfrac{\text{N}}{\text{m}^2}\right)\left(\dfrac{\text{kg}\cdot\text{m}}{\text{s}^2\cdot\text{N}}\right)}{1000\ \dfrac{\text{kg}}{\text{m}^3}}}$$

$$= 1.265\times10^3\ \text{m/s}$$

Δp	instantaneous pressure increase in pipe	N/m^2
Δv	instantaneous water velocity decrease	m/s

Assume the pressure instantaneously increases from zero and the velocity instantaneously decreases from 2.4 m/s to zero.

$$\Delta p = \rho a \Delta\text{v}$$

$$= \left(1000\ \frac{\text{kg}}{\text{m}^3}\right)\left(1.265\times10^3\frac{\text{m}}{\text{s}}\right)$$

$$\times\left(2.4\ \frac{\text{m}}{\text{s}}\right)\left(\frac{\text{N}\cdot\text{s}^2}{\text{kg}\cdot\text{m}}\right)$$

$$= 3.036\times10^6\ \text{N/m}^2$$

g	gravitational acceleration	$9.81\ \text{m/s}^2$
$h_{w,\text{max}}$	maximum head from water hammer	m

$$h_{w,\text{max}} = \frac{\Delta p}{\rho g} = \frac{\left(3.036\times10^6\ \dfrac{\text{N}}{\text{m}^2}\right)\left(\dfrac{\text{kg}\cdot\text{m}}{\text{s}^2\cdot\text{N}}\right)}{\left(1000\ \dfrac{\text{kg}}{\text{m}^3}\right)\left(9.81\ \dfrac{\text{m}}{\text{s}^2}\right)}$$

$$= 309\ \text{m}\quad(310\ \text{m})$$

The answer is (B).

Why Other Options Are Wrong

(A) This incorrect solution fails to take the square root of the elasticity-density term for the speed-of-sound-in-the-pipe calculation. Other assumptions, definitions, and equations are unchanged from the correct solution.

From standard reference tables, 200 mm schedule-40 steel pipe has an inside diameter of 202.7 mm and a wall thickness of 8.18 mm.

$$E = \frac{E_w t_p E_p}{t_p E_p + D_p E_w}$$

$$= \frac{\left(2.07\times10^9\ \dfrac{\text{N}}{\text{m}^2}\right)(8.18\ \text{mm})\left(200\times10^9\ \dfrac{\text{N}}{\text{m}^2}\right)}{\left(\begin{array}{c}(8.18\ \text{mm})\left(200\times10^9\ \dfrac{\text{N}}{\text{m}^2}\right)\\[3mm]+(202.7\ \text{mm})\left(2.07\times10^9\ \dfrac{\text{N}}{\text{m}^2}\right)\end{array}\right)}$$

$$= 1.6\times10^9\ \text{N/m}^2$$

$$a = \frac{\sqrt{E}}{\rho}$$

$$= \frac{\sqrt{\left(1.6 \times 10^9 \; \frac{N}{m^2}\right)\left(\frac{kg \cdot m}{s^2 \cdot N}\right)}}{1000 \; \frac{kg}{m^3}}$$

$$= 40 \text{ m/s}$$

Assume the pressure instantaneously increases from zero and the velocity instantaneously decreases from 2.4 m/s to zero.

$$\Delta p = \rho a \Delta v$$

$$= \left(1000 \; \frac{kg}{m^3}\right)\left(40 \; \frac{m}{s}\right)\left(2.4 \; \frac{m}{s}\right)\left(\frac{N \cdot s^2}{kg \cdot m}\right)$$

$$= 9.6 \times 10^4 \text{ N/m}^2$$

$$h_{w,max} = \frac{\Delta p}{\rho g} = \frac{\left(9.6 \times 10^4 \; \frac{N}{m^2}\right)\left(\frac{kg \cdot m}{s^2 \cdot N}\right)}{\left(1000 \; \frac{kg}{m^3}\right)\left(9.81 \; \frac{m}{s^2}\right)}$$

$$= 9.8 \text{ m}$$

(C) This incorrect solution misreads the values from standard reference tables for the pipe diameter and wall thickness. Other assumptions, definitions, and equations are unchanged from the correct solution.

From standard reference tables, 200 mm schedule-40 steel pipe has an inside diameter of 254.5 mm and a wall thickness of 9.27 mm.

$$E = \frac{E_w t_p E_p}{t_p E_p + D_p E_w}$$

$$= \frac{\left(2.07 \times 10^9 \; \frac{N}{m^2}\right)(12.7 \text{ mm})\left(200 \times 10^9 \; \frac{N}{m^2}\right)}{\left(\begin{array}{c}(12.7 \text{ mm})\left(200 \times 10^9 \; \frac{N}{m^2}\right) \\ + (193.7 \text{ mm})\left(2.07 \times 10^9 \; \frac{N}{m^2}\right)\end{array}\right)}$$

$$= 1.8 \times 10^9 \text{ N/m}^2$$

$$a = \sqrt{\frac{E}{\rho}} = \sqrt{\frac{\left(1.8 \times 10^9 \; \frac{N}{m^2}\right)\left(\frac{kg \cdot m}{s^2 \cdot N}\right)}{1000 \; \frac{kg}{m^3}}}$$

$$= 1.342 \times 10^3 \text{ m/s}$$

Assume the pressure instantaneously increases from zero and the velocity instantaneously decreases from 2.4 m/s to zero.

$$\Delta p = \rho a \Delta v$$

$$= \left(1000 \; \frac{kg}{m^3}\right)\left(1.342 \times 10^3 \; \frac{m}{s}\right)$$

$$\times \left(2.4 \; \frac{m}{s}\right)\left(\frac{N \cdot s^2}{kg \cdot m}\right)$$

$$= 3.22 \times 10^6 \text{ N/m}^2$$

$$h_{w,max} = \frac{\Delta p}{\rho g} = \frac{\left(3.22 \times 10^6 \; \frac{N}{m^2}\right)\left(\frac{kg \cdot m}{s^2 \cdot N}\right)}{\left(1000 \; \frac{kg}{m^2}\right)\left(9.81 \; \frac{m}{s^2}\right)}$$

$$= 328 \text{ m} \quad (330 \text{ m})$$

(D) This incorrect solution misplaces the decimal in the value used for the water bulk modulus. Other assumptions, definitions, and equations are unchanged from the correct solution.

E_w water bulk modulus $20.7 \times 10^9 \text{ N/m}^2$

From standard reference tables, 200 mm schedule-40 steel pipe has an inside diameter of 202.7 mm and a wall thickness of 8.18 mm.

$$E = \frac{E_w t_p E_p}{t_p E_p + D_p E_w}$$

$$= \frac{\left(20.7 \times 10^9 \; \frac{N}{m^2}\right)(8.18 \text{ mm})\left(200 \times 10^9 \; \frac{N}{m^2}\right)}{\left(\begin{array}{c}(8.18 \text{ mm})\left(200 \times 10^9 \; \frac{N}{m^2}\right) \\ + (202.7 \text{ mm})\left(20.7 \times 10^9 \; \frac{N}{m^2}\right)\end{array}\right)}$$

$$= 5.8 \times 10^9 \text{ N/m}^2$$

$$a = \sqrt{\frac{E}{\rho}} = \sqrt{\frac{\left(5.8 \times 10^9 \; \frac{N}{m^2}\right)\left(\frac{kg \cdot m}{s^2 \cdot N}\right)}{1000 \; \frac{kg}{m^3}}}$$

$$= 2.4 \times 10^3 \text{ m/s}$$

Assume the pressure instantaneously increases from zero and the velocity instantaneously decreases from 2.4 m/s to zero.

$$\Delta p = \rho a \Delta v$$

$$= \left(1000 \; \frac{kg}{m^3}\right)\left(2.4 \times 10^3 \; \frac{m}{s}\right)$$

$$\times \left(2.4 \; \frac{m}{s}\right)\left(\frac{N \cdot s^2}{kg \cdot m}\right)$$

$$= 5.8 \times 10^6 \text{ N/m}^2$$

$$h_{w,max} = \frac{\Delta p}{\rho g} = \frac{\left(5.8 \times 10^6 \; \frac{N}{m^2}\right)\left(\frac{kg \cdot m}{s^2 \cdot N}\right)}{\left(1000 \; \frac{kg}{m^2}\right)\left(9.81 \; \frac{m}{s^2}\right)}$$

$$= 589 \text{ m} \quad (590 \text{ m})$$

SOLUTION 52

L_m model length m
L_p prototype length m
L_r scale length ratio

$$L_r = \frac{L_m}{L_p} = \frac{1}{10}$$

v_m model velocity m/s
v_p prototype velocity m/s
ν_m freshwater kinematic
viscosity at 15°C $\quad 1.189 \times 10^{-6}$ m²/s
ν_p seawater kinematic
viscosity at 8∘C $\quad 1.436 \times 10^{-6}$ m²/s

Assume that kinematic viscosity is not significantly influenced by salinity.

Viscous and inertial forces will dominate for a diving sled. The relationship between the model and prototype under the given condition are

$$\frac{L_m \text{v}_m}{\nu_m} = \frac{L_p \text{v}_p}{\nu_p}$$

$$\text{v}_m = \frac{L_p \text{v}_p \nu_m}{L_m \nu_p} = \frac{\text{v}_p \nu_m}{L_r \nu_p}$$

$$= \frac{\left(0.5 \, \frac{\text{m}}{\text{s}}\right)\left(1.189 \times 10^{-6} \, \frac{\text{m}^2}{\text{s}}\right)}{\left(\frac{1}{10}\right)\left(1.436 \times 10^{-6} \, \frac{\text{m}^2}{\text{s}}\right)}$$

$$= 4.1 \text{ m/s}$$

The answer is (C).

Why Other Options Are Wrong

(A) This incorrect solution inverts the scale length ratio. Other assumptions, definitions, and equations are the same as used in the correct solution.

$$L_r = 10$$

$$\text{v}_m = \frac{\text{v}_p \nu_m}{L_r \nu_p} = \frac{\left(0.5 \, \frac{\text{m}}{\text{s}}\right)\left(1.189 \times 10^{-6} \, \frac{\text{m}^2}{\text{s}}\right)}{(10)\left(1.436 \times 10^{-6} \, \frac{\text{m}^2}{\text{s}}\right)}$$

$$= 0.041 \text{ m/s}$$

(B) This incorrect solution confuses velocity and kinematic viscosity terms when setting up the similarity equation in terms of the model velocity and ignores the scientific notation in the viscosity terms. Other assumptions, definitions, and equations are the same as used in the correct solution.

$$L_r = \frac{L_m}{L_p} = 1/10$$

$$\frac{L_m \text{v}_m}{\nu_m} = \frac{L_p \text{v}_p}{\nu_p}$$

$$\text{v}_m = \frac{L_m \nu_m \nu_p}{L_p \text{v}_p} = \frac{L_r \nu_m \nu_p}{\text{v}_p}$$

$$= \frac{\left(\frac{1}{10}\right)\left(1.189 \, \frac{\text{m}^2}{\text{s}}\right)\left(1.436 \, \frac{\text{m}^2}{\text{s}}\right)}{\left(0.5 \, \frac{\text{m}}{\text{s}}\right)}$$

$$= 0.34 \text{ m/s}$$

(D) This incorrect solution reverses the viscosity values for the given temperatures. Other assumptions, definitions, and equations are the same as used in the correct solution.

$$L_r = \frac{L_m}{L_p} = 1/10$$

ν_m freshwater kinematic
viscosity at 15°C $\quad 1.436 \times 10^{-6}$ m²/s
ν_p seawater kinematic
viscosity at 8°C $\quad 1.189 \times 10^{-6}$ m²/s

$$\text{v}_m = \frac{\text{v}_p \nu_m}{L_r \nu_p} = \frac{\left(0.5 \, \frac{\text{m}}{\text{s}}\right)\left(1.436 \times 10^{-6} \, \frac{\text{m}^2}{\text{s}}\right)}{\left(\frac{1}{10}\right)\left(1.189 \times 10^{-6} \, \frac{\text{m}^2}{\text{s}}\right)}$$

$$= 6.0 \text{ m/s}$$

SOLUTION 53

A frontal area m²
C_D coefficient of drag –
F_D drag force N
v velocity m/s
ρ air density at 39°C 1.13 kg/m³

$$F_D = \frac{C_D A \rho \text{v}^2}{2}$$

$$= \frac{\left(\begin{array}{c}(0.24)(0.47 \text{ m}^2)\left(1.13 \, \frac{\text{kg}}{\text{m}^3}\right) \\ \times \left(11.6 \, \frac{\text{m}}{\text{s}}\right)^2 \left(\frac{\text{N·s}^2}{\text{kg·m}}\right)\end{array}\right)}{2}$$

$$= 8.57 \text{ N}$$

f_R rolling resistance coefficient –
F_R rolling resistance force N
g gravitational acceleration m/s²
M rider mass kg

$$F_R = f_R M g$$

$$= \left(\frac{0.05\%}{100\%}\right)(79 \text{ kg})\left(9.81 \, \frac{\text{m}}{\text{s}^2}\right)\left(\frac{\text{N·s}^2}{\text{kg·m}}\right)$$

$$= 0.39 \text{ N}$$

d distance traveled   m
W work   kcal

$$W = (F_D + F_R)d$$
$$= (8.57 \text{ N} + 0.39 \text{ N})(100 \text{ km})$$
$$\times \left(1000 \; \frac{\text{m}}{\text{km}}\right)\left(\frac{1 \text{ kcal}}{4186 \text{ N·m}}\right)$$
$$= 210 \text{ kcal}$$

Note that the calorie used by dieticians is really a kcal. The average healthy daily diet includes about 1500 kcal.

The answer is (D).

Why Other Options Are Wrong

(A) This incorrect solution fails to convert km to m in the work equation, reading 100 km as 100 m. Other assumptions, definitions, and equations are unchanged from the correct solution.

$$F_D = \frac{C_D A \rho v^2}{2}$$
$$= \frac{\left(\begin{array}{c}(0.24)(0.46 \text{ m}^2)\left(1.13 \; \dfrac{\text{kg}}{\text{m}^3}\right) \\ \times \left(11.6 \; \dfrac{\text{m}}{\text{s}}\right)^2 \left(\dfrac{\text{N·s}^2}{\text{kg·m}}\right)\end{array}\right)}{2}$$
$$= 8.57 \text{ N}$$
$$F_R = f_R M g$$
$$= \left(\frac{0.05\%}{100\%}\right)(79 \text{ kg})\left(9.81 \; \frac{\text{m}}{\text{s}^2}\right)\left(\frac{\text{N·s}^2}{\text{kg·m}}\right)$$
$$= 0.39 \text{ N}$$
$$W = (F_D + F_R)d$$
$$= (8.57 \text{ N} + 0.39 \text{ N})(100 \text{ m})\left(\frac{1 \text{ kcal}}{4186 \text{ N·m}}\right)$$
$$= 0.21 \text{ kcal}$$

(B) This incorrect solution squares the entire term in the numerator of the drag force equation. Other assumptions, definitions, and equations are unchanged from the correct solution.

$$F_D = \frac{(C_D A \rho v)^2}{2}$$
$$= \frac{\left(\begin{array}{c}(0.24)(0.46 \text{ m}^2)\left(1.13 \; \dfrac{\text{kg}}{\text{m}^3}\right) \\ \times \left(11.6 \; \dfrac{\text{m}}{\text{s}}\right)\left(\dfrac{\text{N·s}^2}{\text{kg·m}}\right)\end{array}\right)^2}{2}$$
$$= 1.05 \text{ N}^2\text{·s}^2/\text{m}^2$$

The units do not work. Assume units to be newtons for subsequent calculations.

$$F_R = f_R M g$$
$$= \left(\frac{0.05\%}{100\%}\right)(79 \text{ kg})\left(9.81 \; \frac{\text{m}}{\text{s}^2}\right)\left(\frac{\text{N·s}^2}{\text{kg·m}}\right)$$
$$= 0.39 \text{ N}$$
$$W = (F_D + F_R)d$$
$$= (1.05 \text{ N} + 0.39 \text{ N})(100 \text{ km})$$
$$\times \left(1000 \; \frac{\text{m}}{\text{km}}\right)\left(\frac{1 \text{ kcal}}{4186 \text{ N·m}}\right)$$
$$= 34 \text{ kcal}$$

(C) This incorrect solution ignores the force from rolling resistance. Other assumptions, definitions, and equations are unchanged from the correct solution.

$$F_D = \frac{C_D A \rho v^2}{2}$$
$$= \frac{\left(\begin{array}{c}(0.24)(0.47 \text{ m}^2)\left(1.13 \; \dfrac{\text{kg}}{\text{m}^3}\right) \\ \times \left(11.6 \; \dfrac{\text{m}}{\text{s}}\right)^2 \left(\dfrac{\text{N·s}^2}{\text{kg·m}}\right)\end{array}\right)}{2}$$
$$= 8.57 \text{ N}$$
$$W = F_D d$$
$$= (8.57 \text{ N})(100 \text{ km})\left(1000 \; \frac{\text{m}}{\text{km}}\right)\left(\frac{1 \text{ kcal}}{4186 \text{ N·m}}\right)$$
$$= 205 \text{ kcal} \quad (200 \text{ kcal})$$

SOLUTION 54

g gravitational acceleration   9.81 m/s^2
h_s suction head   m
p_o atmospheric pressure at 1370 m
         above mean sea level (MSL)   84 kN/m^2
p_v vapor pressure of water at 13°C   1.5 kN/m^2
TDH   total dynamic head   m
ρ density of water at 13°C   1000 kg/m^3
σ cavitation constant   –

$$\rho g = \left(1000 \; \frac{\text{kg}}{\text{m}^3}\right)\left(9.81 \; \frac{\text{m}}{\text{s}^2}\right)\left(\frac{\text{N·s}^2}{\text{kg·m}}\right)\left(\frac{1 \text{ kN}}{1000 \text{ N}}\right)$$
$$= 9.81 \text{ kN/m}^3$$
$$h_s = \sigma \text{TDH} - \frac{p_o}{\rho g} + \frac{p_v}{\rho g}$$
$$= (0.26)(38 \text{ m}) - \frac{84 \; \dfrac{\text{kN}}{\text{m}^2}}{9.81 \; \dfrac{\text{kN}}{\text{m}^3}} + \frac{1.5 \; \dfrac{\text{kN}}{\text{m}^2}}{9.81 \; \dfrac{\text{kN}}{\text{m}^3}}$$
$$= 1.47 \text{ m} \quad (1.5 \text{ m})$$

The answer is (B).

Why Other Options Are Wrong

(A) This incorrect solution adds instead of subtracts the vapor pressure term in the suction head equation. Other definitions and equations are unchanged from the correct solution.

$$\rho g = \left(1000 \ \frac{kg}{m^3}\right)\left(9.81 \ \frac{m}{s^2}\right)\left(\frac{N{\cdot}s^2}{kg{\cdot}m}\right)\left(\frac{1 \ kN}{1000 \ N}\right)$$

$$= 9.81 \ kN/m^3$$

$$h_s = \sigma TDH - \frac{p_o}{\rho g} - \frac{p_v}{\rho g}$$

$$= (0.26)(38 \ m) - \frac{84 \ \frac{kN}{m^2}}{9.81 \ \frac{kN}{m^3}} - \frac{1.5 \ \frac{kN}{m^2}}{9.81 \ \frac{kN}{m^3}}$$

$$= 1.16 \ m \quad (1.2 \ m)$$

(C) This incorrect solution uses the customary U.S. value for specific weight for water in SI units instead of a density-gravitational acceleration term. Other definitions and equations are unchanged from the correct solution.

γ specific weight of water at 13°C   62.4 kN/m³

Note that the equivalent term in SI units for the specific weight of water is expressed as ρg.

$$h_s = \sigma TDH - \frac{p_o}{\gamma} - \frac{p_v}{\gamma}$$

$$= (0.26)(38 \ m) - \frac{84 \ \frac{kN}{m^2}}{62.4 \ \frac{kN}{m^3}} - \frac{1.5 \ \frac{kN}{m^2}}{62.4 \ \frac{kN}{m^3}}$$

$$= 8.5 \ m$$

(D) This incorrect solution transposes the cavitation constant value. Other definitions and equations are unchanged from the correct solution.

$$\rho g = \left(1000 \ \frac{kg}{m^3}\right)\left(9.81 \ \frac{m}{s^2}\right)\left(\frac{N{\cdot}s^2}{kg{\cdot}m}\right)\left(\frac{1 \ kN}{1000 \ N}\right)$$

$$= 9.81 \ kN/m^3$$

$$h_s = \sigma TDH - \frac{p_o}{\rho g} - \frac{p_v}{\rho g}$$

$$= (0.62)(38 \ m) - \frac{84 \ \frac{kN}{m^2}}{9.81 \ \frac{kN}{m^3}} - \frac{1.5 \ \frac{kN}{m^2}}{9.81 \ \frac{kN}{m^3}}$$

$$= 14.8 \ m \quad (15 \ m)$$

SOLUTION 55

fitting	quantity	unit effective length (ft)	total effective length (ft)
square inlet	1	47	47
gate valve	10	3.2	32
standard radius 90° ell	19	21	399
standard radius 45° ell	37	15	555
straight tee	8	7.2	58
			1091

L total pipe length   m
L_e effective pipe length   m
L_p actual pipe length   m

$$L = L_p + L_e$$

$$= (5 \ km)\left(1000 \ \frac{m}{km}\right) + (1091 \ ft)\left(\frac{1 \ m}{3.28 \ ft}\right)$$

$$= 5333 \ m$$

D pipe diameter   m

$$D = (400 \ mm)\left(\frac{1 \ m}{1000 \ mm}\right)$$

$$= 0.4 \ m$$

A pipe cross-sectional area   m²

$$A = \pi \frac{D^2}{4}$$

$$= \frac{\pi(0.4 \ m)^2}{4}$$

$$= 0.13 \ m^2$$

Assume a Hazen-Williams coefficient of 130 for new welded steel pipe.

C Hazen-Williams coefficient   –
h_f head loss from friction (include minor
    losses as effective pipe length)   m
Q flow rate   m³/s

$$h_f = \frac{10.7Q^{1.85}L}{C^{1.85}D^{4.87}}$$

$$= \frac{(10.7)Q^{1.85}(5333 \ m)}{(130)^{1.85}(0.4 \ m)^{4.87}}$$

$$= 607Q^{1.85}$$

g gravitational acceleration 9.81 m/s²
p pressure   N/m²
z elevation   m
ρ water density   1000 kg/m³

$$\frac{p_1}{\rho g} + z_1 + \frac{Q_1^2}{2gA_1^2} = \frac{p_2}{\rho g} + z_2 + \frac{Q_2^2}{2gA_2^2} + h_f$$

Because the pipeline connects two reservoirs, assume the pressures at the reservoirs are equal and cancel. Assume the flow at the upstream reservoir to be zero.

$$0 + 1100 \text{ m} + 0 = 0 + 835 \text{ m}$$

$$+ \frac{Q_2^2}{(2)\left(9.81 \frac{\text{m}}{\text{s}^2}\right)(0.13 \text{ m}^2)^2}$$

$$+ 607Q^{1.85}$$

$$1100 \text{ m} - 835 \text{ m} = 3.0Q^2 + 607Q^{1.85}$$

Assume the velocity head of $3.0Q^2$ is negligible compared to the friction loss of $607Q^{1.85}$.

$$265 = 607Q^{1.85}$$

$$Q = \left(\frac{265}{607}\right)^{1/1.85}$$

$$= 0.64 \text{ m}^3/\text{s}$$

The answer is (C).

Why Other Options Are Wrong

(A) This incorrect solution uses a Hazen-Williams coefficient of 100 instead of 130. For new pipe, the typical value is 130. Other assumptions, definitions, and equations are the same as used in the correct solution.

fitting	quantity	unit effective length (ft)	total effective length (ft)
square inlet	1	47	47
gate valve	10	3.2	32
standard radius 90° ell	19	21	399
standard radius 45° ell	37	15	555
straight tee	8	7.2	58
			1091

$$L = L_p + L_e = (5 \text{ km})\left(1000 \frac{\text{m}}{\text{km}}\right) + (1091 \text{ m})\left(\frac{1 \text{ m}}{3.28 \text{ ft}}\right)$$

$$= 5333 \text{ m}$$

$$D = (400 \text{ mm})\left(\frac{1 \text{ m}}{1000 \text{ mm}}\right) = 0.4 \text{ m}$$

$$A = \pi\frac{D^2}{4} = \frac{\pi(0.4 \text{ m})^2}{4} = 0.13 \text{ m}^2$$

Assume a Hazen-Williams coefficient of 100.

$$h_f = \frac{10.7Q^{1.85}L}{C^{1.85}D^{4.87}}$$

$$= \frac{(10.7)Q^{1.85}(5333 \text{ m})}{(100)^{1.85}(0.4 \text{ m})^{4.87}}$$

$$= 987Q^{1.85}$$

$$\frac{p_1}{\rho g} + z_1 + \frac{Q_1^2}{2gA_1^2} = \frac{p_2}{\rho g} + z_2 + \frac{Q_2^2}{2gA_2^2} + h_f$$

Because the pipeline connects two reservoirs, assume the pressures at the reservoirs are equal and cancel. Assume the flow at the upstream reservoir to be zero.

$$0 + 1100 \text{ m} + 0 = 0 + 835 \text{ m}$$

$$+ \frac{Q_2^2}{(2)\left(9.81 \frac{\text{m}}{\text{s}^2}\right)(0.13 \text{ m}^2)^2}$$

$$+ 987Q^{1.85}$$

$$1100 \text{ m} - 835 \text{ m} = 3.0Q^2 + 987Q^{1.85}$$

Assume the velocity head of $3.0Q^2$ is negligible compared to the friction loss of $987Q^{1.85}$.

$$265 = 987Q^{1.85}$$

$$Q = \left(\frac{265}{987}\right)^{1/1.85}$$

$$= 0.49 \text{ m}^3/\text{s}$$

(B) This incorrect solution ignores minor losses and uses a Hazen-Williams coefficient of 100. Other assumptions, definitions, and equations are the same as used in the correct solution.

$$L = (5 \text{ km})\left(1000 \frac{\text{m}}{\text{km}}\right) = 5000 \text{ m}$$

$$D = (400 \text{ mm})\left(\frac{1 \text{ m}}{1000 \text{ mm}}\right) = 0.4 \text{ m}$$

$$A = \pi\frac{D^2}{4} = \frac{\pi(0.4 \text{ m})^2}{4} = 0.13 \text{ m}^2$$

Assume a Hazen-Williams coefficient of 100.

$$h_f = \frac{10.7Q^{1.85}L}{C^{1.85}D^{4.87}}$$

$$= \frac{(10.7)Q^{1.85}(5000 \text{ m})}{(100)^{1.85}(0.4 \text{ m})^{4.87}}$$

$$= 925Q^{1.85}$$

$$\frac{p_1}{\rho g} + z_1 + \frac{Q_1^2}{2gA_1^2} = \frac{p_2}{\rho g} + z_2 + \frac{Q_2^2}{2gA_2^2} + h_f$$

Because the pipeline connects two reservoirs, assume the pressures at the reservoirs are equal and cancel. Assume the flow at the upstream reservoir to be zero.

$$0 + 1100 \text{ m} + 0 = 0 + 835 \text{ m}$$

$$+ \frac{Q_2^2}{(2)(9.81 \frac{\text{m}}{\text{s}^2})(0.13 \text{ m}^2)^2}$$

$$+ 925Q^{1.85}$$

$$1100 \text{ m} - 835 \text{ m} = 3.0Q^2 + 925Q^{1.85}$$

Assume the velocity head of $3.0Q^2$ is negligible compared to the friction loss of $925Q^{1.85}$.

$$265 = 925Q^{1.85}$$

$$Q = \left(\frac{265}{925}\right)^{1/1.85}$$

$$= 0.51 \text{ m}^3/\text{s}$$

(D) This incorrect solution ignores minor losses and uses a wrong Hazen-Williams coefficient. Other assumptions, definitions, and equations are the same as used in the correct solution.

$$L = (5 \text{ km}) \left(1000 \ \frac{\text{m}}{\text{km}} \right) = 5000 \text{ m}$$

$$D = (400 \text{ mm}) \left(\frac{1 \text{ m}}{1000 \text{ mm}} \right) = 0.4 \text{ m}$$

$$A = \pi \frac{D^2}{4} = \frac{\pi (0.4 \text{ m})^2}{4} = 0.13 \text{ m}^2$$

Assume a Hazen-Williams coefficient of 150 for new welded steel pipe.

$$
\begin{aligned}
h_f &= \frac{10.7 Q^{1.85} L}{C^{1.85} D^{4.87}} \\
&= \frac{(10.7) Q^{1.85} (5000 \text{ m})}{(150)^{1.85} (0.4 \text{ m})^{4.87}} \\
&= 374 Q^{1.85}
\end{aligned}
$$

$$\frac{p_1}{\rho g} + z_1 + \frac{Q_1^2}{2g A_1^2} = \frac{p_2}{\rho g} + z_2 + \frac{Q_2^2}{2g A_2^2} + h_f$$

Because the pipeline connects two reservoirs, assume the pressures at the reservoirs are equal and cancel. Assume the flow at the upstream reservoir to be zero.

$$0 + 1100 \text{ m} + 0 = 0 + 835 \text{ m}$$

$$+ \frac{Q_2^2}{(2) \left(9.81 \ \frac{\text{m}}{\text{s}^2} \right) (0.13 \text{ m}^2)^2}$$

$$+ 374 Q^{1.85}$$

$$1100 \text{ m} - 835 \text{ m} = 3.0 Q^2 + 374 Q^{1.85}$$

Assume the velocity head of $3.0 Q^2$ is negligible compared to the friction loss of $570 Q^{1.85}$.

$$265 = 374 Q^{1.85}$$

$$Q = \left(\frac{265}{374} \right)^{1/1.85} = 0.83 \text{ m}^3/\text{s}$$

SOLUTION 56

For the channel,

A_1	channel wetted area	ft^2
b	channel base width	ft
d_1	channel water depth	ft
θ	side slope angle measured from the horizontal	degree

For 1-to-1 side slopes, θ is 45°.

$$
\begin{aligned}
A_1 &= \left(b + \frac{d_1}{\tan \theta} \right) d_1 \\
&= \left(8 \text{ ft} + \frac{3 \text{ ft}}{\tan 45°} \right) (3 \text{ ft}) \\
&= 33 \text{ ft}^2
\end{aligned}
$$

R_1 channel hydraulic radius ft

$$
\begin{aligned}
R_1 &= \frac{A_1}{b + \dfrac{2d}{\sin \theta}} = \frac{33 \text{ ft}^2}{8 \text{ ft} + \dfrac{(2)(3 \text{ ft})}{\sin 45°}} \\
&= 2.0 \text{ ft}
\end{aligned}
$$

n	Manning roughness coefficient	–
S	slope	ft/ft
v_1	channel velocity	ft^3/sec

Assume a Manning roughness coefficient of 0.013.

$$
\begin{aligned}
v_1 &= \left(\frac{1.49}{n} \right) R_1^{2/3} \sqrt{S} \\
&= \left(\frac{1.49}{0.013} \right) (2.0)^{2/3} \sqrt{0.02 \ \frac{\text{ft}}{\text{ft}}} \\
&= 25.7 \text{ ft/sec}
\end{aligned}
$$

Q flow rate ft^3/sec

$$
\begin{aligned}
Q = v_1 A_1 &= \left(25.7 \ \frac{\text{ft}}{\text{sec}} \right) (33 \text{ ft}^2) \\
&= 848 \text{ ft}^3/\text{sec}
\end{aligned}
$$

For the culvert,

A_2	culvert wetted area	ft^2
d_2	culvert water depth	ft
w	culvert width	ft

$$A_2 = w d_2 = (8 \text{ ft})(d_2)$$

$$R_2 = \frac{A_2}{w + 2d_2} = \frac{(8 \text{ ft})(d_2)}{8 \text{ ft} + 2d_2} = \frac{(4 \text{ ft})(d_2)}{4 + d_2}$$

$$Q = \left(\frac{1.49}{n} \right) A_2 R_2^{2/3} \sqrt{S} = 848 \text{ ft}^3/\text{sec}$$

Combine equations for velocity, flow, hydraulic radius, and area, and drop units for depth to simplify.

$$848 \ \frac{\text{ft}^3}{\text{sec}} = \frac{(1.49)(8d_2)(4d_2)^{2/3} \sqrt{0.02}}{(0.013)(4 + d_2)^{2/3}}$$

$$6.5 = \frac{d_2 (4d_2)^{2/3}}{(4 + d_2)^{2/3}} = d_2 \left(\frac{4d_2}{4 + d_2} \right)^{2/3}$$

Solve for d_2 by trial and error.

$$d_2 = 4.1 \text{ ft}$$

$$A_2 = (8 \text{ ft})(4.1 \text{ ft}) = 32.8 \text{ ft}^2$$

$$v_2 = \frac{Q}{A_2} = \frac{848 \ \dfrac{\text{ft}^3}{\text{sec}}}{32.8 \text{ ft}^2} = 25.9 \text{ ft/sec}$$

$$R_2 = \frac{(4 \text{ ft})(4.1 \text{ ft})}{4 + 4.1 \text{ ft}} = 2.02 \text{ ft}$$

For head loss from friction,

h_f head loss in culvert from friction   ft
L culvert length   ft

$$h_f = \frac{Ln^2 v_2^2}{2.208 R_2^{4/3}}$$

$$= \frac{(1000 \text{ ft})(0.013)^2 \left(25.9 \dfrac{\text{ft}}{\text{sec}}\right)^2}{(2.208)(2.02 \text{ ft})^{4/3}}$$

$$= 20 \text{ ft}$$

For head loss from contraction at the culvert entrance,

g gravitational acceleration   32.2 ft/sec²
h_c head loss at culvert entrance   ft
k_c contraction constant   –

Assume a contraction constant of 0.5 for an abrupt transition.

$$h_c = k_c \left(\frac{v_2^2}{2g} - \frac{v_1^2}{2g}\right)$$

$$= (0.5) \left(\frac{\left(25.9 \dfrac{\text{ft}}{\text{sec}}\right)^2 - \left(25.7 \dfrac{\text{ft}}{\text{sec}}\right)^2}{(2)\left(32.2 \dfrac{\text{ft}}{\text{sec}^2}\right)}\right)$$

$$= 0.080 \text{ ft}$$

For head loss from expansion at the culvert exit,

h_e head loss at culvert entrance   ft
k_e contraction constant   –

Assume an expansion constant of 1.0 for an abrupt transition and assume the channel velocity is the same at both ends of the culvert.

$$h_e = k_e \left(\frac{v_2^2}{2g} - \frac{v_1^2}{2g}\right)$$

$$= (1.0) \left(\frac{\left(25.9 \dfrac{\text{ft}}{\text{sec}}\right)^2 - \left(25.7 \dfrac{\text{ft}}{\text{sec}}\right)^2}{(2)\left(32.2 \dfrac{\text{ft}}{\text{sec}^2}\right)}\right)$$

$$= 0.16 \text{ ft}$$

h_t total head loss   ft

$$h_t = h_f + h_c + h_e$$

$$= 20 \text{ ft} + 0.08 \text{ ft} + 0.16 \text{ ft}$$

$$= 20.2 \text{ ft} \quad (20 \text{ ft})$$

The answer is (B).

Why Other Options Are Wrong

(A) This incorrect solution makes an error in determining the angle of the channel side slopes and reverses the velocities in expansion head loss calculations. Other assumptions, definitions, and equations are unchanged from the correct solution.

For the channel,

For 1-to-1 side slopes, θ is 90°.

$$A_1 = \left(b + \frac{d_1}{\tan\theta}\right) d_1 = \left(8 \text{ ft} + \frac{3 \text{ ft}}{\tan 90°}\right)(3 \text{ ft})$$

$$= 24 \text{ ft}^2$$

$$R_1 = \frac{A_1}{b + \dfrac{2d}{\sin\theta}} = \frac{24 \text{ ft}^2}{8 \text{ ft} + \dfrac{(2)(3 \text{ ft})}{\sin 90°}}$$

$$= 1.7 \text{ ft}$$

Assume a Manning roughness coefficient of 0.013.

$$v_1 = \left(\frac{1.49}{n}\right) R_1^{2/3} \sqrt{S}$$

$$= \left(\frac{1.49}{0.013}\right)(1.7)^{2/3} \sqrt{0.02 \dfrac{\text{ft}}{\text{ft}}}$$

$$= 23 \text{ ft/sec}$$

Q flow rate   ft³/sec

$$Q = v_1 A_1 = \left(23 \frac{\text{ft}}{\text{sec}}\right)(24 \text{ ft}^2)$$

$$= 552 \text{ ft}^3/\text{sec}$$

For the culvert,

$$A_2 = w d_2 = (8 \text{ ft}) d_2$$

$$R_2 = \frac{A_2}{w + 2d_2} = \frac{(8 \text{ ft}) d_2}{8 \text{ ft} + 2d_2}$$

$$= \frac{(4 \text{ ft}) d_2}{4 + d_2}$$

$$Q = \left(\frac{1.49}{n}\right) A_2 R_2^{2/3} \sqrt{S}$$

$$= 552 \text{ ft}^3/\text{sec}$$

Combine equations for velocity, flow, hydraulic radius, and area, and drop units for depth to simplify.

$$552 \frac{\text{ft}^3}{\text{sec}} = \frac{(1.49)(8d_2)(4d_2)^{2/3}\sqrt{0.02}}{(0.013)(4 + d_2)^{2/3}}$$

$$4.3 = \frac{d_2(4d_2)^{2/3}}{(4 + d_2)^{2/3}}$$

$$= d_2 \left(\frac{4d_2}{4 + d_2}\right)^{2/3}$$

Solve for d_2 by trial and error.

$$d_2 = 3.0 \text{ ft}$$

$$A_2 = (8 \text{ ft})(3.0 \text{ ft}) = 24 \text{ ft}^2$$

$$v_2 = \frac{Q}{A_2} = \frac{552 \frac{\text{ft}^3}{\text{sec}}}{24 \text{ ft}^2} = 23 \text{ ft/sec}$$

$$R_2 = \frac{(4 \text{ ft})(3.0 \text{ ft})}{4 + 3.0 \text{ ft}} = 1.7 \text{ ft}$$

For head loss from friction,

$$h_f = \frac{Ln^2v_2^2}{2.208R_2^{4/3}}$$

$$= \frac{(1000 \text{ ft})(0.013)^2\left(23 \frac{\text{ft}}{\text{sec}}\right)^2}{(2.208)(1.7 \text{ ft})^{4/3}}$$

$$= 20.0 \text{ ft}$$

For head loss from contraction at the culvert entrance, assume a contraction constant of 0.5 for an abrupt transition.

$$h_c = k_c\left(\frac{v_1^2}{2g} - \frac{v_2^2}{2g}\right)$$

$$= (0.5)\left(\frac{\left(23 \frac{\text{ft}}{\text{sec}}\right)^2 - \left(23 \frac{\text{ft}}{\text{sec}}\right)^2}{(2)\left(32.2 \frac{\text{ft}}{\text{sec}^2}\right)}\right)$$

$$= 0 \text{ ft}$$

For head loss from expansion at the culvert exit, assume an expansion constant of 1.0 for an abrupt transition and assume the channel velocity is the same at both ends of the culvert.

$$h_e = k_e\left(\frac{v_1^2}{2g} - \frac{v_2^2}{2g}\right)$$

$$= (1.0)\left(\frac{\left(23 \frac{\text{ft}}{\text{sec}}\right)^2 - \left(23 \frac{\text{ft}}{\text{sec}}\right)^2}{(2)\left(32.2 \frac{\text{ft}}{\text{sec}^2}\right)}\right)$$

$$= 0 \text{ ft}$$

$$h_t = h_f + h_c + h_e$$

$$= 12.2 \text{ ft} + 0 \text{ ft} + 0 \text{ ft}$$

$$= 12.2 \text{ ft} \quad (12 \text{ ft})$$

(C) This incorrect solution fails to apply the exponent to the hydraulic radius term in the friction loss equation. Other assumptions, definitions, and equations are unchanged from the correct solution.

For the channel, for 1-to-1 side slopes, θ is 45°.

$$A_1 = \left(b + \frac{d_1}{\tan\theta}\right)d_1$$

$$= \left(8 \text{ ft} + \frac{3 \text{ ft}}{\tan 45°}\right)(3 \text{ ft})$$

$$= 33 \text{ ft}^2$$

$$R_1 = \frac{A_1}{b + \frac{2d}{\sin\theta}} = \frac{33 \text{ ft}^2}{8 \text{ ft} + \frac{(2)(3 \text{ ft})}{\sin 45°}}$$

$$= 2.0 \text{ ft}$$

Assume a Manning roughness coefficient of 0.013.

$$v_1 = \left(\frac{1.49}{n}\right)R_1^{2/3}\sqrt{S}$$

$$= \left(\frac{1.49}{0.013}\right)(2.0)^{2/3}\sqrt{0.02 \frac{\text{ft}}{\text{ft}}}$$

$$= 25.7 \text{ ft/sec}$$

$$Q = v_1A_1 = \left(25.7 \frac{\text{ft}}{\text{sec}}\right)(33 \text{ ft}^2)$$

$$= 848 \text{ ft}^3/\text{sec}$$

For the culvert,

$$A_2 = wd_2 = (8 \text{ ft})d_2$$

$$R_2 = \frac{A_2}{w + 2d_2} = \frac{(8 \text{ ft})d_2}{8 \text{ ft} + 2d_2}$$

$$= \frac{(4 \text{ ft})d_2}{4 + d_2}$$

$$Q = \left(\frac{1.49}{n}\right)A_2R_2^{2/3}\sqrt{S}$$

$$= 848 \text{ ft}^3/\text{sec}$$

Combine equations for velocity, flow, hydraulic radius, and area, and drop units for depth to simplify.

$$848 \frac{\text{ft}^3}{\text{sec}} = \frac{(1.49)(8d_2)(4d_2)^{2/3}\sqrt{0.02}}{(0.013)(4 + d_2)^{2/3}}$$

$$6.5 = \frac{d_2(4d_2)^{2/3}}{(4 + d_2)^{2/3}} = d_2\left(\frac{4d_2}{4 + d_2}\right)^{2/3}$$

Solve for d_2 by trial and error.

$$d_2 = 4.1 \text{ ft}$$

$$A_2 = (8 \text{ ft})(4.1 \text{ ft}) = 32.8 \text{ ft}^2$$

$$v_2 = \frac{Q}{A_2} = \frac{848 \frac{\text{ft}^3}{\text{sec}}}{32.8 \text{ ft}^2} = 25.9 \text{ ft/sec}$$

$$R_2 = \frac{(4 \text{ ft})(4.1 \text{ ft})}{4 + 4.1 \text{ ft}} = 2.02 \text{ ft}$$

For head loss from friction,

$$h_f = \frac{Ln^2 v_2^2}{2.208R_2}$$

$$= \frac{(1000 \text{ ft})(0.013)^2 \left(25.9 \frac{\text{ft}}{\text{sec}}\right)^2}{(2.208)(2.02 \text{ ft})}$$

$$= 25.4 \text{ ft}$$

For head loss from contraction at the culvert entrance, assume a contraction constant of 0.5 for an abrupt transition.

$$h_c = k_c \left(\frac{v_2^2}{2g} - \frac{v_1^2}{2g}\right)$$

$$= (0.5)\left(\frac{\left(25.9 \frac{\text{ft}}{\text{sec}}\right)^2 - \left(25.7 \frac{\text{ft}}{\text{sec}}\right)^2}{(2)\left(32.2 \frac{\text{ft}}{\text{sec}^2}\right)}\right)$$

$$= 0.080 \text{ ft}$$

For head loss from expansion at the culvert exit, assume an expansion constant of 1.0 for an abrupt transition and assume the channel velocity is the same at both ends of the culvert.

$$h_e = k_e \left(\frac{v_2^2}{2g} - \frac{v_1^2}{2g}\right)$$

$$= (1.0)\left(\frac{\left(25.9 \frac{\text{ft}}{\text{sec}}\right)^2 - \left(25.7 \frac{\text{ft}}{\text{sec}}\right)^2}{(2)\left(32.2 \frac{\text{ft}}{\text{sec}^2}\right)}\right)$$

$$= 0.16 \text{ ft}$$

$$h_t = h_f + h_c + h_e$$

$$= 25.4 \text{ ft} + 0.08 \text{ ft} + 0.16 \text{ ft}$$

$$= 25.6 \text{ ft} \quad (26 \text{ ft})$$

(D) This incorrect solution makes an error when applying the exponent in the velocity equation for the culvert. Other assumptions, definitions, and equations are unchanged from the correct solution.

For the channel, for 1-to-1 side slopes, θ is 45°.

$$A_1 = \left(b + \frac{d_1}{\tan\theta}\right)d_1 = \left(8 \text{ ft} + \frac{3 \text{ ft}}{\tan 45°}\right)(3 \text{ ft})$$

$$= 33 \text{ ft}^2$$

$$R_1 = \frac{A_1}{b + \frac{2d}{\sin\theta}} = \frac{33 \text{ ft}^2}{8 \text{ ft} + \frac{(2)(3 \text{ ft})}{\sin 45°}}$$

$$= 2.0 \text{ ft}$$

Assume a Manning roughness coefficient of 0.013.

$$v_1 = \left(\frac{1.49}{n}\right) R_1^{2/3}\sqrt{S}$$

$$= \left(\frac{1.49}{0.013}\right)(2.0)^{2/3}\sqrt{0.02 \frac{\text{ft}}{\text{ft}}}$$

$$= 25.7 \text{ ft/sec}$$

$$Q = v_1 A_1$$

$$= \left(25.7 \frac{\text{ft}}{\text{sec}}\right)(33 \text{ ft}^2)$$

$$= 848 \text{ ft}^3/\text{sec}$$

For the culvert,

$$A_2 = wd_2 = (8 \text{ ft})d_2$$

$$R_2 = \frac{A_2}{w + 2d_2} = \frac{(8 \text{ ft})d_2}{8 \text{ ft} + 2d_2}$$

$$= \frac{(4 \text{ ft})d_2}{4 + d_2}$$

$$Q = \left(\frac{1.49}{n}\right) A_2 R_2^{2/3}\sqrt{S}$$

$$= 848 \text{ ft}^3/\text{sec}$$

Combine equations for velocity, flow, hydraulic radius, and area, and drop units for depth to simplify.

$$848 \frac{\text{ft}^3}{\text{sec}} = \frac{(1.49)(8d_2)(4)(d_2)^{2/3}\sqrt{0.02}}{(0.013)(4 + d_2)^{2/3}}$$

$$1.63 = \frac{d_2(d_2)^{2/3}}{(4 + d_2)^{2/3}}$$

$$= d_2 \left(\frac{d_2}{4 + d_2}\right)^{2/3}$$

Solve for d_2 by trial and error.

$$d_2 = 2.9 \text{ ft}$$

$$A_2 = (8 \text{ ft})(2.9 \text{ ft}) = 23.2 \text{ ft}^2$$

$$v_2 = \frac{Q}{A_2} = \frac{848 \frac{\text{ft}^3}{\text{sec}}}{23.2 \text{ ft}^2}$$

$$= 36.6 \text{ ft/sec}$$

$$R_2 = \frac{(4 \text{ ft})(2.9 \text{ ft})}{4 + 2.9 \text{ ft}}$$

$$= 1.68 \text{ ft}$$

For head loss from friction,

$$h_f = \frac{Ln^2 v_2^2}{2.208R_2^{4/3}}$$

$$= \frac{(1000 \text{ ft})(0.013)^2 \left(36.6 \frac{\text{ft}}{\text{sec}}\right)^2}{(2.208)(1.68 \text{ ft})^{4/3}}$$

$$= 51 \text{ ft}$$

For head loss from contraction at the culvert entrance, assume a contraction constant of 0.5 for an abrupt transition.

$$h_c = k_c \left(\frac{v_2^2}{2g} - \frac{v_1^2}{2g} \right)$$

$$= (0.5) \left(\frac{\left(36.6 \ \frac{\text{ft}}{\text{sec}} \right)^2 - \left(25.7 \ \frac{\text{ft}}{\text{sec}} \right)^2}{(2) \left(32.2 \ \frac{\text{ft}}{\text{sec}^2} \right)} \right)$$

$$= 5.3 \ \text{ft}$$

For head loss from expansion at the culvert exit, assume an expansion constant of 1.0 for an abrupt transition and assume the channel velocity is the same at both ends of the culvert.

$$h_e = k_e \left(\frac{v_2^2}{2g} - \frac{v_1^2}{2g} \right)$$

$$= (1.0) \left(\frac{\left(36.6 \ \frac{\text{ft}}{\text{sec}} \right)^2 - \left(25.7 \ \frac{\text{ft}}{\text{sec}} \right)^2}{(2) \left(32.2 \ \frac{\text{ft}}{\text{sec}^2} \right)} \right)$$

$$= 10.5 \ \text{ft}$$

$$h_t = h_f + h_c + h_e = 51 \ \text{ft} + 5.3 \ \text{ft} + 10.5 \ \text{ft}$$

$$= 66.8 \ \text{ft} \quad (67 \ \text{ft})$$

SOLUTION 57

Assume that the only head loss in the pipe is from friction.

h_f head loss in the pipe   ft
z_1 upstream manometer reading   ft
z_2 downstream manometer reading   ft

$$h_f = z_2 - z_1 = (37.1 \ \text{in} - 9.2 \ \text{in}) \left(\frac{1 \ \text{ft}}{12 \ \text{in}} \right)$$

$$= 2.33 \ \text{ft}$$

A pipe cross-sectional area   ft^2
D pipe inside diameter   ft

$$A = \pi \frac{D^2}{4} = \left(\frac{\pi (0.75 \ \text{in})^2}{4} \right) \left(\frac{1 \ \text{ft}^2}{144 \ \text{in}^2} \right) = 0.0031 \ \text{ft}^2$$

Q flow rate   ft^3/sec
v flow velocity   ft/sec

$$v = \frac{Q}{A} = \left(\frac{\left(8 \ \frac{\text{gal}}{\text{min}} \right) \left(0.134 \ \frac{\text{ft}^3}{\text{gal}} \right)}{0.0031 \ \text{ft}^2} \right) \left(\frac{1 \ \text{min}}{60 \ \text{sec}} \right)$$

$$= 5.8 \ \text{ft/sec}$$

f friction factor   –
g gravitational acceleration   32.2 ft/sec^2
L pipe length   ft

$$f = \frac{2 h_f D g}{L v^2}$$

$$= \frac{(2)(2.33 \ \text{ft})(0.75 \ \text{in}) \left(\frac{1 \ \text{ft}}{12 \ \text{in}} \right) \left(32.2 \ \frac{\text{ft}}{\text{sec}^2} \right)}{(20 \ \text{ft}) \left(5.8 \ \frac{\text{ft}}{\text{sec}} \right)^2}$$

$$= 0.014$$

Assume that the flow is turbulent, so the von Karman-Nikuradse smooth pipe equation applies.

ε specific roughness   ft
ε/D relative roughness   –

$$\frac{1}{\sqrt{f}} = 1.74 - 2 \log \left(\frac{2\varepsilon}{D} \right)$$

$$\frac{1}{\sqrt{0.014}} = 1.74 - 2 \log \left(\frac{2\varepsilon}{D} \right)$$

$$\frac{10^{-3.36}}{2} = \frac{\varepsilon}{D}$$

$$= 0.00022$$

$$\varepsilon = \left(\frac{\varepsilon}{D} \right) D$$

$$= (0.00022)(0.75 \ \text{in}) \left(\frac{1 \ \text{ft}}{12 \ \text{in}} \right)$$

$$= 0.000014 \ \text{ft}$$

The answer is (A).

Why Other Options Are Wrong

(B) This incorrect solution calculates the relative roughness instead of the specific roughness. Other assumptions, definitions, and equations are the same as used in the correct solution.

Assume that the only head loss in the pipe is from friction.

$$h_f = z_2 - z_1$$

$$= (37.1 \ \text{in} - 9.2 \ \text{in}) \left(\frac{1 \ \text{ft}}{12 \ \text{in}} \right)$$

$$= 2.33 \ \text{ft}$$

$$A = \left(\frac{\pi (0.75 \ \text{in})^2}{4} \right) \left(\frac{1 \ \text{ft}^2}{144 \ \text{in}^2} \right)$$

$$= 0.0031 \ \text{ft}^2$$

$$v = \frac{Q}{A}$$

$$= \left(\frac{\left(8\ \frac{\text{gal}}{\text{min}}\right)\left(0.134\ \frac{\text{ft}^3}{\text{gal}}\right)}{0.0031\ \text{ft}^2}\right)\left(\frac{1\ \text{min}}{60\ \text{sec}}\right)$$

$$= 5.8\ \text{ft/sec}$$

$$f = \frac{h_f 2Dg}{Lv^2}$$

$$= \frac{(2)(2.33\ \text{ft})(0.75\ \text{in})\left(\frac{1\ \text{ft}}{12\ \text{in}}\right)\left(32.2\ \frac{\text{ft}}{\text{sec}^2}\right)}{(20\ \text{ft})\left(5.8\ \frac{\text{ft}}{\text{sec}}\right)^2}$$

$$= 0.014$$

Assume that the flow is turbulent, so the von Karman-Nikuradse smooth pipe equation applies.

ε/D specific roughness –

$$\frac{1}{\sqrt{f}} = 1.74 - 2\log\left(\frac{2\varepsilon}{D}\right)$$

$$\frac{1}{\sqrt{0.014}} = 1.74 - 2\log\left(\frac{2\varepsilon}{D}\right)$$

$$\frac{10^{-3.36}}{2} = \frac{\varepsilon}{D} = 0.00022$$

(C) This incorrect solution fails to convert units for friction loss from inches to feet. Other assumptions, definitions, and equations are the same as used in the correct solution.

Assume that the only head loss in the pipe is from friction.

$$h_f = z_2 - z_1 = 37.1\ \text{ft} - 9.2\ \text{ft} = 27.9\ \text{ft}$$

$$A = \pi\frac{D^2}{4} = \left(\frac{\pi(0.75\ \text{in})^2}{4}\right)\left(\frac{1\ \text{ft}^2}{144\ \text{in}^2}\right)$$

$$= 0.0031\ \text{ft}^2$$

$$v = \frac{Q}{A} = \frac{\left(8\ \frac{\text{gal}}{\text{min}}\right)\left(0.134\ \frac{\text{ft}^3}{\text{gal}}\right)\left(\frac{1\ \text{min}}{60\ \text{sec}}\right)}{0.0031\ \text{ft}^2}$$

$$= 5.8\ \text{ft/sec}$$

$$f = \frac{2h_f Dg}{Lv^2}$$

$$= \frac{(2)(27.9\ \text{ft})(0.75\ \text{in})\left(\frac{1\ \text{ft}}{12\ \text{in}}\right)\left(32.2\ \frac{\text{ft}}{\text{sec}^2}\right)}{(20\ \text{ft})\left(5.8\ \frac{\text{ft}}{\text{sec}}\right)^2}$$

$$= 0.17$$

Assume that the flow is turbulent, so the von Karman-Nikuradse smooth pipe equation applies.

$$\frac{1}{\sqrt{f}} = 1.74 - 2\log\left(\frac{2\varepsilon}{D}\right)$$

$$\frac{1}{\sqrt{0.17}} = 1.74 - 2\log\left(\frac{2\varepsilon}{D}\right)$$

$$\frac{10^{-0.34}}{2} = \frac{\varepsilon}{D} = 0.23$$

$$\varepsilon = \left(\frac{\varepsilon}{D}\right)D = (0.23)(0.75\ \text{in})\left(\frac{1\ \text{ft}}{12\ \text{in}}\right)$$

$$= 0.014\ \text{ft}$$

(D) This incorrect solution uses the wrong conversion factor for gal/min to ft^3/sec. Other assumptions, definitions, and equations are the same as used in the correct solution.

Assume that the only head loss in the pipe is from friction.

$$h_f = z_2 - z_1 = (37.1\ \text{in} - 9.2\ \text{in})\left(\frac{1\ \text{ft}}{12\ \text{in}}\right)$$

$$= 2.33\ \text{ft}$$

$$A = \pi\frac{D^2}{4} = \left(\frac{\pi(0.75\ \text{in})^2}{4}\right)\left(\frac{1\ \text{ft}^2}{144\ \text{in}^2}\right)$$

$$= 0.0031\ \text{ft}^2$$

$$v = \frac{Q}{A} = \frac{\left(8\ \frac{\text{gal}}{\text{min}}\right)\left(\frac{1\ \text{min}}{60\ \text{sec}}\right)}{\left(28.32\ \frac{\text{gal}}{\text{ft}^3}\right)(0.0031\ \text{ft}^2)}$$

$$= 1.5\ \text{ft/sec}$$

$$f = \frac{2h_f Dg}{Lv^2}$$

$$= \frac{(2)(2.33\ \text{ft})(0.75\ \text{in})\left(\frac{1\ \text{ft}}{12\ \text{in}}\right)\left(32.2\ \frac{\text{ft}}{\text{sec}^2}\right)}{(20\ \text{ft})\left(1.5\ \frac{\text{ft}}{\text{sec}}\right)^2}$$

$$= 0.21$$

Assume that the flow is turbulent, so the von Karman-Nikuradse smooth pipe equation applies.

$$\frac{1}{\sqrt{f}} = 1.74 - 2\log\left(\frac{2\varepsilon}{D}\right)$$

$$\frac{1}{\sqrt{0.21}} = 1.74 - 2\log\left(\frac{2\varepsilon}{D}\right)$$

$$\frac{10^{-0.22}}{2} = \frac{\varepsilon}{D} = 0.30$$

$$\varepsilon = \left(\frac{\varepsilon}{D}\right)D$$

$$= (0.30)(0.75\ \text{in})\left(\frac{1\ \text{ft}}{12\ \text{in}}\right)$$

$$= 0.019\ \text{ft}$$

SOLUTION 58

h_f friction head loss for straight pipe   m/m

$$h_f = \frac{14.2 \text{ cm} - 12.9 \text{ cm}}{10 \text{ cm}}$$
$$= 0.13 \frac{\text{cm}}{\text{cm}}$$
$$= 0.13 \text{ m/m}$$

h_m total head loss for straight pipe and valve   m

$$h_m = (18 \text{ cm} - 3.7 \text{ cm}) \left(\frac{1 \text{ m}}{100 \text{ cm}}\right) = 0.143 \text{ m}$$

h_v minor loss through valve   m
L pipe length   m

$$h_v = h_m - Lh_f$$
$$= 0.143 \text{ m} - (2)(20 \text{ cm}) \left(\frac{1 \text{ m}}{100 \text{ cm}}\right) \left(0.13 \frac{\text{m}}{\text{m}}\right)$$
$$= 0.091 \text{ m}$$

g gravitational acceleration   9.81 m/s^2
k loss coefficient   –
v flow velocity   m/s

$$h_v = \frac{kv^2}{2g}$$
$$k = \frac{h_v 2g}{v^2} = \frac{(0.091 \text{ m})(2) \left(9.81 \frac{\text{m}}{\text{s}^2}\right)}{\left(3.2 \frac{\text{m}}{\text{s}}\right)^2}$$
$$= 0.17$$

The answer is (A).

Why Other Options Are Wrong

(B) This incorrect solution accounts for only one of the two straight pipe lengths on either side of the valve. Other definitions and equations are unchanged from the correct solution.

$$h_f = \frac{14.2 \text{ cm} - 12.9 \text{ cm}}{10 \text{ cm}} = 0.13 \frac{\text{cm}}{\text{cm}}$$
$$= 0.13 \text{ m/m}$$
$$h_m = (18 \text{ cm} - 3.7 \text{ cm}) \left(\frac{1 \text{ m}}{100 \text{ cm}}\right)$$
$$= 0.143 \text{ m}$$
$$h_v = h_m - Lh_f$$
$$= 0.143 \text{ m} - (20 \text{ cm}) \left(\frac{1 \text{ m}}{100 \text{ cm}}\right) \left(0.13 \frac{\text{m}}{\text{m}}\right)$$
$$= 0.117 \text{ m}$$

$$h_v = \frac{kv^2}{2g}$$
$$k = \frac{h_v 2g}{v^2} = \frac{(0.117 \text{ m})(2) \left(9.81 \frac{\text{m}}{\text{s}^2}\right)}{\left(3.2 \frac{\text{m}}{\text{s}}\right)^2}$$
$$= 0.22$$

(C) This incorrect solution does not account for the differences in straight pipe length between each set of manometers. Other definitions and equations are unchanged from the correct solution.

h_f friction head loss for straight pipe   m

$$h_f = (14.2 \text{ cm} - 12.9 \text{ cm}) \left(\frac{1 \text{ m}}{100 \text{ cm}}\right) = 0.013 \text{ m}$$
$$h_m = (18 \text{ cm} - 3.7 \text{ cm}) \left(\frac{1 \text{ m}}{100 \text{ cm}}\right) = 0.143 \text{ m}$$
$$h_v = h_m - h_f = 0.143 \text{ m} - 0.13 \text{ m} = 0.13 \text{ m}$$
$$= \frac{kv^2}{2g}$$
$$k = \frac{h_v 2g}{v^2} = \frac{(0.13 \text{ m})(2) \left(9.81 \frac{\text{m}}{\text{s}^2}\right)}{\left(3.2 \frac{\text{m}}{\text{s}}\right)^2} = 0.25$$

(D) This incorrect solution ignores the set of manometers with the straight pipe only and uses the head loss for the straight pipe and valve to calculate the loss coefficient. Other definitions and equations are unchanged from the correct solution.

$$h_m = (18 \text{ cm} - 3.7 \text{ cm}) \left(\frac{1 \text{ m}}{100 \text{ cm}}\right) = 0.143 \text{ m}$$
$$h_v = \frac{kv^2}{2g}$$
$$k = \frac{h_v 2g}{v^2} = \frac{(0.143 \text{ m})(2) \left(9.81 \frac{\text{m}}{\text{s}^2}\right)}{\left(3.2 \frac{\text{m}}{\text{s}}\right)^2} = 0.27$$

SOLUTION 59

This problem can be solved using either conjugate depths or the specific force equation. This solution uses the specific force equation.

A cross-sectional flow area   m^2
b channel base width   m
d water depth   m
θ side slope angle measured from the horizontal   –
1, 2 subscripts   upstream and downstream, respectively

For a trapezoidal channel,

$$A = bd + \frac{d^2}{\tan \theta}$$

For 1-to-1 side slopes, the side slope angle is $45°$.

$$A_1 = (4.2 \text{ m})(0.74 \text{ m}) + \frac{(0.74 \text{ m})^2}{\tan 45°} = 3.66 \text{ m}^2$$

$$A_2 = (4.2 \text{ m})d_2 + d_2^2$$

$\overline{d}$ distance from centroid to water surface   m

$$A\overline{d} = \frac{d^2 \left(3b + \dfrac{2d}{\tan \theta} \right)}{6}$$

$$A_1\overline{d}_1 = \frac{(0.74 \text{ m})^2 \left((3)(4.2 \text{ m}) + \dfrac{(2)(0.74 \text{ m})}{\tan 45°} \right)}{6}$$

$$= 1.29 \text{ m}^3$$

$$A_2\overline{d}_2 = \frac{d_2^2 \left((3)(4.2 \text{ m}) + \dfrac{(2)d_2}{\tan 45°} \right)}{6}$$

$$= d_2^2(2.1 \text{ m} + 0.33d_2)$$

g gravitational acceleration 9.81 m/s^2
Q flow rate m^3/s

$$A_1\overline{d}_1 + \frac{Q^2}{gA_1} = A_2\overline{d}_2 + \frac{Q^2}{gA_2}$$

$$1.29 \text{ m}^3 + \frac{\left(38 \dfrac{\text{m}^3}{\text{s}} \right)^2}{\left(9.81 \dfrac{\text{m}}{\text{s}^2} \right)(3.66 \text{ m}^2)}$$

$$= d_2^2(2.1 \text{ m} + 0.33d_2)$$

$$+ \frac{\left(38 \dfrac{\text{m}^3}{\text{s}} \right)^2}{\left(9.81 \dfrac{\text{m}}{\text{s}^2} \right)\left((4.2 \text{ m})d_2 + d_2^2 \right)}$$

$$41.5 = 2.1d_2^2 + 0.33d_2^3 + \frac{147}{4.2d_2 + d_2^2}$$

$$d_2 = 3.3 \text{ m}$$

The answer is (D).

Why Other Options Are Wrong

(A) This incorrect solution squares both the flow rate and area terms in the specific force equation calculation. Other assumptions, definitions, and equations are unchanged from the correct solution.

For a trapezoidal channel,

$$A = bd + \frac{d^2}{\tan \theta}$$

For 1-to-1 side slopes, the side slope angle is $45°$.

$$A_1 = (4.2 \text{ m})(0.74 \text{ m}) + \frac{(0.74 \text{ m})^2}{\tan 45°}$$

$$= 3.66 \text{ m}^2$$

$$A_2 = (4.2 \text{ m})d_2 + d_2^2$$

$$A\overline{d} = \frac{d^2 \left(3b + \dfrac{2d}{\tan \theta} \right)}{6}$$

$$A_1\overline{d}_1 = \frac{(0.74 \text{ m})^2 \left((3)(4.2 \text{ m}) + \dfrac{2d_2}{\tan 45°} \right)}{6}$$

$$= 1.29 \text{ m}^3$$

$$A_2\overline{d}_2 = \frac{d_2^2 \left((3)(4.2 \text{ m}) + \dfrac{2d_2}{\tan 45°} \right)}{6}$$

$$= d_2^2(2.1 \text{ m} + 0.33d_2)$$

$$A_1\overline{d}_1 + \frac{Q^2}{gA_1{}^2} = A_2\overline{d}_2 + \frac{Q^2}{gA_2{}^2}$$

$$1.29 \text{ m}^3 + \frac{\left(38 \dfrac{\text{m}^3}{\text{s}} \right)^2}{\left(9.81 \dfrac{\text{m}}{\text{s}^2} \right)(3.66 \text{ m}^2)^2}$$

$$= d_2^2(2.1 \text{ m} + 0.33d_2)$$

$$+ \frac{\left(38 \dfrac{\text{m}^3}{\text{s}} \right)^2}{\left(9.81 \dfrac{\text{m}}{\text{s}^2} \right)\left((4.2 \text{ m})d_2 + d_2^2 \right)^2}$$

$$14.18 = 2.1d_2^2 + 0.33d_2^3 + \frac{147}{(4.2d_2 + d_2^2)^2}$$

Possible values for d_2 are 2.18 m and 0.69 m. Assume the reasonable solution is $d_2 = 0.69 \text{ m}$.

(B) This incorrect solution uses the geometry and equations for a rectangular channel and assumes that the centroid depth and water depth are equal. Other assumptions, definitions, and equations are unchanged from the correct solution.

For a rectangular channel,

$$A = bd$$

$$A_1 = (4.2 \text{ m})(0.74 \text{ m}) = 3.11 \text{ m}^2$$

$$A_2 = (4.2 \text{ m})d_2$$

$$A_1d_1 + \frac{Q^2}{gA_1} = A_2d_2 + \frac{Q^2}{gA_2}$$

$$(3.11 \text{ m}^2)(0.74 \text{ m}) + \frac{\left(38 \frac{\text{m}^3}{\text{s}}\right)^2}{\left(9.81 \frac{\text{m}}{\text{s}^2}\right)(3.11 \text{ m}^2)}$$

$$= (4.2 \text{ m})d_2^2 + \frac{\left(38 \frac{\text{m}^3}{\text{s}}\right)^2}{\left(9.81 \frac{\text{m}}{\text{s}^2}\right)(4.2 \text{ m})d_2}$$

$$4.2d_2^3 - 50d_2 + 35 = 0$$

Possible values for d_2 are 3 m and 0.74 m. Assume the reasonable solution is

$$d_2 = 0.74 \text{ m}$$

(C) This incorrect solution assumes that the centroid depth and the water depth are equal. Other assumptions, definitions, and equations are unchanged from the correct solution.

For a trapezoidal channel,

$$A = bd + \frac{d^2}{\tan \theta}$$

For 1-to-1 side slopes, the side slope angle is 45°.

$$A_1 = (4.2 \text{ m})(0.74 \text{ m}) + \frac{(0.74 \text{ m})^2}{\tan 45°}$$

$$= 3.66 \text{ m}^2$$

$$A_2 = (4.2 \text{ m})d_2 + d_2^2$$

$$A_1d_1 + \frac{Q^2}{gA_1} = A_2d_2 + \frac{Q^2}{gA_2}$$

$$(3.66 \text{ m}^2)(0.74 \text{ m}) + \frac{\left(38 \frac{\text{m}^3}{\text{s}}\right)^2}{\left(9.81 \frac{\text{m}}{\text{s}^2}\right)(3.66 \text{ m}^2)}$$

$$= d_2\left((4.2 \text{ m})d_2 + d_2^2\right)$$

$$+ \frac{\left(38 \frac{\text{m}^3}{\text{s}}\right)^2}{\left(9.81 \frac{\text{m}}{\text{s}^2}\right)\left((4.2 \text{ m})d_2 + d_2^2\right)}$$

$$42.9 = 4.2d_2^2 + d_2^3 + \frac{147}{4.2d_2 + d_2^2}$$

Solve for $d_2 = 2.3 \text{ m}$

SOLUTION 60

d	water depth	m
v	flow velocity	m/s

1, 2 subscripts   upstream and downstream, respectively, of the hydraulic jump

$$d_2 = -0.5d_1 + \sqrt{\frac{2v_1^2 d_1}{g} + \frac{d_1^2}{4}}$$

$$= -(0.5)(0.65 \text{ m}) + \sqrt{\frac{(2)\left(8 \frac{\text{m}}{\text{s}}\right)^2 (0.65 \text{ m})}{9.81 \frac{\text{m}}{\text{s}^2}} + \frac{(0.65 \text{ m})^2}{4}}$$

$$= 2.6 \text{ m}$$

q flow rate per unit of channel width   m³/s·m

$$q = v_1d_1 = v_2d_2 = \left(8 \frac{\text{m}}{\text{s}}\right)(0.65 \text{ m})$$

$$= 5.2 \text{ m}^3/\text{s·m}$$

$$v_2 = \frac{q}{d_2} = \frac{5.2 \frac{\text{m}^3}{\text{s·m}}}{2.6 \text{ m}}$$

$$= 2.0 \text{ m/s}$$

ΔE change in specific energy   m

$$\Delta E = d_1 + \frac{v_1^2}{2g} - d_2 - \frac{v_2^2}{2g}$$

$$= 0.65 \text{ m} + \frac{\left(8 \frac{\text{m}}{\text{s}}\right)^2}{(2)\left(9.81 \frac{\text{m}}{\text{s}^2}\right)} - 2.6 \text{ m}$$

$$- \frac{\left(2 \frac{\text{m}}{\text{s}}\right)^2}{(2)\left(9.81 \frac{\text{m}}{\text{s}^2}\right)}$$

$$= 1.11 \text{ m}$$

q_m	mass flow rate per unit of channel width	kg/s·m
ρ	water density	1000 kg/m³

$$q_m = q\rho = \left(5.2 \frac{\text{m}^3}{\text{s·m}}\right)\left(1000 \frac{\text{kg}}{\text{m}^3}\right)$$

$$= 5200 \text{ kg/s·m}$$

g	gravitational acceleration	9.81 m/s²
P	power dissipated	kW/m

$$P = q_m g \Delta E$$

$$= \left(5200 \ \frac{\text{kg}}{\text{s·m}}\right)\left(9.81 \ \frac{\text{m}}{\text{s}^2}\right)(1.11 \ \text{m})$$

$$\times \left(\frac{1 \ \text{W·s}^3}{\text{kg·m}^2}\right)\left(\frac{1 \ \text{kW}}{1000 \ \text{W}}\right)$$

$$= 56.6 \ \text{kW/m} \quad (57 \ \text{kW/m})$$

The answer is (B).

Why Other Options Are Wrong

(A) This incorrect solution fails to square the velocity term when calculating the water depth after the hydraulic jump. Other assumptions, definition, and equations are the same as used in the correct solution.

$$d_2 = -0.5d_1 + \sqrt{\frac{2v_1 d_1}{g} + \frac{d_1^2}{4}}$$

$$= -(0.5)(0.65 \ \text{m}) + \sqrt{\begin{array}{c}\dfrac{(2)\left(8 \ \frac{\text{m}}{\text{s}}\right)(0.65 \ \text{m})}{9.81 \ \frac{\text{m}}{\text{s}^2}} \\ + \dfrac{(0.65 \ \text{m})^2}{4}\end{array}}$$

$$= 0.75 \ \text{m}$$

$$q = v_1 d_1 = v_2 d_2 = \left(8 \ \frac{\text{m}}{\text{s}}\right)(0.65 \ \text{m})$$

$$= 5.2 \ \text{m}^3/\text{s·m}$$

$$v_2 = \frac{q}{d_2} = \frac{5.2 \ \frac{\text{m}^3}{\text{s·m}}}{0.75 \ \text{m}} = 6.9 \ \text{m/s}$$

$$\Delta E = d_1 + \frac{v_1^2}{2g} - d_2 - \frac{v_2^2}{2g}$$

$$= 0.65 \ \text{m} + \frac{\left(8 \ \frac{\text{m}}{\text{s}}\right)^2}{(2)\left(9.81 \ \frac{\text{m}}{\text{s}^2}\right)} - 0.75 \ \text{m}$$

$$- \frac{\left(6.9 \ \frac{\text{m}}{\text{s}}\right)^2}{(2)\left(9.81 \ \frac{\text{m}}{\text{s}^2}\right)}$$

$$= 0.74 \ \text{m}$$

$$q_m = q\rho = \left(5.2 \ \frac{\text{m}^3}{\text{s·m}}\right)\left(1000 \ \frac{\text{kg}}{\text{m}^3}\right)$$

$$= 5200 \ \text{kg/s·m}$$

$$P = q_m g \Delta E$$

$$= \left(5200 \ \frac{\text{kg}}{\text{s·m}}\right)\left(9.81 \ \frac{\text{m}}{\text{s}^2}\right)(0.74 \ \text{m})$$

$$\times \left(\frac{1 \ \text{W·s}^3}{\text{kg·m}^2}\right)\left(\frac{1 \ \text{kW}}{1000 \ \text{W}}\right)$$

$$= 37.7 \ \text{kW/m} \quad (38 \ \text{kW/m})$$

(C) This incorrect solution fails to divide the velocity terms by two in the specific energy equation. Other

assumptions, definition, and equations are the same as used in the correct solution.

$$d_2 = -0.5d_1 + \sqrt{\frac{2v_1^2 d_1}{g} + \frac{d_1^2}{4}}$$

$$= -(0.5)(0.65 \ \text{m}) + \sqrt{\begin{array}{c}\dfrac{(2)\left(8 \ \frac{\text{m}}{\text{s}}\right)^2 (0.65 \ \text{m})^2}{9.81 \ \frac{\text{m}}{\text{s}^2}} \\ + \dfrac{(0.65 \ \text{m})^2}{4}\end{array}}$$

$$= 2.6 \ \text{m}$$

$$q = v_1 d_1 = v_2 d_2 = \left(8 \ \frac{\text{m}}{\text{s}}\right)(0.65 \ \text{m})$$

$$= 5.2 \ \text{m}^3/\text{s·m}$$

$$v_2 = \frac{q}{d_2} = \frac{5.2 \ \frac{\text{m}^3}{\text{s·m}}}{2.6 \ \text{m}} = 2.0 \ \text{m/s}$$

$$\Delta E = d_1 + \frac{v_1^2}{g} - d_2 - \frac{v_2^2}{g}$$

$$= 0.65 \ \text{m} + \frac{\left(8 \ \frac{\text{m}}{\text{s}}\right)^2}{9.81 \ \frac{\text{m}}{\text{s}^2}} - 2.6 \ \text{m} - \frac{\left(2 \ \frac{\text{m}}{\text{s}}\right)^2}{9.81 \ \frac{\text{m}}{\text{s}^2}}$$

$$= 4.17 \ \text{m}$$

$$q_m = q\rho = \left(5.2 \ \frac{\text{m}^3}{\text{s·m}}\right)\left(1000 \ \frac{\text{kg}}{\text{m}^3}\right)$$

$$= 5200 \ \text{kg/s·m}$$

$$P = q_m g \Delta E$$

$$= \left(5200 \ \frac{\text{kg}}{\text{s·m}}\right)\left(9.81 \ \frac{\text{m}}{\text{s}^2}\right)(4.17 \ \text{m})$$

$$\times \left(\frac{1 \ \text{W·s}^3}{\text{kg·m}^2}\right)\left(\frac{1 \ \text{kW}}{1000 \ \text{W}}\right)$$

$$= 213 \ \text{kW/m} \quad (210 \ \text{kW/m})$$

(D) This incorrect solution ignores the square root when calculating the water depth after the hydraulic jump. Other assumptions, definition, and equations are the same as used in the correct solution.

$$d_2 = -0.5d_1 + \frac{2v_1^2 d_1}{g} + \frac{d_1^2}{4}$$

$$= -(0.5)(0.65 \ \text{m}) + \frac{(2)\left(8 \ \frac{\text{m}}{\text{s}}\right)^2 (0.65 \ \text{m})}{9.81 \ \frac{\text{m}}{\text{s}^2}}$$

$$+ \frac{(0.65 \ \text{m})^2}{4}$$

$$= 8.3 \ \text{m}$$

$$q = v_1 d_1 = v_2 d_2 = \left(8 \ \frac{\text{m}}{\text{s}}\right)(0.65 \ \text{m})$$

$$= 5.2 \ \text{m}^3/\text{s·m}$$

$$v_2 = \frac{q}{d_2} = \frac{5.2 \ \frac{\text{m}^3}{\text{s·m}}}{8.3 \ \text{m}} = 0.63 \ \text{m/s}$$

$$\Delta E = d_1 + \frac{v_1^2}{2g} - d_2 - \frac{v_2^2}{2g}$$

$$= 0.65 \text{ m} + \frac{\left(8 \, \frac{\text{m}}{\text{s}}\right)^2}{(2)\left(9.81 \, \frac{\text{m}}{\text{s}^2}\right)} - 8.3 \text{ m}$$

$$- \frac{\left(0.63 \, \frac{\text{m}}{\text{s}}\right)^2}{(2)\left(9.81 \, \frac{\text{m}}{\text{s}^2}\right)}$$

$$= -4.41 \text{ m}$$

Ignore the negative sign and use a positive value in subsequent calculations.

$$q_m = q\rho = \left(5.2 \, \frac{\text{m}^3}{\text{s·m}}\right)\left(1000 \, \frac{\text{kg}}{\text{m}^3}\right)$$

$$= 5200 \text{ kg/s·m}$$

$$P = q_m g \Delta E$$

$$= \left(5200 \, \frac{\text{kg}}{\text{s·m}}\right)\left(9.81 \, \frac{\text{m}}{\text{s}^2}\right)(4.41 \text{ m})$$

$$\times \left(\frac{1 \text{ W·s}^3}{\text{kg·m}^2}\right)\left(\frac{1 \text{ kW}}{1000 \text{ W}}\right)$$

$$= 225 \text{ kW/m} \quad (230 \text{ kW/m})$$

SOLUTION 61

For a slope of 5% with 20 cm diameter logs placed at 4.0 m intervals, the conditions depicted in the following illustration will result, and the maximum water velocity will occur as the water flows over each log.

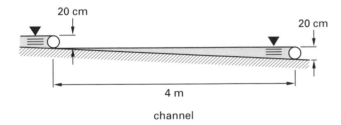

channel

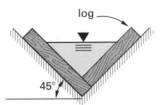

channel cross section

A flow area through the V-notch formed by the logs m^2
d water depth over the log m
θ side slope angle measure from the horizontal   degree

$$A = \frac{d^2}{\tan\theta} = \frac{d^2}{\tan 45°} = d^2$$

Q flow rate m^3/s
v allowable scour velocity m/s

With the logs placed perpendicular to the flow in a 90° V-shaped cross section channel, they will act as 90° V-notch weirs, and the flow over the weir can be approximated by

$$Q = Av = 1.4d^{2.5}$$

$$d^2\left(0.76 \, \frac{\text{m}}{\text{s}}\right) = 1.4d^{2.5}$$

$$\sqrt{d} = \frac{0.76 \, \frac{\text{m}}{\text{s}}}{1.4} = 0.54$$

$$d = 0.29 \text{ m}$$

$$A = (0.29 \text{ m})^2 = 0.084 \text{ m}^2$$

$$Q = (0.84 \text{ m}^2)\left(0.76 \, \frac{\text{m}}{\text{s}}\right) = 0.064 \text{ m}^3/\text{s}$$

The answer is (C).

Why Other Options Are Wrong

(A) This incorrect solution ignores the influence of the logs. Other assumptions, definitions, and equations are the same as used in the correct solution.

d depth of flow m
R channel hydraulic radius m

For a 90° V-shaped cross section, the side slope angle is 45°.

$$R = \frac{d\cos\theta}{2} = \frac{d\cos 45°}{2} = 0.354d$$

S channel slope m/m
n Manning roughness coefficient   –

Assume Manning roughness coefficient of 0.018 for ordinary firm loam.

$$v = \left(\frac{1}{n}\right) R^{2/3}\sqrt{S}$$

$$= \left(\frac{1}{0.018}\right)(0.354d)^{2/3}\sqrt{\frac{5}{100}}$$

$$= 0.76 \, \frac{\text{m}}{\text{s}}$$

Solve for $d = 0.043 \text{ m}$

A channel cross-sectional area of flow m^2

$$A = \frac{d^2}{\tan\theta} = \frac{(0.043 \text{ m})^2}{\tan 45°}$$
$$= 0.0018 \text{ m}^2$$

$$Q = Av = (0.0018 \text{ m}^2)\left(0.76 \frac{\text{m}}{\text{s}}\right)$$
$$= 0.0014 \text{ m}^3/\text{s}$$

(B) This incorrect solution ignores the influence of the logs and does not square the depth term in the area calculation. Other assumptions, definitions, and equations are the same as used in the correct solution.

$$R = \frac{d \cos 45°}{2} = 0.354 \text{ d}$$

$$v = \left(\frac{1}{n}\right) R^{2/3}\sqrt{S}$$
$$= \left(\frac{1}{0.018}\right)(0.354d)^{2/3}\sqrt{\frac{5}{100}}$$
$$= 0.76 \frac{\text{m}}{\text{s}}$$

Solve for $d = 0.043$ m

$$A = \frac{d}{\tan\theta} = \frac{0.043 \text{ m}}{\tan 45°}$$
$$= 0.043 \text{ m}^2$$

$$Q = Av = (0.043 \text{ m}^2)\left(0.76 \frac{\text{m}}{\text{s}}\right)$$
$$= 0.033 \text{ m}^3/\text{s}$$

(D) This incorrect solution makes an error in the calculation of the flow depth through the logs. Other assumptions, definitions, and equations, and the figure are the same as used in the correct solution.

$$A = \frac{d^2}{\tan\theta} = \frac{d^2}{\tan 45°}$$
$$= d^2$$

$$Q = Av = 1.4d^{2.5}$$

$$d^2\left(0.76 \frac{\text{m}}{\text{s}}\right) = 1.4d^{2.5}$$

$$d^{1.25} = \frac{0.76 \frac{\text{m}}{\text{s}}}{1.4} = 0.54$$

$$d = 0.61 \text{ m}$$

$$A = (0.61 \text{ m})^2 = 0.37 \text{ m}^2$$

$$Q = Av = (0.37 \text{ m}^2)\left(0.76 \frac{\text{m}}{\text{s}}\right)$$
$$= 0.28 \text{ m}^3/\text{s}$$

SOLUTION 62

β beta ratio   –
D_1 upstream pipe diameter   in or ft
D_2 throat diameter   in or ft

$$\beta = \frac{D_2}{D_1} = \frac{10}{16} = 0.625$$

C_v velocity coefficient   –
g gravitational acceleration   32.2 ft/sec^2
h manometer reading   ft
ρ water density   62.3 lbm/ft^3 at 70°F
ρ_m manometer fluid density   848 lbm/ft^3 at 70°F
v_2 throat velocity   ft/sec

Assume that the velocity coefficient is 1.0.

$$v_2 = \left(\frac{C_v}{\sqrt{1-\beta^4}}\right)\sqrt{\left(\frac{2g}{\rho}\right)(\rho_m - \rho)h}$$

$$= \left(\frac{1}{\sqrt{1-0.625^4}}\right)$$
$$\times\sqrt{\frac{(2)\left(32.2 \frac{\text{ft}}{\text{sec}}^2\right)\times\left(848 \frac{\text{lbm}}{\text{ft}^3} - 62.3 \frac{\text{lbm}}{\text{ft}^3}\right)(9.4 \text{ in})}{\left(62.3 \frac{\text{lbm}}{\text{ft}^3}\right)\left(\frac{12 \text{ in}}{\text{ft}}\right)}}$$

$$= 27.4 \text{ ft/sec}$$

A_2 throat cross-sectional area   ft^2

$$A_2 = \pi\frac{D_2^2}{4} = \left(\frac{\pi(10 \text{ in})^2}{4}\right)\left(\frac{1 \text{ ft}^2}{144 \text{ in}^2}\right)$$
$$= 0.55 \text{ ft}^2$$

Q flow rate   ft^3/sec

$$Q = A_2v_2 = (0.55 \text{ ft}^2)\left(27.4 \frac{\text{ft}}{\text{sec}}\right)$$
$$= 15 \text{ ft}^3/\text{sec}$$

The answer is (B).

Why Other Options Are Wrong

(A) This incorrect solution inverts the beta ratio. Other assumptions, definitions, and equations are unchanged from the correct solution.

$$\beta = \frac{D_1}{D_2} = \frac{16}{10} = 1.6$$

Assume that the velocity coefficient is 1.0.

$$v_2 = \left(\frac{C_v}{\sqrt{1-\beta^4}}\right)\sqrt{\left(\frac{2g}{\rho}\right)(\rho_m - \rho)h}$$

$$= \left(\frac{1}{\sqrt{1-1.6^4}}\right)$$

$$\times \sqrt{\frac{(2)\left(32.2\,\dfrac{\text{ft}}{\text{sec}^2}\right) \times \left(848\,\dfrac{\text{lbm}}{\text{ft}^3} - 62.3\,\dfrac{\text{lbm}}{\text{ft}^3}\right)(9.4\text{ in})}{\left(62.3\,\dfrac{\text{lbm}}{\text{ft}^3}\right)\left(\dfrac{12\text{ in}}{\text{ft}}\right)}}$$

$$= 10.7\text{ ft/sec}$$

The negative sign in the beta ratio term is ignored.

$$A_2 = \pi\frac{D_2^2}{4} = \frac{\pi(10\text{ in})^2\left(\dfrac{1\text{ ft}^2}{144\text{ in}^2}\right)}{4}$$

$$= 0.55\text{ ft}^2$$

$$Q = A_2 v_2 = \left(0.55\text{ ft}^2\right)\left(10.7\,\frac{\text{ft}}{\text{sec}}\right)$$

$$= 5.9\text{ ft}^3/\text{sec}$$

(C) This incorrect solution fails to include the exponent in the beta ratio term in the velocity equation. Other assumptions, definitions, and equations are unchanged from the correct solution.

$$\beta = \frac{D_2}{D_1} = \frac{10}{16} = 0.625$$

Assume that the velocity coefficient is 1.0.

$$v_2 = \left(\frac{C_v}{\sqrt{1-\beta^4}}\right)\sqrt{\left(\frac{2g}{\rho}\right)(\rho_m - \rho)h}$$

$$= \left(\frac{1}{\sqrt{1-0.625}}\right)$$

$$\times \sqrt{\frac{(2)\left(32.2\,\dfrac{\text{ft}}{\text{sec}^2}\right) \times \left(848\,\dfrac{\text{lbm}}{\text{ft}^3} - 62.3\,\dfrac{\text{lbm}}{\text{ft}^3}\right)(9.4\text{ in})}{\left(62.3\,\dfrac{\text{lbm}}{\text{ft}^3}\right)\left(\dfrac{12\text{ in}}{\text{ft}}\right)}}$$

$$= 41.2\text{ft/sec}$$

$$A_2 = \pi\frac{D_2^2}{4} = \frac{\pi(10\text{ in})^2\left(\dfrac{1\text{ ft}^2}{144\text{ in}^2}\right)}{4} = 0.55\text{ ft}^2$$

$$Q = A_2 v_2 = \left(0.55\text{ ft}^2\right)\left(41.2\,\frac{\text{ft}}{\text{sec}}\right) = 23\text{ ft}^3/\text{sec}$$

(D) This incorrect solution applies the area of the pipe instead of the area of the throat in the velocity equation. Other assumptions, definitions, and equations are unchanged from the correct solution.

$$\beta = \frac{D_2}{D_1} = \frac{10}{16} = 0.625$$

Assume that the velocity coefficient is 1.0.

$$v_2 = \left(\frac{C_v}{\sqrt{1-\beta^4}}\right)\sqrt{\left(\frac{2g}{\rho}\right)(\rho_m - \rho)h}$$

$$= \left(\frac{1}{\sqrt{1-0.625^4}}\right)$$

$$\times \sqrt{\frac{(2)\left(32.2\,\dfrac{\text{ft}}{\text{sec}^2}\right) \times \left(848\,\dfrac{\text{lbm}}{\text{ft}^3} - 62.3\,\dfrac{\text{lbm}}{\text{ft}^3}\right)(9.4\text{ in})}{\left(62.3\,\dfrac{\text{lbm}}{\text{ft}^3}\right)\left(\dfrac{12\text{ in}}{\text{ft}}\right)}}$$

$$= 27.4\text{ ft/sec}$$

$$A_2 = \pi\frac{D_2^2}{4} = \frac{\pi(16\text{ in})^2\left(\dfrac{1\text{ ft}^2}{144\text{ in}^2}\right)}{4} = 1.4\text{ ft}^2$$

$$Q = A_2 v_2 = \left(1.4\text{ ft}^2\right)\left(27.4\,\frac{\text{ft}}{\text{sec}}\right) = 38\text{ ft}^3/\text{sec}$$

SOLUTION 63

Assume that the average depth of flow between the two manholes is correlated to the average flow.

d average water depth between manholes   in

$$d = \frac{3.8\text{ in} + 5.2\text{ in}}{2} = 4.5\text{ in}$$

D pipe diameter   in
d/D partial to full flow ratio   –

$$\frac{d}{D} = \frac{4.5\text{ in}}{16\text{ in}} = 0.28$$

A average cross sectional area of flow   ft^2

From reference tables for areas of partially full pipes, based on the partial-to-full flow ratio,

$$A = 0.32\text{ ft}^2$$

t average travel time between manholes   sec

$$t = \frac{398\text{ sec} + 512\text{ sec}}{2} = 455\text{ sec}$$

Q flow rate $\quad\quad\quad\quad\quad$ ft^3/sec
x distance between manholes $\quad$ ft

$$Q = \frac{Ax}{t} = \frac{(0.32 \text{ ft}^2)(1076 \text{ ft})}{455 \text{ sec}}$$
$$= 0.76 \text{ ft}^3/\text{sec}$$

The answer is (B).

Why Other Options Are Wrong

(A) This incorrect solution makes an error in the conversion from in to ft in the average depth equation. Other assumptions, definitions, and equations are unchanged from the correct solution.

d average water depth between manholes $\quad$ ft

$$d = \frac{(3.8 \text{ in} + 5.2 \text{ in})\left(\dfrac{1 \text{ ft}}{144 \text{ in}}\right)}{2}$$
$$= 0.0313 \text{ ft}$$

D pipe diameter $\quad$ ft

$$\frac{d}{D} = \frac{0.0313 \text{ ft}}{(16 \text{ in})\left(\dfrac{1 \text{ ft}}{12 \text{ in}}\right)} = 0.023$$

From reference tables for areas of partially full pipes, based on the partial-to-full flow ratio,

$$A = 0.009 \text{ ft}^2$$
$$t = \frac{398 \text{ sec} + 512 \text{ sec}}{2}$$
$$= 455 \text{ sec}$$
$$Q = \frac{Ax}{t} = \frac{(0.009 \text{ ft}^2)(1076 \text{ ft})}{455 \text{ sec}}$$
$$= 0.021 \text{ ft}^3/\text{sec}$$

(C) This incorrect solution uses the deeper of the two depths instead of the average depth and uses the longer of the two times instead of the average time. Other assumptions, definitions, and equations are unchanged from the correct solution.

d maximum water depth between manholes $\quad$ in

$$d = 5.2 \text{ in}$$
$$\frac{d}{D} = \frac{5.2 \text{ in}}{16 \text{ in}}$$
$$= 0.325$$

From reference tables for areas of partially full pipes, based on the partial-to-full flow ratio,

$$A = 0.39 \text{ ft}^2$$

t maximum travel time between manholes $\quad$ sec

$$t = 512 \text{ sec}$$
$$Q = \frac{Ax}{t} = \frac{(0.39 \text{ ft}^2)(1076 \text{ ft})}{512 \text{ sec}}$$
$$= 0.82 \text{ ft}^3/\text{sec}$$

(D) This incorrect solution uses the ratio of depth to diameter to estimate the cross-sectional area of flow. Other assumptions, definitions, and equations are unchanged from the correct solution.

$$d = \frac{3.8 \text{ in} + 5.2 \text{ in}}{2} = 4.5 \text{ in}$$
$$\frac{d}{D} = \frac{4.5 \text{ in}}{16 \text{ in}} = 0.28$$

A_f cross-sectional area of full flow $\quad$ ft^2

$$A_f = \frac{\pi D^2}{4}$$
$$= \frac{\pi (16 \text{ in})^2 \left(\dfrac{1 \text{ ft}^2}{144 \text{ in}^2}\right)}{4}$$
$$= 1.4 \text{ ft}^2$$

A_p cross-sectional area of partial flow $\quad$ ft^2

$$A_p = A_f\left(\frac{d}{D}\right) = (1.4 \text{ ft}^2)(0.28)$$
$$= 0.392 \text{ ft}^2$$
$$t = \frac{398 \text{ sec} + 512 \text{ sec}}{2}$$
$$= 455 \text{ sec}$$
$$Q = \frac{A_p x}{t} = \frac{(0.392 \text{ ft}^2)(1076 \text{ ft})}{455 \text{ sec}}$$
$$= 0.93 \text{ ft}^3/\text{sec}$$

SOLUTION 64

D culvert diameter $\quad\quad\quad\quad$ ft
P culvert wetted perimeter $\quad$ ft

$$P = \pi D = \pi (6.0 \text{ ft}) = 18.85 \text{ ft}$$

A culvert cross-sectional area $\quad$ ft^2

$$A = \pi \frac{D^2}{4} = \frac{\pi (6.0 \text{ ft})^2}{4} = 28.3 \text{ ft}^2$$

R culvert hydraulic radius $\quad$ ft

$$R = \frac{A}{P} = \frac{28.3 \text{ ft}^2}{18.85 \text{ ft}} = 1.5 \text{ ft}$$

C_d discharge coefficient   –
f_c friction loss correction   –
L culvert length   ft
n Manning roughness coefficient   –

For concrete culverts, the Manning roughness coefficient is 0.013. For square flush openings at the culvert inlet and outlet, the discharge coefficient is assumed to be equal to the orifice coefficient for a sharp-edged opening with a value 0.62.

$$f_c = 1 + \frac{29C_d^2 n^2 L}{R^{4/3}}$$
$$= 1 + \frac{(29)(0.62)^2(0.013)^2(276 \text{ ft})}{(1.5 \text{ ft})^{4/3}}$$
$$= 1.3$$

$d_{\text{CL},i}$ water depth above the culvert
        centerline at the inlet   ft
d_i inlet elevation head relative to
        the outlet invert   ft
S culvert slope   ft/ft

$$d_i = 0.5D + d_{\text{CL},i} + LS$$
$$= (0.5)(6.0 \text{ ft}) + 12.5 \text{ ft} + (270 \text{ ft})\left(\frac{1 \text{ ft}}{83 \text{ ft}}\right)$$
$$= 18.8 \text{ ft}$$

d_o outlet elevation head relative to
        the outlet invert   ft
$d_{\text{CL},o}$ water depth above the
        culvert centerline at the outlet   ft

$$d_o = 0.5D + d_{\text{CL},o}$$
$$= (0.5)(6.0 \text{ ft}) + 4.6 \text{ ft}$$
$$= 7.6 \text{ ft}$$

For the culvert depicted by the illustration, the flow is submerged outlet.

Q discharge   ft³/sec
g acceleration of gravity   32.2 ft/sec²

The maximum discharge for submerged outlet flow can be approximated by

$$Q = C_d A \sqrt{\frac{2g(d_i - d_o)}{f_c}}$$
$$= (0.62)(28.3 \text{ ft}^2)$$
$$\times \sqrt{\frac{(2)\left(32.2 \dfrac{\text{ft}}{\text{sec}^2}\right)(18.8 \text{ ft} - 7.6 \text{ ft})}{1.3}}$$
$$= 413 \text{ ft}^3/\text{sec} \quad (410 \text{ ft}^3/\text{sec})$$

The answer is (B).

Why Other Options Are Wrong

(A) This incorrect solution uses the water depth above the culvert invert as the inlet and outlet elevation heads. Other assumptions, definitions, and equations are unchanged from the correct solution.

$$P = \pi D = \pi(6.0 \text{ ft}) = 18.85 \text{ ft}$$
$$A = \frac{\pi D^2}{4} = \frac{\pi(6.0 \text{ ft})^2}{4} = 28.3 \text{ ft}^2$$
$$R = \frac{A}{P} = \frac{28.3 \text{ ft}^2}{18.85 \text{ ft}} = 1.5 \text{ ft}$$

For concrete culverts, the Manning roughness coefficient is 0.013. For square flush openings at the culvert inlet and outlet, the discharge coefficient is assumed to be equal to the orifice coefficient for a sharp-edged opening with a value 0.62.

$$f_c = 1 + \frac{29C_d^2 n^2 L}{R^{4/3}}$$
$$= 1 + \frac{(29)(0.62)^2(0.013)^2(276 \text{ ft})}{(1.5 \text{ ft})^{4/3}}$$
$$= 1.3$$

d_i inlet water depth   ft
$$d_i = 0.5D + d_{\text{CL},i} = (0.5)(6.0 \text{ ft}) + 12.5 \text{ ft} = 15.5 \text{ ft}$$
d_o outlet water depth   ft
$$d_o = 0.5D + d_{\text{CL},o} = (0.5)(6.0 \text{ ft}) + 4.6 \text{ ft} = 7.6 \text{ ft}$$

For the culvert depicted by the figure, the flow is submerged outlet. The maximum discharge for submerged outlet flow can be approximated by

$$Q = C_d A \sqrt{\frac{2g(d_i - d_o)}{f_c}}$$
$$= (0.62)(28.3 \text{ ft}^2)$$
$$\times \sqrt{\frac{(2)\left(32.2 \dfrac{\text{ft}}{\text{sec}^2}\right)(15.5 \text{ ft} - 7.6 \text{ ft})}{1.3}}$$
$$= 347 \text{ ft}^3/\text{sec} \quad (350 \text{ ft}^3/\text{sec})$$

(C) This incorrect solution neglects the friction loss correction term. Other assumptions, definitions, and equations are unchanged from the correct solution.

$$A = \frac{\pi D^2}{4} = \frac{\pi(6.0 \text{ ft}^2)}{4}$$
$$= 28.3 \text{ ft}^2$$
$$d_i = 0.5D + d_{\text{CL},i} + LS$$
$$= (0.5)(6.0 \text{ ft}) + 12.5 \text{ ft} + (270 \text{ ft})\left(\frac{1 \text{ ft}}{83 \text{ ft}}\right)$$
$$= 18.8 \text{ ft}$$
$$d_o = 0.5D + d_{\text{CL},o}$$
$$= (0.5)(6.0 \text{ ft}) + 4.6 \text{ ft}$$
$$= 7.6 \text{ ft}$$

For square flush openings at the culvert inlet and outlet, the discharge coefficient is assumed to be equal to the orifice coefficient for a sharp-edged opening with a value 0.62. For the culvert depicted by the figure, the flow is submerged outlet. The maximum discharge for submerged outlet flow can be approximated by

$$Q = C_d A \sqrt{2g(d_i - d_o)}$$
$$= (0.62)(28.3\ \text{ft}^2)$$
$$\times \sqrt{(2)\left(32.2\ \frac{\text{ft}}{\text{sec}^2}\right)(18.8\ \text{ft} - 7.6\ \text{ft})}$$
$$= 471\ \text{ft}^3/\text{sec} \quad (470\ \text{ft}^3/\text{sec})$$

(D) This incorrect solution uses the discharge equation for conditions of rapid flow at the inlet, and neglects the friction loss correction term. Other assumptions, definitions, and equations are unchanged from the correct solution.

$$A = \pi \frac{D^2}{4} = \frac{\pi(6.0\ \text{ft}^2)}{4} = 28.3\ \text{ft}^2$$
$$d_i = 0.5D + d_{\text{CL},i} + LS$$
$$= (0.5)(6.0\ \text{ft}) + 12.5\ \text{ft} + (270\ \text{ft})\left(\frac{1\ \text{ft}}{83\ \text{ft}}\right)$$
$$= 18.8\ \text{ft}$$

s inlet invert elevation relative to
    outlet invert elevation   ft

$$s = LS = (270\ \text{ft})\left(\frac{1\ \text{ft}}{83\ \text{ft}}\right) = 3.25\ \text{ft}$$

For square flush openings at the culvert inlet and outlet, the discharge coefficient is assumed to be equal to the orifice coefficient for a sharp-edged opening with a value 0.62. For the culvert depicted by the figure, the flow is submerged outlet. The maximum discharge for submerged outlet flow can be approximated by

$$Q = C_d A \sqrt{2g(d_i - s)}$$
$$= (0.62)(28.3\ \text{ft}^2)$$
$$\times \sqrt{(2)\left(32.2\ \frac{\text{ft}}{\text{sec}^2}\right)(18.8\ \text{ft} - 3.25\ \text{ft})}$$
$$= 555\ \text{ft}^3/\text{sec} \quad (560\ \text{ft}^3/\text{sec})$$

HYDROLOGY

SOLUTION 65

The form of the data provided suggests the isohyctal method for determining the areal average precipitation for the region.

area	isohyet (cm)	enclosed area (km²)	net area (km²)	average precipitation (cm)	precipitation volume (cm·km²)
I	>22	84	84	23	1932
II	20	252	168	21	3528
III	18	578	326	19	6194
IV	16	892	314	17	5338
V	14	1136	244	15	3660
VI	<14	1294	158	13	2054
					22 706

A total area   km²
P_m average precipitation   cm
$\sum P$ cumulative precipitation volume   cm·km²

$$P_m = \frac{\sum P}{A} = \frac{22\,706\ \text{cm·km}^2}{1294\ \text{km}^2}$$
$$= 17.5\ \text{cm}$$

The answer is (C).

Why Other Options Are Wrong

(A) This incorrect solution takes the arithmetic average of the isohyets as the average precipitation. Other definitions are unchanged from the correct solution.

$\sum_d$ sum of the isohyets   cm
n number of sub-regions   –

$$P_m = \frac{\sum d}{n}$$
$$= \frac{22\ \text{cm} + 20\ \text{cm} + 18\ \text{cm} + 16\ \text{cm} + 14\ \text{cm}}{6}$$
$$= 15.0\ \text{cm}$$

(B) This incorrect solution calculates the average precipitation using the isohyet instead of the average precipitation for each subregion. Other definitions and equations are unchanged from the correct solution.

area	isohyet (cm)	enclosed area (km²)	net area (km²)	precipitation volume (cm·km²)
I	>22	84	84	1848
II	20	252	168	3360
III	18	578	326	5868
IV	16	892	314	5024
V	14	1136	244	3416
VI	<14	1294	158	2212
				21 728

$$P_m = \frac{\sum P}{A} = \frac{21\,728\ \text{cm·km}^2}{1294\ \text{km}^2}$$
$$= 16.8\ \text{cm}$$

(D) This incorrect solution uses the values for each isohyet instead of the average isohyet for each sub-region and uses the enclosed area instead of the net area. Other definitions and equations are unchanged from the correct solution.

area	isohyet (cm)	average precipitation (cm)	enclosed area (km²)	precipitation volume (cm·km²)
I	>22	23	84	1932
II	20	21	252	5292
III	18	19	578	10 982
IV	16	17	892	15 164
V	14	15	1136	17 040
VI	<14	13	1294	16 822
				67 232

$$P_m = \frac{\sum P}{A} = \frac{67\,232 \text{ cm·km}^2}{1294 \text{ km}^2} = 52.0 \text{ cm}$$

SOLUTION 66

I floods creating economic impact   %
n_e number of events
n_f number of floods

$$I = \left(\frac{n_f}{n_e}\right) \times 100\% = \left(\frac{18}{112}\right) \times 100\% = 16\%$$

Entering the figure at 16% on the y-axis and extending horizontally to intersect the curve, the peak flow is 1600 m³/s.

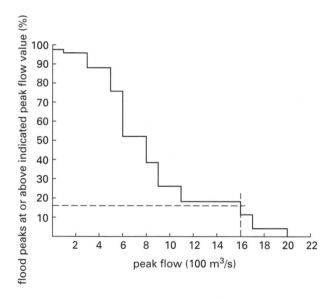

The answer is (C).

Why Other Options Are Wrong

(A) This incorrect solution determines the peak flow for the percentage of floods that did not result in significant economic impact. Other definitions are unchanged from the correct solution.

$$I = \frac{(n_e - n_f) \times 100\%}{n_f} = \frac{(112 - 18) \times 100\%}{112} = 84\%$$

Entering the figure at 84% on the y-axis and extending horizontally to intersect the curve, the peak flow is 500 m³/s.

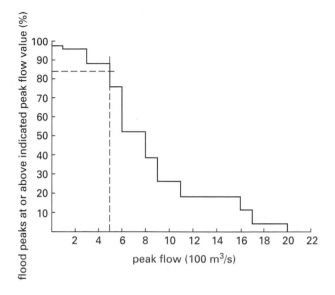

(B) This incorrect solution reads the number of events instead of the percentage of these events resulting in economic impact.

I floods creating economic impact

$$I = 18$$

Entering the figure at 18 on the y-axis and extending horizontally to intersect the curve, the peak flow is 1100 m³/s.

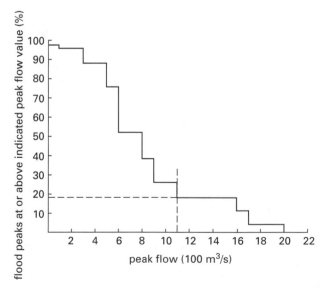

(D) This incorrect solution reads the x-axis scale as 1000 m³/s instead of 100 m³/s. The figure and other definitions and equations are unchanged from the correct solution.

$$I = \left(\frac{18}{112}\right) \times 100\% = 16\%$$

Entering the figure at 16% on the y-axis and extending horizontally to intersect the curve, the peak flow is 16 000 m^3/s.

SOLUTION 67

From the illustration, 40 recorded events experienced a flow exceeding 40,000 ft^3/sec.

m number of events of interest
     that exceed peak flow   –
N period of record   yr
t_p recurrence interval   yr

$$t_p = \frac{N+1}{m} = \frac{112 \text{ yr} + 1}{40} = 2.8 \text{ yr}$$

The answer is (C).

Why Other Options Are Wrong

(A) This incorrect solution calculates the probability of recurrence. Other definitions are unchanged from the correct solution.

From the illustration, 40 events of the record experience a flow exceeding 40,000 ft^3/sec.

$$\frac{1}{t_p} = \frac{N+1}{m} = \frac{112 \text{ yr} + 1}{40} = 2.8 \text{ yr}$$
$$t_p = 0.35 \text{ yr}^{-1}$$

The units do not work.

(B) This incorrect solution calculates recurrence based on the number of events, taken from the y-axis of the figure, instead of the years of record. Other definitions and equations are unchanged from the correct solution.

From the illustration, there are 86 events and 40 of these events experience a flow exceeding 40,000 ft^3/sec.

N number of total events   –
m number of events of interest
     that exceed peak flow   –

$$t_p = \frac{N+1}{m} = \frac{86+1}{40} = 2.2 \text{ yr}$$

The units are assumed.

(D) This incorrect solution calculates the recurrence based on the median storm magnitude instead of the given storm.

From the illustration, there are 86 events covering a range from 0 to 100,000 ft^3/sec. The median event peak flow is 50,000 ft^3/sec. Thirty-three events of record experience a flow exceeding 50,000 ft^3/sec.

$$t_p = \frac{N+1}{m} = \frac{112 \text{ yr} + 1}{33} = 3.4 \text{ yr}$$

SOLUTION 68

The given information requires use of the log Pearson type III distribution to find the flood magnitude.

N number of years of record   –
X flood magnitude   ft^3/sec
$\bar{X}$ mean flood magnitude   ft^3/sec
σ standard deviation   ft^3/sec

$$\sigma = \sqrt{\frac{\sum \left(\log X - \log \bar{X}\right)^2}{N-1}} = \sqrt{\frac{3.894}{97-1}} = 0.20$$

g skew coefficient –

$$g = \frac{N\sum\left(\log X - \log \bar{X}\right)^3}{(N-1)(N-2)\sigma^3}$$
$$= \frac{(97)(0.181)}{(97-1)(97-2)(0.20)^3}$$
$$= 0.24$$

K log Pearson type III distribution coefficient –

From a standard reference table for the log Pearson type III distribution with a recurrence interval of 25 yr and skew coefficient of 0.24, the log Pearson type III distribution coefficient is

$$K = 1.833$$
$$\log X = \log \bar{X} + K\sigma = 3.571 + (1.833)(0.20) = 3.94$$
$$X = 8662 \text{ ft}^3/\text{sec} \quad (8700 \text{ ft}^3/\text{sec})$$

The answer is (C).

Why Other Options Are Wrong

(A) This incorrect solution misses a decimal place in the standard deviation value when calculating the flood magnitude. Other assumptions, definitions, and equations are the same as used in the correct solution.

$$\sigma = \sqrt{\frac{\sum \left(\log X - \log \bar{X}\right)^2}{N-1}} = \sqrt{\frac{3.894}{97-1}} = 0.20$$
$$g = \frac{N\sum\left(\log X - \log \bar{X}\right)^3}{(N-1)(N-2)\sigma^3}$$
$$= \frac{(97)(0.181)}{(97-1)(97-2)(0.20)^3} = 0.24$$

From a standard reference table for the log Pearson type III distribution with a recurrence interval of 25 yr and skew coefficient of 0.24, the log Pearson type III distribution coefficient is

$$K = 1.833$$
$$\log X = \log \bar{X} + K\sigma$$
$$= 3.571 + (1.833)(0.020)$$
$$= 3.61$$
$$X = 4073 \text{ ft}^3/\text{sec} \quad (4100 \text{ ft}^3/\text{sec})$$

(B) This incorrect solution fails to cube the standard deviation when calculating the skew coefficient. Other assumptions, definitions, and equations are the same as used in the correct solution.

$$\sigma = \sqrt{\frac{\sum \left(\log X - \log \bar{X}\right)^2}{N-1}} = \sqrt{\frac{3.894}{97-1}} = 0.20$$

$$g = \frac{N \sum \left(\log X - \log \bar{X}\right)^3}{(N-1)(N-2)\,\sigma}$$

$$= \frac{(97)\,(0.181)}{(97-1)(97-2)\,(0.20)}$$

$$= 0.0096$$

From a standard reference table for the log Pearson type III distribution with a recurrence interval of 25 yr and skew coefficient of 0.0096, the log Pearson type III distribution coefficient is

$$K = 1.751$$

$$\log X = \log \bar{X} + K\sigma$$

$$= 3.571 + (1.751)\,(0.20)$$

$$= 3.92$$

$$X = 8318 \text{ ft}^3/\text{sec} \quad (8300 \text{ ft}^3/\text{sec})$$

(D) This incorrect solution misreads the log Pearson type III distribution coefficient for a 50-year instead of 25-year event. Other assumptions, definitions, and equations are the same as used in the correct solution.

$$\sigma = \sqrt{\frac{\sum \left(\log X - \log \bar{X}\right)^2}{N-1}} = \sqrt{\frac{3.894}{97-1}} = 0.20$$

$$g = \frac{N \sum \left(\log X - \log \bar{X}\right)^3}{(N-1)(N-2)\,\sigma^3}$$

$$= \frac{(97)\,(0.181)}{(97-1)(97-2)\,(0.20)^3} = 0.24$$

From a standard reference table for the log Pearson type III distribution with a recurrence interval of 25 yr and skew coefficient of 0.24, the log Pearson type III distribution coefficient is

$$K = 2.185$$

$$\log X = \log \bar{X} + K\sigma$$

$$= 3.571 + (2.185)\,(0.20)$$

$$= 4.01$$

$$X = 10{,}233 \text{ ft}^3/\text{sec} \quad (10{,}000 \text{ ft}^3/\text{sec})$$

SOLUTION 69

d	depth of rainfall	cm
i	intensity	cm/h
t	storm duration	h

From the illustration, for a 50-year recurrence interval and 2 h duration storm, the rainfall depth is 9 cm.

$$i = \frac{d}{t} = \frac{9 \text{ cm}}{2 \text{ h}} = 4.5 \text{ cm/h}$$

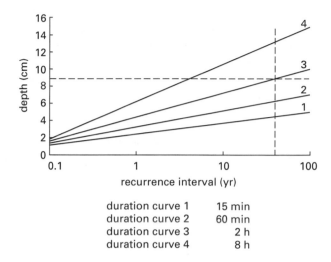

duration curve 1	15 min
duration curve 2	60 min
duration curve 3	2 h
duration curve 4	8 h

For the Steel formula, intensity must be in units of in/hr.

$$i = \left(4.5 \ \frac{\text{cm}}{\text{h}}\right) \left(\frac{1 \text{ in}}{2.54 \text{ cm}}\right) = 1.8 \text{ in/hr}$$

b	constant	min
K	constant	in-min/hr
t_c	time of concentration	min

The Steel formula is

$$i = \frac{K}{t_c + b}$$

$$t_c = \frac{K}{i} - b$$

$$= \frac{180 \ \dfrac{\text{in-min}}{\text{hr}}}{1.8 \ \dfrac{\text{in}}{\text{hr}}} - 25 \text{ min}$$

$$= 75 \text{ min}$$

The answer is (D).

Why Other Options Are Wrong

(A) This incorrect solution uses SI units for all calculations instead of converting to customary U.S. units for the Steel formula. The illustration and other definitions and equations are unchanged from the correct solution.

From the illustration, for a 50-year recurrence interval and 2 h duration storm, the rainfall is 9 cm.

$$i = \frac{d}{t} = \frac{9 \text{ cm}}{2 \text{ h}}$$

$$= 4.5 \text{ cm/h}$$

Assume units for the constant, K, should be cm·min/h.

$$t_c = \frac{K}{i} - b = \frac{180 \ \frac{\text{cm·min}}{\text{h}}}{4.5 \ \frac{\text{cm}}{\text{h}}} - 25 \ \text{min}$$

$$= 15 \ \text{min}$$

(B) This incorrect solution takes the rainfall as the intensity instead of calculating intensity from rainfall and duration. The illustration is unchanged from the correct solution.

From the illustration,

$$i = 9.0 \ \frac{\text{cm}}{\text{h}} \left(\frac{1 \ \text{in}}{2.54 \ \text{cm}} \right)$$

$$= 3.5 \ \text{in/hr}$$

$$t_c = \frac{K}{i} - b$$

$$= \frac{180 \ \frac{\text{in·min}}{\text{hr}}}{3.5 \ \frac{\text{in}}{\text{hr}}} - 25 \ \text{min}$$

$$= 26 \ \text{min}$$

(C) This incorrect solution misreads the illustration, mistaking the 8 h duration for the 2 h duration storm. Other definitions and equations are unchanged from the correct solution.

From the illustration, for a 50-year recurrence interval and 2 h duration storm, the rainfall is 13.5 cm.

$$i = \frac{d}{t} = \frac{13.5 \ \text{cm}}{2 \ \text{h}}$$

$$= 6.8 \ \text{cm/h}$$

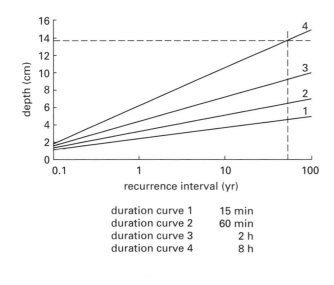

duration curve 1	15 min
duration curve 2	60 min
duration curve 3	2 h
duration curve 4	8 h

$$i = \left(6.8 \ \frac{\text{cm}}{\text{h}} \right) \left(\frac{1 \ \text{in}}{2.54 \ \text{cm}} \right)$$

$$= 2.68 \ \text{in/hr}$$

$$t_c = \frac{K}{i} - b$$

$$= \frac{180 \ \frac{\text{in·min}}{\text{hr}}}{2.68 \ \frac{\text{in}}{\text{hr}}} - 25 \ \text{min}$$

$$= 42.2 \ \text{min} \quad (42 \ \text{min})$$

SOLUTION 70

Because the normal annual precipitation between station C and the other three stations varies by more than 10%, the normal-ratio method should be used for computing an estimate of the missing record.

N normal annual precipitation   cm
P annual precipitation for year of interest   cm

Subscripts A, B, C, and D refer to the precipitation stations.

$$P_C = \left(\frac{N_C}{3} \right) \left(\frac{P_A}{N_A} + \frac{P_B}{N_B} + \frac{P_D}{N_D} \right)$$

$$= \left(\frac{42 \ \text{cm}}{3} \right) \left(\frac{34 \ \text{cm}}{39 \ \text{cm}} + \frac{28 \ \text{cm}}{31 \ \text{cm}} + \frac{32 \ \text{cm}}{37 \ \text{cm}} \right)$$

$$= 37 \ \text{cm}$$

The answer is (B).

Why Other Options Are Wrong

(A) This incorrect solution takes the arithmetic average of the other three stations for 1991 as the missing record for station C in 1991. Definitions are unchanged from the correct solution.

$$P_C = \frac{P_A + P_B + P_D}{3}$$

$$= \frac{34 \ \text{cm} + 28 \ \text{cm} + 32 \ \text{cm}}{3}$$

$$= 31.3 \ \text{cm} \quad (31 \ \text{cm})$$

(C) This incorrect solution misreads one of the values entered in the normal-ratio equation. Definitions and the equation are unchanged from the correct solution.

$$P_C = \left(\frac{N_C}{3} \right) \left(\frac{P_A}{N_A} + \frac{P_B}{N_B} + \frac{P_D}{N_D} \right)$$

$$= \left(\frac{42 \ \text{cm}}{3} \right) \left(\frac{42 \ \text{cm}}{39 \ \text{cm}} + \frac{28 \ \text{cm}}{31 \ \text{cm}} + \frac{32 \ \text{cm}}{37 \ \text{cm}} \right)$$

$$= 39.8 \ \text{cm} \quad (40 \ \text{cm})$$

(D) This incorrect solution reverses the values for the normal annual precipitation and the annual precipitation in the year of interest. Other definitions and the equation are unchanged from the correct solution.

$$P_C = \left(\frac{N_C}{3}\right)\left(\frac{P_A}{N_A} + \frac{P_B}{N_B} + \frac{P_D}{N_D}\right)$$
$$= \left(\frac{42 \text{ cm}}{3}\right)\left(\frac{39 \text{ cm}}{34 \text{ cm}} + \frac{31 \text{ cm}}{28 \text{ cm}} + \frac{37 \text{ cm}}{32 \text{ cm}}\right)$$
$$= 47.7 \text{ cm} \quad (48 \text{ cm})$$

SOLUTION 71

i rainfall intensity   in/hr
L overland flow distance   ft

$$iL = \left(0.89 \frac{\text{in}}{\text{hr}}\right)(150 \text{ ft}) = 134 < 500$$

Because the site is characterized by sheet flow and the product of rainfall intensity in units of in/hr and the flow distance in units of ft is less than 500, the Izzard method for calculating time to concentration applies. The rainfall data is not provided in a format that will allow use of the Manning kinematic equation.

c overland flow retardance coefficient   –
k Izzard coefficient   –
S ground surface slope   ft/ft

For smooth asphalt, the value of the overland flow retardance coefficient is 0.007.

$$k = \frac{0.0007i + c}{S^{1/3}}$$

$$k_A = \frac{(0.0007)\left(0.89 \frac{\text{in}}{\text{hr}}\right) + 0.007}{(0.001)^{1/3}} = 0.076$$

$$k_B = \frac{(0.0007)\left(0.89 \frac{\text{in}}{\text{hr}}\right) + 0.007}{(0.0006)^{1/3}} = 0.090$$

t_c time to concentration   min

$$t_c = \frac{41kL^{1/3}}{i^{2/3}}$$

$$t_{c,A} = \frac{(41)(0.076)(150 \text{ ft})^{1/3}}{(0.89)^{2/3} \frac{\text{in}}{\text{hr}}} = 17.9 \text{ min}$$

$$t_{c,B} = \frac{(41)(0.090)(90 \text{ ft})^{1/3}}{(0.89)^{2/3} \frac{\text{in}}{\text{hr}}} = 17.9 \text{ min}$$

$$t_{c,A} = t_{c,B} = 17.9 \text{ min} \quad (18 \text{ min})$$

The answer is (B).

Why Other Options Are Wrong

(A) This incorrect solution uses the Manning kinematic equation and manipulates the rainfall intensity for application to the equation. The Manning equation stipulates specific conditions for rainfall. Other assumptions, definitions, and equations are the same as used in the correct solution.

Because both the overland flow distances are less than 300 ft, the Manning kinematic equation applies.

t rainfall duration   hr
P rainfall   in

$$P = it = \left(0.89 \frac{\text{in}}{\text{hr}}\right)(1 \text{ hr})$$
$$= 0.89 \text{ in}$$

n Manning roughness coefficient   –

For smooth asphalt, the value of the Manning roughness coefficient is 0.011.

$$t_c = \frac{0.007(nL)^{0.8}}{\sqrt{P}S^{0.4}}$$

$$t_{c,A} = \frac{(0.007)\left((0.011)(150 \text{ ft})\right)^{0.8}}{\sqrt{0.89 \text{ in}}\,(0.001)^{0.4}} = 0.18 \text{ min}$$

$$t_{c,B} = \frac{(0.007)\left((0.011)(90 \text{ ft})\right)^{0.8}}{\sqrt{0.89 \text{ in}}\,(0.0006)^{0.4}} = 0.14 \text{ min}$$

Use the lesser of the two values,

$$t_c = 0.14 \text{ min}$$

(C) This incorrect solution uses the Manning roughness coefficient for the overland flow retardance coefficient. Other assumptions, definitions, and equations are the same as used in the correct solution.

$$iL = \left(0.89 \frac{\text{in}}{\text{hr}}\right)(150 \text{ ft})$$
$$= 134 < 500$$

For smooth asphalt, the value of the Manning roughness coefficient, n, is 0.011.

$$k = \frac{0.0007i + n}{S^{1/3}}$$

$$k_A = \frac{(0.0007)\left(0.89 \frac{\text{in}}{\text{hr}}\right) + 0.011}{(0.001)^{1/3}}$$
$$= 0.12$$

$$k_B = \frac{(0.0007)\left(0.89 \frac{\text{in}}{\text{hr}}\right) + 0.011}{(0.0006)^{1/3}}$$
$$= 0.14$$

$$t_c = \frac{41kL^{1/3}}{i^{2/3}}$$

$$t_{c,A} = \frac{(41)(0.12)(150 \text{ ft})^{1/3}}{(0.89)^{2/3} \frac{\text{in}}{\text{hr}}}$$

$$= 28.3 \text{ min} \quad (28 \text{ min})$$

$$t_{c,B} = \frac{(41)(0.14)(90 \text{ ft})^{1/3}}{(0.89)^{2/3} \frac{\text{in}}{\text{hr}}}$$

$$= 27.8 \text{ min} \quad (28 \text{ min})$$

$$t_{c,A} \approx t_{c,B} = 28 \text{ min}$$

(D) This incorrect solution adds the time to concentration for the two lots. Other assumptions, definitions, and equations are the same as used in the correct solution.

$$iL = \left(0.89 \frac{\text{in}}{\text{hr}}\right)(150 \text{ ft}) = 134 < 500$$

$$k = \frac{0.0007i + c}{S^{1/3}}$$

$$k_A = \frac{(0.0007)\left(0.89 \frac{\text{in}}{\text{hr}}\right) + 0.007}{(0.001)^{1/3}}$$

$$= 0.076$$

$$k_B = \frac{(0.0007)\left(0.89 \frac{\text{in}}{\text{hr}}\right) + 0.007}{(0.0006)^{1/3}}$$

$$= 0.090$$

$$t_c = \frac{41kL^{1/3}}{i^{2/3}}$$

$$t_{c,A} = \frac{(41)(0.076)(150 \text{ ft})^{1/3}}{(0.89)^{2/3} \frac{\text{in}}{\text{hr}}} = 17.9 \text{ min}$$

$$t_{c,B} = \frac{(41)(0.090)(90 \text{ ft})^{1/3}}{(0.89)^{2/3} \frac{\text{in}}{\text{hr}}} = 17.9 \text{ min}$$

$t_{c,T}$ total time to concentration   min

$$t_{c,T} = t_{c,A} + t_{c,B}$$

$$= 17.9 \text{ min} + 17.9 \text{ min}$$

$$= 35.8 \text{ min} \quad (36 \text{ min})$$

SOLUTION 72

The hydrograph for each storm is produced by multiplying the unit hydrograph by the rainfall for each storm. The peak discharge is determined by offsetting the hydrographs by 3 h and then adding them. This is shown in the following table and illustration.

storm discharge (m³/s)	storm runoff (cm)			combined discharge (m³/s)	duration (h)
	1.6	3.1	2.7		
0	0	–	–	0	0
25	40	–	–	40	1
125	200	–	–	200	2
368	588.8	0	–	588.8	3
445	712	77.5	–	789.5	4
425	680	387.5	–	1067.5	5
300	480	1140.8	0	1620.8	6
220	352	1379.5	67.5	1799	7
159	254.4	1317.5	337.5	1909.4	8
118	188.8	930	993.6	2112.4	9
84	134.4	682	1201.5	2017.9	10
60	96	492.9	1147.5	1736.4	11
35	56	365.8	810	1231.8	12
22	35.2	260.4	594	889.6	13
14	22.4	186	429.3	637.7	14
11	17.6	108.5	318.6	444.7	15
8	12.8	68.2	226.8	307.8	16
4	6.4	43.4	162	211.8	17
–	–	34.1	94.5	128.6	18
–	–	24.8	59.4	84.2	19
–	–	12.4	37.8	50.2	20
–	–	–	29.7	29.7	21
–	–	–	21.6	21.6	22
–	–	–	10.8	10.8	23

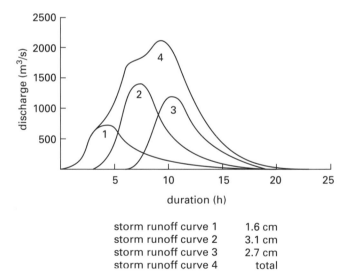

storm runoff curve 1	1.6 cm
storm runoff curve 2	3.1 cm
storm runoff curve 3	2.7 cm
storm runoff curve 4	total

From the illustration, the peak discharge occurs at 9 h after the beginning of the first storm and is equal to 2100 m³/s.

The answer is (B).

Why Other Options Are Wrong

(A) This incorrect solution calculates the hydrograph for the largest storm (3.1 cm) only and then takes the peak for that storm for the peak discharge.

The hydrograph for the largest storm (3.1 cm) is produced by multiplying the unit hydrograph by the rainfall for that storm. The peak discharge is determined by adding the storm hydrographs.

From the illustration, the peak discharge occurs at about 4.3 h after the beginning of the storm and is equal to 1400 m³/s.

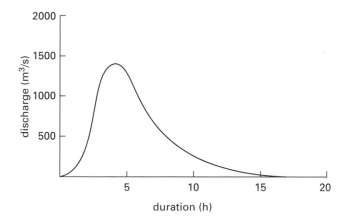

(C) This incorrect solution adds the rainfall for the three storms, multiplies it by the area of the drainage area, and then divides by the storm duration of 3 h. The unit hydrograph is ignored.

A drainage area   km²
d storm depth   cm
Q peak discharge   m³/s
t storm duration   h

$$Q = \frac{A \sum d}{t}$$

$$= \frac{(420 \text{ km}^2) \left(\dfrac{1000 \text{ m}}{1 \text{ km}}\right)^2 \left(\begin{matrix} 1.6 \text{ cm} + 3.1 \text{ cm} \\ +2.7 \text{ cm} \end{matrix}\right)}{(3 \text{ h}) \left(\dfrac{3600 \text{ s}}{1 \text{ h}}\right) \left(\dfrac{100 \text{ cm}}{1 \text{ m}}\right)}$$

$$= 2878 \text{ m}^3/\text{s} \quad (2900 \text{ m}^3/\text{s})$$

(D) This incorrect solution does not offset the hydrographs before adding them to produce the peak hydrograph.

The hydrograph for each storm is produced by multiplying the unit hydrograph by the rainfall for each storm. The peak discharge is determined by adding the storm hydrographs.

From the illustration, the peak discharge occurs at about 4.3 h after the beginning of the storm and is equal to 3300 m³/s.

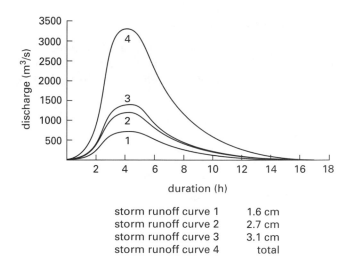

storm runoff curve 1	1.6 cm
storm runoff curve 2	2.7 cm
storm runoff curve 3	3.1 cm
storm runoff curve 4	total

SOLUTION 73

The rainfall volume in surface detention is defined by the area to the left of the curve and under a horizontal line extended from the y-axis to the peak of the curve as shown in the following illustration.

Each segment of grid area on the illustration defines a volume of

$$\left(25 \ \frac{\text{m}^3}{\text{s}}\right) (1 \text{ h}) \left(3600 \ \frac{\text{s}}{\text{h}}\right) = 90\,000 \text{ m}^3$$

time interval (h)	segments
0–1	15.5
1–2	11.5
2–3	3.0
3–4	0.5
4–5	0.0
	30.5

I surface detention   m³

$$I = (30.5)(90\,000 \text{ m}^3) = 2.7 \times 10^6 \text{ m}^3$$

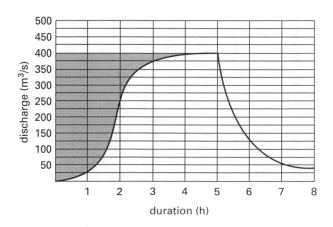

The answer is (B).

Why Other Options Are Wrong

(A) This incorrect solution calculates the volume as the area under the recession portion of the curve beginning at 5 h when rainfall ends. Definitions are unchanged from the correct solution.

The rainfall volume in surface detention is defined by the area under the recession portion of the curve as shown in the following illustration.

Each segment of grid area on the illustration defines a volume of

$$\left(25 \ \frac{\text{m}^3}{\text{s}}\right) (1 \ \text{h}) \left(3600 \ \frac{\text{s}}{\text{h}}\right) = 90\,000 \ \text{m}^3$$

time interval (h)	segments
5–6	8.5
6–7	3.0
7–8	2.0
	13.5

$$I = (13.5) \left(90\,000 \ \text{m}^3\right) = 1.2 \times 10^6 \ \text{m}^3$$

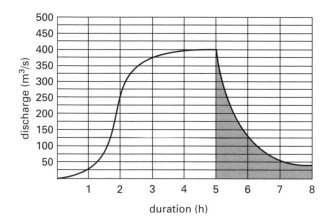

(C) This incorrect solution calculates the volume as the area under the curve until the end of rainfall at 5 h. Definitions are unchanged from the correct solution.

The rainfall volume in surface detention is defined by the area under the curve until the end of rainfall at 5 h as shown in the following illustration.

Each segment of grid area on the illustration defines a volume of

$$\left(25 \ \frac{\text{m}^3}{\text{s}}\right) (1 \ \text{h}) \left(3600 \ \frac{\text{s}}{\text{h}}\right) = 90\,000 \ \text{m}^3$$

time interval (h)	segments
0–1	0.5
1–2	4.5
2–3	13
3–4	15.5
4–5	16
	49.5

$$I = (49.5) \left(90\,000 \ \text{m}^3\right) = 4.5 \times 10^6 \ \text{m}^3$$

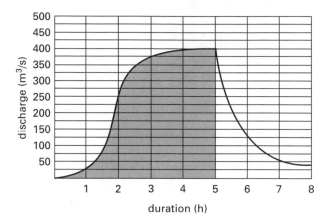

(D) This incorrect solution calculates the volume based on the total area under the curve instead of to the left of it. Definitions are unchanged from the correct solution.

The rainfall volume in surface detention is defined by the total area under the curve as shown in the following illustration.

Each segment of grid area on the illustration defines a volume of

$$\left(25 \ \frac{\text{m}^3}{\text{s}}\right) (1 \ \text{h}) \left(3600 \ \frac{\text{s}}{\text{h}}\right) = 90\,000 \ \text{m}^3$$

time interval (h)	segments
0–1	0.5
1–2	4.5
2–3	13
3–4	15.5
4–5	16
5–6	8.5
6–7	3.0
7–8	2.0
	63

$$I = (63) \left(90\,000 \ \text{m}^3\right) = 5.7 \times 10^6 \ \text{m}^3$$

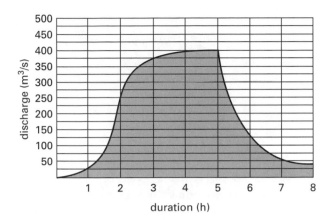

$$Q = \frac{(116.5)(41{,}322 \text{ ac-ft})\left(12 \dfrac{\text{in}}{\text{ft}}\right)}{(5480 \text{ mi}^2)\left(640 \dfrac{\text{ac}}{\text{mi}^2}\right)}$$

$$= 16.47 \text{ in} \quad (16 \text{ in})$$

SOLUTION 74

Separate groundwater from direct runoff as shown in the following illustration. With groundwater separated, graphically integrate the illustration to find direct runoff.

Each grid segment on the graph represents a discharge volume of

$$\left(250{,}000 \ \frac{\text{ft}^3}{\text{sec}}\right)(2 \text{ hr})$$

$$\times \left(3600 \ \frac{\text{sec}}{\text{hr}}\right)\left(\frac{1 \text{ ac}}{43{,}560 \text{ ft}^2}\right) = 41{,}322 \text{ ac-ft}$$

duration interval (hr)	segments
4–8	8.5
8–12	37
12–16	32.5
16–20	20
20–24	11.5
24–28	5.5
28–32	1.5
	116.5

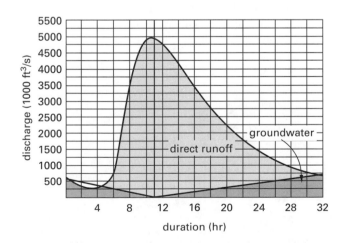

Q total discharge   in

The answer is (D).

Other options are incorrect because

(A) This incorrect solution miscalculates the grid segment volume. The illustration and other assumptions, definitions, and equations are the same as used in the correct solution.

Each grid segment on the graph represents a discharge volume of

$$\left(250 \ \frac{\text{ft}^3}{\text{sec}}\right)(2 \text{ hr})$$

$$\times \left(3600 \ \frac{\text{sec}}{\text{hr}}\right)\left(\frac{1 \text{ ac}}{43{,}560 \text{ ft}^2}\right) = 41.3 \text{ ac-ft}$$

duration interval (hr)	segments
4–8	8.5
8–12	37
12–16	32.5
16–20	20
20–24	11.5
24–28	5.5
28–32	1.5
	116.5

$$Q = \frac{(116.5)(41.3 \text{ ac-ft})\left(12 \dfrac{\text{in}}{\text{ft}}\right)}{(5480 \text{ mi}^2)\left(640 \dfrac{\text{ac}}{\text{mi}^2}\right)} = 0.016 \text{ in}$$

(B) This incorrect solution uses the conversion from m^2 to ac instead of mi^2 to ac. The illustration and other assumptions, definitions, and equations are the same as used in the correct solution.

Each grid segment on the graph represents a discharge volume of

$$\left(250{,}000 \ \frac{\text{ft}^3}{\text{sec}}\right)(2 \text{ hr})$$

$$\times \left(3600 \ \frac{\text{sec}}{\text{hr}}\right)\left(\frac{1 \text{ ac}}{43{,}560 \text{ ft}^2}\right) = 41{,}322 \text{ ac-ft}$$

duration interval (hr)	segments
4–8	8.5
8–12	37
12–16	32.5
16–20	20
20–24	11.5
24–28	5.5
28–32	1.5
	116.5

$$Q = \frac{(116.5)\,(41{,}322 \text{ ac-ft})\left(12\,\dfrac{\text{in}}{\text{ft}}\right)}{(5480 \text{ mi}^2)\left(4047\,\dfrac{\text{ac}}{\text{mi}^2}\right)} = 2.6 \text{ in}$$

(C) This incorrect solution improperly separates the groundwater flow from direct runoff. Other assumptions, definitions, and equations are the same as used in the correct solution.

Each grid segment on the graph represents a discharge volume of

$$\left(250{,}000\,\frac{\text{ft}^3}{\text{sec}}\right)(2 \text{ hr})$$

$$\times \left(3600\,\frac{\text{sec}}{\text{hr}}\right)\left(\frac{1 \text{ ac}}{43{,}560 \text{ ft}^2}\right) = 41{,}322 \text{ ac-ft}$$

duration interval (hr)	segments
4–8	5.5
8–12	31
12–16	26.5
16–20	15.5
20–24	8.0
24–28	3.0
28–32	0.5
	90

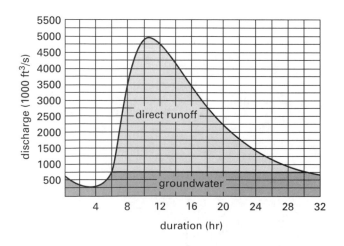

$$Q = \frac{(90)\,(41{,}322 \text{ ac-ft})\left(12\,\dfrac{\text{in}}{\text{ft}}\right)}{(5480 \text{ mi}^2)\left(640\,\dfrac{\text{ac}}{\text{mi}^2}\right)}$$

$$= 12.7 \text{ in} \quad (13 \text{ in})$$

SOLUTION 75

The wilting point is defined as the moisture content of the soil corresponding to a moisture tension of 15 atm during the drying cycle. From the illustration, the moisture content corresponding to a moisture tension of 15 atm is 23%.

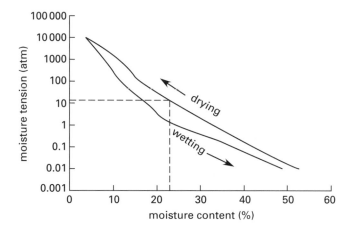

The answer is (C).

Why Other Options Are Wrong

(A) This incorrect solution determined the wilting point using the wetting cycle part of the curve.

The wilting point is defined as the moisture content of the soil corresponding to a moisture tension of 15 atm during the drying cycle. From the illustration, the moisture content corresponding to a moisture tension of 15 atm is 17%.

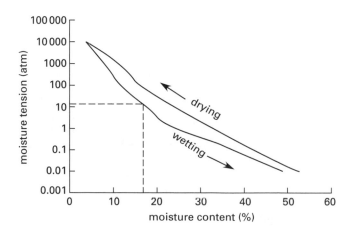

(B) This incorrect solution determined the wilting point as the average moisture content between the wetting and drying cycles.

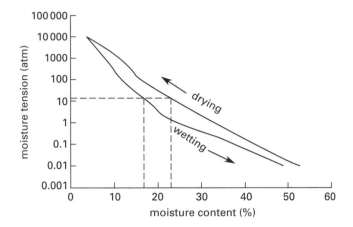

The wilting point is defined as the average moisture content of the soil corresponding to a moisture tension of 15 atm during the wetting and drying cycle. From the figure, the moisture content corresponding to a moisture tension of 15 atm on the wetting cycle is 17% and on the drying cycle is 23%. The average of 17% and 23% is 20%.

(D) This incorrect solution misread the plot, taking the moisture tension at 1.5 atm instead of 15 atm. The wilting point is 32%.

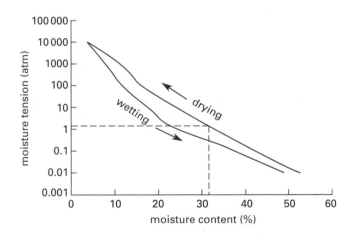

SOLUTION 76

T_w temperature at the weather station °C
z_f freezing elevation m
z_w weather station elevation m
Γ ambient lapse rate °C/m

$$z_f = z_w + \frac{0°\text{C} - T_w}{\Gamma} = 2036 \text{ m} + \frac{0°\text{C} - 12°\text{C}}{-0.016 \ \dfrac{°\text{C}}{\text{m}}}$$

$$= 2786 \text{ m}$$

The area between the snow line elevation and the freezing point elevation is the melting area. This area is determined from the figure.

A_m melting area m²

$$A_m = \left(1850 \text{ km}^2 - 755 \text{ km}^2\right)\left(1000 \ \frac{\text{m}}{\text{km}}\right)^2$$

$$= 1.1 \times 10^9 \text{ m}^2$$

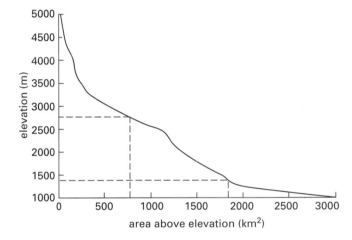

T_s temperature at the snow line °C
z_s snow line elevation m

$$T_s = T_w + \Gamma\left(z_s - z_w\right)$$

$$= 12°\text{C} + \left(-0.016 \ \frac{°\text{C}}{\text{m}}\right)(1370 \text{ m} - 2036 \text{ m})$$

$$= 22.7°\text{C}$$

T_m average temperature over the melting area °C

$$T_m = 0.5T_s = (0.5)(22.7°\text{C}) = 11.3°\text{C}$$

d_m average degree days °C·d

$$d_m = 11.3°\text{C·d}$$

V_s snow melt volume m³
d_f degree day factor mm/°C·d

$$V_s = d_m d_f A_m$$

$$= (11.3°\text{C·d})\left(3 \ \frac{\text{mm}}{°\text{C·d}}\right)\left(1.1 \times 10^9 \text{m}^2\right)\left(\frac{1 \text{ m}}{10^3 \text{ mm}}\right)$$

$$= 3.73 \times 10^7 \text{ m}^3 \quad \left(3.7 \times 10^7 \text{ m}^3\right)$$

The answer is (B).

Why Other Options Are Wrong

(A) This incorrect solution miscalculates the freezing elevation and then determines the melting area between the freezing elevation and the weather station. Other assumptions, definitions, and equations are unchanged from the correct solution.

$$z_f = z_w + \frac{T_w - 0°C}{\Gamma} = 2036 \text{ m} + \frac{12°C - 0°C}{-0.016 \dfrac{°C}{m}}$$

$$= 1286 \text{ m}$$

The area between the snow line elevation and the freezing point elevation is the melting area. This area is determined from the figure.

$$A_m = \left(1850 \text{ km}^2 - 1300 \text{ km}^2\right)\left(1000 \ \frac{m}{km}\right)^2$$

$$= 5.5 \times 10^8 \text{ m}^2$$

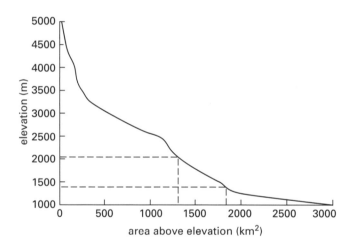

$$T_s = T_w + \Gamma\left(z_s - z_w\right)$$

$$= 12°C + \left(-0.016 \ \frac{°C}{m}\right)(1286 \text{ m} - 2036 \text{ m})$$

$$= 24°C$$

$$T_m = 0.5 Ts = (0.5)(24°C) = 12°C$$

$$d_m = 12°C\cdot d$$

$$V_s = d_m d_f A_m$$

$$= (12°C\cdot d)\left(3 \ \frac{mm}{°C\cdot d}\right)\left(5.5 \times 10^8 \text{ m}^2\right)\left(\frac{1 \text{ m}}{10^3 \text{ mm}}\right)$$

$$= 1.98 \times 10^7 \text{ m}^3 \quad (2.0 \times 10^7 \text{ m}^3)$$

(C) This incorrect solution defines the melt area from the snow line to 5000 m. Other assumptions, definitions, and equations are unchanged from the correct solution.

The area between the snow line elevation and the freezing point elevation is the melting area. This area is determined from the figure.

$$A_m = \left(1850 \text{ km}^2\right)\left(1000 \ \frac{m}{km}\right)^2 = 1.85 \times 10^9 \text{ m}^2$$

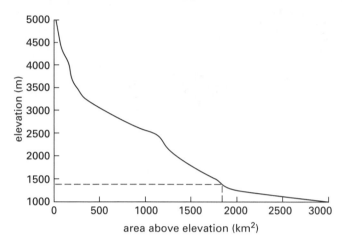

$$T_s = T_w + \Gamma\left(z_s - z_w\right)$$

$$= 12°C + \left(-0.016 \ \frac{°C}{m}\right)(1370 \text{ m} - 2036 \text{ m})$$

$$= 22.7°C$$

$$T_m = 0.5 T_s = (0.5)(22.7°C) = 11.3°C$$

$$d_m = 11.3°C\cdot d$$

$$V_s = d_m d_f A_m$$

$$= (11.3°C\cdot d)\left(3 \ \frac{mm}{°C\cdot d}\right)\left(1.85 \times 10^9 \text{ m}^2\right)\left(\frac{1 \text{ m}}{10^3 \text{ mm}}\right)$$

$$= 6.27 \times 10^7 \text{ m}^3 \quad (6.3 \times 10^7 \text{ m}^3)$$

(D) This incorrect solution miscalculates the temperature at the snow line and the average temperature over the melt area and fails to convert from mm to m in the final calculation. The figure and other assumptions, definitions, and equations are unchanged from the correct solution.

$$z_f = z_w + \frac{0°C - T_w}{\Gamma}$$

$$= 2036 \text{ m} + \frac{0°C - 12°C}{-0.016 \dfrac{°C}{m}}$$

$$= 2786 \text{ m}$$

$$A_m = \left(1850 \text{ km}^2 - 755 \text{ km}^2\right)\left(1000 \ \frac{m}{km}\right)^2$$

$$= 1.1 \times 10^9 \text{ m}^2$$

$$T_s = T_w + \Gamma\left(z_s - z_w\right)$$

$$= 12°C + \left(-0.016 \ \frac{°C}{m}\right)(2036 \text{ m} - 1370 \text{ m})$$

$$= 1.3°C$$

$$T_m = (0.5)(Ts + Tw)$$
$$= (0.5)(1.3°\text{C} + 24°\text{C})$$
$$= 12.65°\text{C}$$
$$d_m = 12.65°\text{C·d}$$
$$V_s = d_m d_f A_m$$
$$= (12.65°\text{C·d}) \left(3 \; \frac{\text{mm}}{°\text{C·d}}\right) (1.1 \times 10^9 \; \text{m}^2)$$
$$= 4.2 \times 10^{10} \; \text{mm·m}^2$$

The units do not work.

SOLUTION 77

f relative humidity   %

p_{a2} water vapor pressure in air at
     specified temperature 2 m
     above lake surface   millibar

p_{s0} water saturation vapor pressure
     in air at lake surface temperature   millibar

p_{s2} water saturation vapor pressure
     in air at specified temperature 2 m
     above lake surface   millibar

The saturation vapor pressure of air at 18°C is 20.63 millibar and at 24°C is 29.83 millibar.

$$p_{a2} = \frac{f p_{s2}}{100}$$
$$= \frac{(16\%)(29.83 \; \text{millibar})}{100\%}$$
$$= 4.77 \; \text{millibar}$$

E evaporation   mm/d
v_4 wind speed at 4 m above the
     water surface   m/s

The information provided suggests one of the empirical formulas available that are based on vapor pressure and wind speed. One of these that has been applied to Lake Mead and may be considered to be generally applicable, especially in the southwestern United States, is

$$E = 0.122 \, (p_{s0} - p_{a2}) \, v_4$$
$$= (0.122)(20.63 \; \text{millibar} - 4.77 \; \text{millibar}) \left(3.6 \; \frac{\text{m}}{\text{s}}\right)$$
$$= 7.0 \; \text{mm/d}$$

The answer is (D).

Why Other Options Are Wrong

(A) This incorrect solution uses the evaporation equation for customary U.S. units instead of SI units. Other assumptions, definitions, and equations are unchanged from the correct solution.

$$p_{a2} = \frac{f p_{s2}}{100} = \frac{(16\%)(29.83 \; \text{millibar})}{100\%}$$
$$= 4.77 \; \text{millibar}$$
$$E = 0.00304 \, (p_{s0} - p_{a2}) \, v_4$$
$$= (0.00304)(20.63 \; \text{millibar} - 4.77 \; \text{millibar})$$
$$\times \left(3.6 \; \frac{\text{m}}{\text{s}}\right)$$
$$= 0.17 \; \text{mm/d}$$

(B) This incorrect solution uses the saturation vapor pressure for both vapor pressure entries. Other assumptions, definitions, and equations are unchanged from the correct solution.

p_{a2} water saturation vapor pressure
     in air at specified temperature
     2 m above lake surface   millibar

$$E = 0.122 \, (p_{s0} - p_{a2}) \, v_4$$
$$= (0.122)(20.63 \; \text{millibar} - 29.83 \; \text{millibar}) \left(3.6 \; \frac{\text{m}}{\text{s}}\right)$$
$$= 4.0 \; \text{mm/d}$$

The negative sign is ignored.

(C) This incorrect solution uses the vapor pressures in units of mm Hg instead of millibars. Other assumptions, definitions, and equations are unchanged from the correct solution.

p_{s0} water saturation vapor pressure in
     air at lake surface temperature   mm Hg

p_{a2} water vapor pressure in air at
     specified temperature 2 m above
     lake surface   mm Hg

p_{s2} water saturation vapor in air at
     specified temperature 2 m above
     lake surface   mm Hg

The saturation vapor pressure of air at 18°C is 15.46 mm Hg and at 24°C is 22.38 mm Hg.

$$p_{a2} = \frac{f p_{s2}}{100} = \frac{(16\%)(22.38 \; \text{mmHg})}{100\%} = 3.58 \; \text{mmHg}$$
$$E = 0.122 \, (p_{s0} - p_{a2}) \, v_4$$
$$= (0.122)(15.46 \; \text{mmHg} - 3.58 \; \text{mmHg}) \left(3.6 \; \frac{\text{m}}{\text{s}}\right)$$
$$= 5.2 \; \text{mm/d}$$

SOLUTION 78

L_s total sediment load   lbm/day

From the figure for a stream flow of 20 ft³/sec, the sediment load is about

$$L_s = 100\,000 \; \text{lbm/day}$$

V reservoir original volume ft^3
V_s sediment volume ft^3
x reservoir volume fraction occupied
     by sediment

$$V_s = Vx = (3600 \text{ ac-ft}) (0.25) \left(43,560 \ \frac{ft^3}{\text{ac-ft}} \right)$$

$$= 3.9 \times 10^7 \ ft^3$$

Assume typical specific weight values of sediment continuously submerged for 50 years are 50 lbm/ft^3 for clay and 65 lbm/ft^3 for silt.

f_i fraction of soil type in sediment
γ_i sediment specific weight for soil
     type fraction lbm/ft^3
γ_s total sediment specific weight
     at time t lbm/ft^3

$$\gamma_s = \sum (f_i \gamma_i) = (0.36) \left(65 \ \frac{lbm}{ft^3} \right) + (0.64) \left(50 \ \frac{lbm}{ft^3} \right)$$

$$= 55.4 \ lbm/ft^3$$

t reservoir useful life   yr

$$t = \frac{V_s \gamma_s}{L_s} = \frac{(3.9 \times 10^7 \ ft^3) \left(55.4 \ \frac{lbm}{ft^3} \right)}{\left(100,000 \ \frac{lbm}{day} \right) \left(365 \ \frac{day}{yr} \right)}$$

$$= 59 \ yr$$

The answer is (B).

Why Other Options Are Wrong

(A) This incorrect solution uses the wrong conversion factor from ft^3 to ac-ft in the sediment volume calculation. Other assumptions, definitions, and equations are unchanged from the correct solution.

$$L_s = 100,000 \ lbm/day$$

$$V_s = Vx = (3600 \text{ ac-ft}) (0.25) \left(10,000 \ \frac{ft^3}{\text{ac-ft}} \right)$$

$$= 9.0 \times 10^6 \ ft^3$$

$$\gamma_s = \sum (f_i \gamma_i) = (0.36) \left(65 \ \frac{lbm}{ft^3} \right) + (0.64) \left(50 \ \frac{lbm}{ft^3} \right)$$

$$= 55.4 \ lbm/ft^3$$

$$t = \frac{V_s \gamma_s}{L_s} = \frac{(9.0 \times 10^6 \ ft^3) \left(55.4 \ \frac{lbm}{ft^3} \right)}{\left(100,000 \ \frac{lbm}{day} \right) \left(365 \ \frac{day}{yr} \right)}$$

$$= 14 \ yr$$

(C) This incorrect solution does not correct the reservoir volume for the fraction occupied by sediment. Other assumptions, definitions, and equations are unchanged from the correct solution.

$$L_s = 100,000 \ lbm/day$$

$$V_s = (3600 \text{ ac-ft}) \left(43,560 \ \frac{ft^3}{\text{ac-ft}} \right)$$

$$= 1.57 \times 10^8 \ ft^3$$

$$\gamma_s = \sum (f_i \gamma_i)$$

$$= (0.36) \left(65 \ \frac{lbm}{ft^3} \right) + (0.64) \left(50 \ \frac{lbm}{ft^3} \right)$$

$$= 55.4 \ lbm/ft^3$$

$$t = \frac{V_s \gamma_s}{L_s}$$

$$= \frac{(1.57 \times 10^8 \ ft^3) \left(55.4 \ \frac{lbm}{ft^3} \right)}{\left(100,000 \ \frac{lbm}{day} \right) \left(365 \ \frac{day}{yr} \right)}$$

$$= 238 \ yr \quad (240 \ yr)$$

(D) This incorrect solution divides by instead of multiplying by the fraction of reservoir volume occupied by sediment. Other assumptions, definitions, and equations are unchanged from the correct solution.

$$L_s = 100,000 \ lbm/day$$

$$V_s = \frac{V}{x}$$

$$= \frac{(3600 \text{ ac-ft}) \left(43,560 \ \frac{ft^3}{\text{ac-ft}} \right)}{0.25}$$

$$= 6.3 \times 10^8 \ ft^3$$

$$\gamma_s = \sum (f_i \gamma_i)$$

$$= (0.36) \left(65 \ \frac{lbm}{ft^3} \right) + (0.64) \left(50 \ \frac{lbm}{ft^3} \right)$$

$$= 55.4 \ lbm/ft^3$$

$$t = \frac{V_s \gamma_s}{L_s}$$

$$= \frac{(6.3 \times 10^8 \ ft^3) \left(55.4 \ \frac{lbm}{ft^3} \right)}{\left(100,000 \ \frac{lbm}{day} \right) \left(365 \ \frac{day}{yr} \right)}$$

$$= 956 \ yr \quad (960 \ yr)$$

SOLUTION 79

Typical runoff coefficients are selected from standard reference tables. Where a range of typical values are listed in reference tables, an average of the range of values is used.

land use	area (ac)	typical runoff coefficient	area weighted runoff coefficient
shingle roof	0.8	0.85	0.68
concrete surface	1.1	0.88	0.97
asphalt surface	2.6	0.83	2.2
poorly drained lawn	10	0.15	1.5
	14.5		5.4

A watershed total area   ac
C_m watershed average weighted
         runoff coefficient   –
C_w area weighted runoff coefficient   –

$$C_m = \frac{\sum C_w}{A} = \frac{5.4}{14.5} = 0.37$$

L longest runoff flow path   ft
S average ground surface slope   %
t_c time of concentration   min

A commonly used equation for runoff from urban areas is the FAA formula.

$$t_c = \frac{(1.8)(1.1 - C_m)\sqrt{L}}{S^{1/3}}$$

$$= \frac{(1.8)(1.1 - 0.37)\sqrt{(337 \text{ yd})\left(3 \frac{\text{ft}}{\text{yd}}\right)}}{(0.013 \times 100\%)^{1/3}}$$

$$= 38 \text{ min}$$

b Steel formula constant   min
K Steel formula constant   in-min/hr

The rainfall intensity can be estimated from the Steel formula with values for coefficients K and b taken for generalized rainfall regions and storm frequency. The typical coefficient values for the western United States and a 10-year storm are

$$K = 60 \text{ in-min/hr}$$
$$b = 13 \text{ min}$$

i rainfall intensity   in/hr

$$i = \frac{K}{t_c + b} = \frac{60 \frac{\text{in-min}}{\text{hr}}}{38 \text{ min} + 13 \text{ min}} = 1.18 \text{ in/hr}$$

Q runoff flow rate   ft^3/sec

$$Q = CiA = (0.37)\left(1.18 \frac{\text{in}}{\text{hr}}\right)(14.5 \text{ ac})\left(1 \frac{\text{ft}^3\text{-hr}}{\text{ac-in-sec}}\right)$$

$$= 6.3 \text{ ft}^3/\text{sec}$$

The answer is (B).

Why Other Options Are Wrong

(A) This incorrect solution uses the slope with units of foot per foot instead of as a percent. Other assumptions, definitions, and equations are unchanged from the correct solution.

land use	area (ac)	typical runoff coefficient	area weighted runoff coefficient
shingle roof	0.8	0.85	0.68
concrete surface	1.1	0.88	0.97
asphalt surface	2.6	0.83	2.2
poorly drained lawn	10	0.15	1.5
	14.5		5.4

$$C_m = \frac{\sum C_w}{A} = \frac{5.4}{14.5}$$
$$= 0.37$$

$$t_c = \frac{(1.8)(1.1 - C_m)\sqrt{L}}{S^{1/3}}$$

$$= \frac{(1.8)(1.1 - 0.37)\sqrt{(337 \text{ yd})\left(3 \frac{\text{ft}}{\text{yd}}\right)}}{(0.013)^{1/3}}$$

$$= 178 \text{ min}$$

$$K = 60 \text{ in-min/hr}$$
$$b = 13 \text{ min}$$

$$i = \frac{K}{t_c + b} = \frac{60 \frac{\text{in-min}}{\text{hr}}}{178 \text{ min} + 13 \text{ min}}$$
$$= 0.31 \text{ in/hr}$$

$$Q = CiA$$
$$= (0.37)\left(0.31 \frac{\text{in}}{\text{hr}}\right)(14.5 \text{ ac})\left(1 \frac{\text{ft}^3\text{-hr}}{\text{ac-in-sec}}\right)$$
$$= 1.7 \text{ ft}^3/\text{sec}$$

(C) This incorrect solution reads the flow distance figure in feet instead of yards. Other assumptions, definitions, and equations are unchanged from the correct solution.

land use	area (ac)	typical runoff coefficient	area weighted runoff coefficient
shingle roof	0.8	0.85	0.68
concrete surface	1.1	0.88	0.97
asphalt surface	2.6	0.83	2.2
poorly drained lawn	10	0.15	1.5
	14.5		5.4

$$C_m = \frac{\sum C_w}{A} = \frac{5.4}{14.5}$$
$$= 0.37$$

$$t_c = \frac{(1.8)(1.1 - C_m)\sqrt{L}}{S^{1/3}}$$

$$= \frac{(1.8)(1.1 - 0.37)\sqrt{(337 \text{ yd})}}{((0.013)(100))^{1/3}}$$

$$= 22 \text{ min}$$

$$K = 60 \text{ in-min/hr}$$

$$b = 13 \text{ min}$$

$$i = \frac{K}{t_c + b} = \frac{60 \frac{\text{in-min}}{\text{hr}}}{22 \text{ min} + 13 \text{ min}}$$

$$= 1.71 \text{ in/hr}$$

$$Q = CiA$$

$$= (0.37)\left(1.71 \frac{\text{in}}{\text{hr}}\right)(14.5 \text{ ac})\left(1 \frac{\text{ft}^3\text{-hr}}{\text{ac-in-sec}}\right)$$

$$= 9.2 \text{ ft}^3/\text{sec}$$

(D) This incorrect solution selects values for the Steel formula coefficients of 100 for K and 10 for b, assuming these to be reasonable approximate values. Other assumptions, definitions, and equations are unchanged from the correct solution.

land use	area (ac)	typical runoff coefficient	area weighted runoff coefficient
shingle roof	0.8	0.85	0.68
concrete surface	1.1	0.88	0.97
asphalt surface	2.6	0.83	2.2
poorly drained lawn	10	0.15	1.5
	14.5		5.4

$$C_m = \frac{\sum C_w}{A} = \frac{5.4}{14.5}$$

$$= 0.37$$

$$t_c = \frac{(1.8)(1.1 - C_m)\sqrt{L}}{S^{1/3}}$$

$$= \frac{(1.8)(1.1 - 0.37)\sqrt{(337 \text{ yd})\left(3 \frac{\text{ft}}{\text{yd}}\right)}}{((0.013)(100))^{1/3}}$$

$$= 38 \text{ min}$$

$$K = 100 \text{ in-min/hr}$$

$$b = 10 \text{ min}$$

$$i = \frac{K}{t_c + b} = \frac{100 \frac{\text{in-min}}{\text{hr}}}{38 \text{ min} + 10 \text{ min}}$$

$$= 2.08 \text{ in/hr}$$

$$Q = CiA$$

$$= (0.37)\left(2.08 \frac{\text{in}}{\text{hr}}\right)(14.5 \text{ ac})\left(1 \frac{\text{ft}^3\text{-hr}}{\text{ac-in-sec}}\right)$$

$$= 11 \text{ ft}^3/\text{sec}$$

SOLUTION 80

Because the watershed is small and rainfall intensity, ground surface slope, and flow distance are known, the appropriate equation to calculate time of concentration would be either the Izzard equation or the kinematic wave formula. Because well-defined drainage channels characterize the watershed, the kinematic wave formula would be the better choice. The kinematic wave formula requires customary U.S. units, so conversion from SI to customary U.S. units is necessary. For manicured sod, the Manning roughness coefficient for overland flow is approximately 0.30.

i	rainfall intensity	in/hr
L	length of flow path	ft
n	Manning roughness coefficient for overland flow	–
S	ground surface slope	ft/ft
t_c	time of concentration	min

$$L = (83 \text{ m})\left(3.28 \frac{\text{ft}}{\text{m}}\right)$$

$$= 272 \text{ ft}$$

$$i = \left(5.3 \frac{\text{cm}}{\text{h}}\right)\left(\frac{1 \text{ in}}{2.54 \text{ cm}}\right)$$

$$= 2.09 \text{ in/hr}$$

$$t_c = \frac{0.94(nL)^{0.6}}{i^{0.4}S^{0.3}}$$

$$= \frac{(0.94)(0.30)^{0.6}(272 \text{ ft})^{0.6}}{\left(2.09 \frac{\text{in}}{\text{hr}}\right)^{0.4}(0.011)^{0.3}}$$

$$= 38 \text{ min}$$

The answer is (D).

Why Other Options Are Wrong

(A) This incorrect solution assumes the channel flow occurs along the longest flow path through the watershed. With this assumption, the flow velocity and time of concentration are computed from velocity-time-distance relationships.

Assume that channel flow velocity can be defined by the NRCS equation. The NRCS equation requires customary U.S. units, so conversion from SI to customary U.S. units is necessary.

v   flow velocity   ft/sec

$$v = 16.1345\sqrt{S} = (16.1345)\sqrt{0.011}$$

$$= 1.69 \text{ ft/sec}$$

$$L = (83 \text{ m})\left(3.28 \frac{\text{ft}}{\text{m}}\right) = 272 \text{ ft}$$

$$t_c = \frac{L}{v} = \frac{(272 \text{ ft}) \left(\dfrac{1 \text{ min}}{60 \text{ sec}}\right)}{1.69 \dfrac{\text{ft}}{\text{sec}}}$$

$$= 2.7 \text{ min}$$

(B) This incorrect solution uses the FAA formula. The FAA formula does not account for rainfall intensity, and this solution ignores the rainfall intensity data. Other definitions are the same as used in the correct solution.

Assume the FAA formula applies. The FAA formula is for English units, so conversion from SI to customary U.S. units is required. The rational method runoff coefficient for manicured sod is about 0.2.

C rational method runoff
     coefficient   –
S ground surface slope $(0.011) \times 100\% = 1.1\%$

$$t_c = \frac{1.8\,(1.1 - C)\,\sqrt{L}}{S^{1/3}}$$

$$= \frac{(1.8)\,(1.1 - 0.2)\,\sqrt{272 \text{ ft}}}{(1.1\%)^{1/3}}$$

$$= 26 \text{ min}$$

(C) This incorrect option uses the Izzard formula. The Izzard formula is applicable to small watersheds with undefined or poorly defined flow channels. Other definitions are the same as used in the correct solution.

Assume the Izzard equation applies. For manicured sod, assume a retardance coefficient of 0.046.

b Izzard coefficient   –
c retardance coefficient   –
i rainfall intensity   mm/h
S ground surface slope   m/m

$$i = \left(5.3 \frac{\text{cm}}{\text{h}}\right) \left(10 \frac{\text{mm}}{\text{cm}}\right)$$

$$= 53 \text{ mm/h}$$

$$b = \frac{0.000028i + c}{S^{1/3}}$$

$$= \frac{(0.000028) \left(53 \dfrac{\text{mm}}{\text{h}}\right) + 0.046}{(0.011)^{0.3}}$$

$$= 0.18$$

t_c time of concentration   min
L length of flow path   m

$$t_c = \frac{526 b L^{1/3}}{i^{2/3}} = \frac{(526)\,(0.18)\,(83 \text{ m})^{1/3}}{\left(53 \dfrac{\text{mm}}{\text{hr}}\right)^{2/3}}$$

$$= 29 \text{ min}$$

SOLUTION 81

s reservoir storage volume m^3
z water surface elevation   m

$$s = 9.4z^2 + 3854z + 14\,470$$

inflow (m^3/s)	outflow (m^3/s)	water surface elevation (m)	reservoir volume (10^6 m^3)
0.67	0.67	371.1	2.74
1.2	0.70	373.2	2.76
2.8	2.9	379.1	2.83
2.5	3.8	377.8	2.81

I inflow m^3/s
O outflow m^3/s
Δ_t elapsed time increment   s
$1, 2$ subscripts   successive time steps

For discharge from deep uncontrolled reservoirs, the following equation applies.

$$I_1 + I_2 + \frac{2s_1}{\Delta t} - O_1 = \frac{2s_2}{\Delta t} + O_2$$

$$\Delta t = \frac{2\,(s_2 - s_1)}{I_1 + I_2 - O_1 - O_2}$$

$$\Delta t_1 = \frac{(2)\,\left(2.76 \times 10^6 \text{ m}^3 - 2.74 \times 10^6 \text{ m}^3\right) \left(\dfrac{1 \text{ h}}{3600 \text{ s}}\right)}{0.67 \dfrac{\text{m}^3}{\text{s}} + 1.2 \dfrac{\text{m}^3}{\text{s}} - 0.67 \dfrac{\text{m}^3}{\text{s}} - 0.70 \dfrac{\text{m}^3}{\text{s}}}$$

$$= 22 \text{ h}$$

$$\Delta t_2 = \frac{(2)\,\left(2.83 \times 10^6 \text{ m}^3 - 2.76 \times 10^6 \text{ m}^3\right) \left(\dfrac{1 \text{ h}}{3600 \text{ s}}\right)}{1.2 \dfrac{\text{m}^3}{\text{s}} + 2.8 \dfrac{\text{m}^3}{\text{s}} - 0.70 \dfrac{\text{m}^3}{\text{s}} - 2.9 \dfrac{\text{m}^3}{\text{s}}}$$

$$= 97 \text{ h}$$

$$\Delta t_3 = \frac{(2)\,\left(2.81 \times 10^6 \text{ m}^3 - 2.83 \times 10^6 \text{ m}^3\right) \left(\dfrac{1 \text{ h}}{3600 \text{ s}}\right)}{2.8 \dfrac{\text{m}^3}{\text{s}} + 2.5 \dfrac{\text{m}^3}{\text{s}} - 2.9 \dfrac{\text{m}^3}{\text{s}} - 3.8 \dfrac{\text{m}^3}{\text{s}}}$$

$$= 8 \text{ h}$$

t total time to maximum water elevation   h

$$t = \Delta t_1 + \Delta t_2 + \Delta t_3$$

$$= 22 \text{ h} + 97 \text{ h} + 8 \text{ h}$$

$$= 127 \text{ h} \quad (130 \text{ h})$$

The answer is (D).

Why Other Options Are Wrong

(A) This incorrect solution fails to multiply the numerator of the time equation by two and adds instead of subtracts outflow. Other assumptions, definitions, and equations are unchanged from the correct solution.

inflow (m^3/s)	outflow (m^3/s)	water surface elevation (m)	reservoir volume (10^6 m^3)
0.67	0.67	371.1	2.74
1.2	0.70	373.2	2.76
2.8	2.9	379.1	2.83
2.5	3.8	377.8	2.81

$$\Delta t = \frac{(s_2 - s_1)}{I_1 + I_2 + O_1 + O_2}$$

$$\Delta t_1 = \frac{\left(2.76 \times 10^6 \text{ m}^3 - 2.74 \times 10^6 \text{ m}^3\right)\left(\dfrac{1 \text{ h}}{3600 \text{ s}}\right)}{0.67 \, \dfrac{\text{m}^3}{\text{s}} + 1.2 \, \dfrac{\text{m}^3}{\text{s}} + 0.67 \, \dfrac{\text{m}^3}{\text{s}} + 0.70 \, \dfrac{\text{m}^3}{\text{s}}}$$
$$= 1.7 \text{ h}$$

$$\Delta t_2 = \frac{\left(2.83 \times 10^6 \text{ m}^3 - 2.76 \times 10^6 \text{ m}^3\right)\left(\dfrac{1 \text{ h}}{3600 \text{ s}}\right)}{1.2 \, \dfrac{\text{m}^3}{\text{s}} + 2.8 \, \dfrac{\text{m}^3}{\text{s}} + 0.70 \, \dfrac{\text{m}^3}{\text{s}} + 2.9 \, \dfrac{\text{m}^3}{\text{s}}}$$
$$= 2.6 \text{ h}$$

$$\Delta t_3 = \frac{\left(2.81 \times 10^6 \text{ m}^3 - 2.83 \times 10^6 \text{ m}^3\right)\left(\dfrac{1 \text{ h}}{3600 \text{ s}}\right)}{2.8 \, \dfrac{\text{m}^3}{\text{s}} + 2.5 \, \dfrac{\text{m}^3}{\text{s}} + 2.9 \, \dfrac{\text{m}^3}{\text{s}} + 3.8 \, \dfrac{\text{m}^3}{\text{s}}}$$
$$= -0.5 \text{ h}$$
$$t = \Delta t_1 + \Delta t_2 + \Delta t_3 = 1.7 \text{ h} + 2.6 \text{ h} - 0.5 \text{ h}$$
$$= 3.8 \text{ h}$$

(B) This incorrect solution adds instead of subtracts the outflow terms in the time increment calculations. Other assumptions, definitions, and equations are unchanged from the correct solution.

inflow (m^3/s)	outflow (m^3/s)	water surface elevation (m)	reservoir volume (10^6 m^3)
0.67	0.67	371.1	2.74
1.2	0.70	373.2	2.76
2.8	2.9	379.1	2.83
2.5	3.8	377.8	2.81

$$\Delta t = \frac{2(s_2 - s_1)}{I_1 + I_2 + O_1 + O_2}$$

$$\Delta t_1 = \frac{(2)\left(2.76 \times 10^6 \text{ m}^3 - 2.74 \times 10^6 \text{ m}^3\right)\left(\dfrac{1 \text{ h}}{3600 \text{ s}}\right)}{0.67 \, \dfrac{\text{m}^3}{\text{s}} + 1.2 \, \dfrac{\text{m}^3}{\text{s}} + 0.67 \, \dfrac{\text{m}^3}{\text{s}} + 0.70 \, \dfrac{\text{m}^3}{\text{s}}}$$
$$= 3.4 \text{ h}$$

$$\Delta t_2 = \frac{(2)\left(2.83 \times 10^6 \text{ m}^3 - 2.76 \times 10^6 \text{ m}^3\right)\left(\dfrac{1 \text{ h}}{3600 \text{ s}}\right)}{1.2 \, \dfrac{\text{m}^3}{\text{s}} + 2.8 \, \dfrac{\text{m}^3}{\text{s}} + 0.70 \, \dfrac{\text{m}^3}{\text{s}} + 2.9 \, \dfrac{\text{m}^3}{\text{s}}}$$
$$= 5.1 \text{ h}$$

$$\Delta t_3 = \frac{(2)\left(2.81 \times 10^6 \text{ m}^3 - 2.83 \times 10^6 \text{ m}^3\right)\left(\dfrac{1 \text{ h}}{3600 \text{ s}}\right)}{2.8 \, \dfrac{\text{m}^3}{\text{s}} + 2.5 \, \dfrac{\text{m}^3}{\text{s}} + 2.9 \, \dfrac{\text{m}^3}{\text{s}} + 3.8 \, \dfrac{\text{m}^3}{\text{s}}}$$
$$= -0.9 \text{ h}$$
$$t = \Delta t_1 + \Delta t_2 + \Delta t_3 = 3.4 \text{ h} + 5.1 \text{ h} - 0.9 \text{ h}$$
$$= 7.6 \text{ h}$$

(C) This incorrect solution calculates the time based on the initial and final values instead of the incremental data values and reverses the storage terms in the equation. Other assumptions, definitions, and equations are unchanged from the correct solution.

inflow (m^3/s)	outflow (m^3/s)	water surface elevation (m)	reservoir volume (10^6 m^3)
0.67	0.67	371.1	2.74
2.5	3.8	377.8	2.81

$$\Delta t = \frac{2(s_2 - s_1)}{I_1 + I_2 - O_1 - O_2}$$

$$= \frac{(2)\left(2.74 \times 10^6 \text{ m}^3 - 2.81 \times 10^6 \text{ m}^3\right)\left(\dfrac{1 \text{ h}}{3600 \text{ s}}\right)}{0.67 \, \dfrac{\text{m}^3}{\text{s}} + 2.5 \, \dfrac{\text{m}^3}{\text{s}} - 0.67 \, \dfrac{\text{m}^3}{\text{s}} - 3.8 \, \dfrac{\text{m}^3}{\text{s}}}$$
$$= 29.9 \text{ h} \quad (30 \text{ h})$$

SOLUTION 82

For a paved site where rainfall intensity data is available, the kinematic wave formula can be used for calculating time of concentration. This is a variation of the Manning kinematic equation and uses the Manning roughness coefficient.

i rainfall intensity   in/hr

From the first illustration, the rainfall intensity for a 45 min duration 10-year storm is 3.5 in/hr and for a 45 min 25-year storm is 5.0 in/hr.

L maximum flow path   ft

From the second illustration, the maximum flow path is calculated by

$$L = \sqrt{(1230 \text{ ft})^2 + (390 \text{ ft})^2} = 1290 \text{ ft}$$

n Manning roughness coefficient   –

For smooth impervious surfaces such as paved asphalt surfaces, the Manning roughness coefficient is 0.011.

S average ground surface slope   ft/ft
t_c time of concentration   min

$$t_c = \frac{0.94\,(nL)^{0.6}}{i^{0.4} S^{0.3}}$$

$$t_{c,10} = \frac{(0.94)(0.011)^{0.6}\,(1290\ \text{ft})^{0.6}}{\left(3.5\ \dfrac{\text{in}}{\text{hr}}\right)^{0.4} \left(\dfrac{1.2}{100}\right)^{0.3}}$$

$$= 10.5\ \text{min}$$

$$t_{c,25} = \frac{(0.94)(0.011)^{0.6}\,(1290\ \text{ft})^{0.6}}{\left(5.0\ \dfrac{\text{in}}{\text{hr}}\right)^{0.4} \left(\dfrac{1.2}{100}\right)^{0.3}}$$

$$= 9.14\ \text{min}$$

$$\Delta t_c = t_{c,10} - t_{c,25}$$
$$= 10.5\ \text{min} - 9.14\ \text{min}$$
$$= 1.4\ \text{min}$$

The answer is (B).

Why Other Options Are Wrong

(A) This incorrect solution uses the slope as a percent instead of as a fraction. Other assumptions, definitions, and equations are the same as used in the correct solution.

From the first illustration, the rainfall intensity for a 45 min duration 10-year storm is 3.5 in/hr and for a 45 min 25-year storm is 5.0 in/hr.

$$L = \sqrt{(1230\ \text{ft})^2 + (390\ \text{ft})^2}$$
$$= 1290\ \text{ft}$$

S average ground surface slope   %

$$t_c = \frac{0.94\,(nL)^{0.6}}{i^{0.4} S^{0.3}}$$

$$t_{c,10} = \frac{(0.94)(0.035)^{0.6}\,(1290\ \text{ft})^{0.6}}{\left(3.5\ \dfrac{\text{in}}{\text{hr}}\right)^{0.4} (1.2)^{0.3}}$$

$$= 5.3\ \text{min}$$

$$t_{c,25} = \frac{(0.94)(0.035)^{0.6}\,(1290\ \text{ft})^{0.6}}{\left(5.0\ \dfrac{\text{in}}{\text{hr}}\right)^{0.4} (1.2)^{0.3}}$$

$$= 4.6\ \text{min}$$

$$\Delta t_c = t_{c,10} - t_{c,25}$$
$$= 5.3\ \text{min} - 4.6\ \text{min}$$
$$= 0.70\ \text{min}$$

(C) This incorrect solution misread the illustration for obtaining rainfall intensity values for the 10-year and 25-year storms. Other assumptions, definitions, and equations are the same as used in the correct solution.

From the first illustration, the rainfall intensity for a 45 min duration 10-year storm is 5.0 in/hr and for a 45 min 25-year storm is 7.0 in/hr.

$$L = \sqrt{(1230\ \text{ft})^2 + (390\ \text{ft})^2}$$
$$= 1290\ \text{ft}$$

$$t_c = \frac{0.94\,(nL)^{0.6}}{i^{0.4} S^{0.3}}$$

$$t_{c,10} = \frac{(0.94)(0.035)^{0.6}\,(1290\ \text{ft})^{0.6}}{\left(5.0\ \dfrac{\text{in}}{\text{hr}}\right)^{0.4} \left(\dfrac{1.2}{100}\right)^{0.3}}$$

$$= 18.3\ \text{min}$$

$$t_{c,25} = \frac{(0.94)(0.035)^{0.6}\,(1290\ \text{ft})^{0.6}}{\left(7.0\ \dfrac{\text{in}}{\text{hr}}\right)^{0.4} \left(\dfrac{1.2}{100}\right)^{0.3}}$$

$$= 16.0\ \text{min}$$

$$\Delta t_c = t_{c,10} - t_{c,25}$$
$$= 18.3\ \text{min} - 16.0\ \text{min}$$
$$= 2.3\ \text{min}$$

(D) This incorrect solution takes the difference of the rainfall intensities and then applies the time of concentration equation. Other assumptions, definitions, and equations are the same as used in the correct solution.

From the first illustration, the rainfall intensity for a 45 min duration 10-year storm is 3.5 in/hr and for a 45 min 25-year storm is 5.0 in/hr.

Δi difference between the rainfall intensities for the two storms being compared   in/hr

$$\Delta i = i_{25} - i_{10}$$
$$= 5.0\ \frac{\text{in}}{\text{hr}} - 3.5\ \frac{\text{in}}{\text{hr}}$$
$$= 1.5\ \frac{\text{in}}{\text{hr}}$$

$$L = \sqrt{(1230\ \text{ft})^2 + (390\ \text{ft})^2}$$
$$= 1290\ \text{ft}$$

$$t_c = \frac{0.94\,(nL)^{0.6}}{\Delta i^{0.4} S^{0.3}}$$

$$= \frac{(0.94)(0.035)^{0.6}\,(1290\ \text{ft})^{0.6}}{\left(1.5\ \dfrac{\text{in}}{\text{hr}}\right)^{0.4} \left(\dfrac{1.2}{100}\right)^{0.3}}$$

$$= 29.6\ \text{min} \quad (30\ \text{min})$$

SOLUTION 83

Q corrected stream flow   m^3/s
Q_o uncorrected stream flow   m^3/s

From the following illustration for an observed fall of 0.3 m,

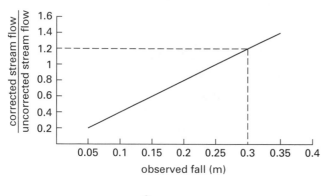

$$\frac{Q}{Q_o} = 1.2$$

From the following illustration for a 2.3 m stage,

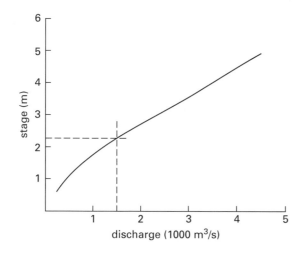

$$Q_o = 1500 \text{ m}^3/\text{s}$$

$$Q = \left(\frac{Q}{Q_o}\right) Q_o = (1.2)\left(1500 \ \frac{\text{m}^3}{\text{s}}\right)$$

$$= 1800 \text{ m}^3/\text{s}$$

The answer is (D).

Why Other Options Are Wrong

(A) This incorrect option does not account for the $(10)^3$ factor on the y-axis of the stage-discharge plot. The illustrations and other definitions and equations are unchanged from the correct solution.

$$\frac{Q}{Q_o} = 1.2$$

$$Q_o = 1.5 \text{ m}^3/\text{s}$$

$$Q = \left(\frac{Q}{Q_o}\right) Q_o = (1.2)\left(1.5 \ \frac{\text{m}^3}{\text{s}}\right)$$

$$= 1.8 \text{ m}^3/\text{s}$$

(B) This incorrect option multiplies the stage by the observed fall to get the discharge. Other definitions are unchanged from the correct solution.

F observed fall   m
s corrected stage   m
s_o uncorrected stage   m

$$s = s_o F = (2.3 \text{ m})(0.3 \text{ m})$$

$$= 0.69$$

The units are ignored.

From the following illustration for a corrected stage of 0.69,

$$Q = 300 \text{ m}^3/\text{s}$$

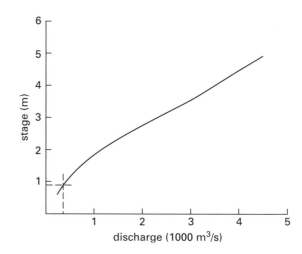

(C) This incorrect option reverses the symbols used for corrected and uncorrected flow. The illustrations and other definitions and equations are unchanged from the correct solution.

Q uncorrected stream flow   m^3/s
Q_o corrected stream flow   m^3/s

$$\frac{Q}{Q_o} = 1.2$$

$$Q = 1500 \text{ m}^3/\text{s}$$

$$Q_o = \frac{Q}{\dfrac{Q}{Q_o}} = \frac{1500 \ \dfrac{\text{m}^3}{\text{s}}}{1.2}$$

$$= 1250 \text{ m}^3/\text{s} \quad (1200 \text{ m}^3/\text{s})$$

SOLUTION 84

The Gumbel distribution can be used with the given information.

K_{50} Gumbel distribution coefficient for
       50 yr record and 50-year flood   –
K_{83} Gumbel distribution coefficient for
       83 yr record and 50-year flood   –
K_{100} Gumbel distribution coefficient for
       100 yr record and 50-year flood   –
n years of record   –

From reference tables, the Gumbel distribution coefficient is available for record periods of 50 yr and 100 yr. Interpolation is required to find the Gumbel distribution coefficient for 83 yr.

$$K_{83} = K_{50} + \frac{(K_{100} - K_{50})(n - 50)}{100 - 50}$$
$$= 2.89 + \frac{(2.77 - 2.89)(83 - 50)}{100 - 50}$$
$$= 2.81$$

σ standard deviation ft^3/sec
Q_m average flow rate over period of record ft^3/sec
Q_p flow rate for the period of interest ft^3/sec

$$Q_p = Q_m + K_{83}\sigma = 1947\,\frac{\text{ft}^3}{\text{sec}} + (2.81)\left(613\,\frac{\text{ft}^3}{\text{sec}}\right)$$
$$= 3670\ \text{ft}^3/\text{sec} \quad (3700\ \text{ft}^3/\text{sec})$$

The answer is (B).

Why Other Options Are Wrong

(A) This incorrect solution takes the simple ratio of the average flow for the period of record to a 50 yr period. Other definitions are unchanged from the correct solution.

$$\frac{Q_p}{50} = \frac{1947\,\frac{\text{ft}^3}{\text{sec}}}{83}$$
$$Q_p = \frac{\left(1947\,\frac{\text{ft}^3}{\text{sec}}\right)(50)}{83}$$
$$= 1173\ \text{ft}^3/\text{sec} \quad (1200\ \text{ft}^3/\text{sec})$$

(C) This incorrect solution reads the value for the Gumbel coefficient for the return period instead of the record of length from the reference table. Other assumptions, definitions, and equations are unchanged from the correct solution.

$$K_{83} = K_{50} + \frac{(K_{100} - K_{50})(n - 50)}{100 - 50}$$
$$= 2.89 + \frac{(3.49 - 2.89)(83 - 50)}{100 - 50}$$
$$= 3.29$$
$$Q_p = Q_m + K_{83}\sigma$$
$$= 1947\,\frac{\text{ft}^3}{\text{sec}} + (3.29)\left(613\,\frac{\text{ft}^3}{\text{sec}}\right)$$
$$= 3964\ \text{ft}^3/\text{sec} \quad (4000\ \text{ft}^3/\text{sec})$$

(D) This incorrect solution multiplies the average flow instead of the standard deviation by the Gumbel coefficient. Other assumptions, definitions, and equations are unchanged from the correct solution.

$$K_{83} = K_{50} + \frac{(K_{100} - K_{50})(n - 50)}{100 - 50}$$
$$= 2.89 + \frac{(2.77 - 2.89)(83 - 50)}{100 - 50}$$
$$= 2.81$$
$$Q_p = Q_m K_{83} + \sigma$$
$$= \left(1947\,\frac{\text{ft}^3}{\text{sec}}\right)(2.81) + \left(613\,\frac{\text{ft}^3}{\text{sec}}\right)$$
$$= 6084\ \text{ft}^3/\text{sec} \quad (6100\ \text{ft}^3/\text{sec})$$

WATER TREATMENT

SOLUTION 85

Typical average water demand for residential use varies between about 75 to 130 gal/person day. These uses are augmented by public and commercial activities that support the community and by losses. Because of these augmenting activities, the average annual daily demand is generally taken to be about 160 to 165 gal/person-day. During summer months, the demand may increase by 20% to 30% and daily maximums may add another 50% to 100%. Fire demand contributes to increased water demand, with 500 gal/min as the acceptable minimum. Based on population, fire demand can be estimated by

P population $(10)^3$ people
Q_f fire demand   gal/min

$$Q_f = 1020\sqrt{P}\left(1 - 0.01\sqrt{P}\right)$$
$$= 1020\sqrt{3.2}\left(1 - \left(0.01\sqrt{3.2}\right)\right)$$
$$= 1792\ \text{gal/min}$$

Accept the fire demand because 1792 gal/min is greater than 500 gal/min.

f_p daily peak correction factor   %
f_s seasonal correction factor   %
Q total maximum daily demand gal/min
Q_m average annual daily demand gal/min

Assume that the following are representative.

parameter	value
Q_m	165 gal/person-day
f_s	25% or 1.25
f_p	75% or 1.75

$$Q = Q_m P f_s f_p + Q_f$$
$$= \left(165 \ \frac{\text{gal}}{\text{person-day}}\right)(3200 \text{ people})\left(\frac{1 \text{ day}}{1440 \text{ min}}\right)$$
$$\times (1.25)(1.75) + 1792 \ \frac{\text{gal}}{\text{min}}$$
$$= 2594 \text{ gal/min} \quad (2600 \text{ gal/min})$$

The answer is (D).

Why Other Options Are Wrong

(A) This incorrect solution uses the typical residential demand instead of the annual average demand, does not include the seasonal and daily peak multipliers, and uses the minimum fire demand instead of calculating fire demand on the basis of population. Other assumptions, definitions, and equations are unchanged from the correct solution.

Typical average water demand for residential use varies between about 75 to 130 gal/person-day. Use 130 gal/person-day as the maximum. Fire demand contributes to increased demand with 500 gal/min as the acceptable minimum.

$$Q = Q_m P f_s f_p + Q_f$$
$$= \left(130 \ \frac{\text{gal}}{\text{person-day}}\right)(3200 \text{ people})\left(\frac{1 \text{ day}}{1440 \text{ min}}\right)$$
$$+ 500 \ \frac{\text{gal}}{\text{min}}$$
$$= 789 \text{ gal/min} \quad (790 \text{ gal/min})$$

(B) This incorrect solution uses the minimum fire demand and multiplies the sum of the per capita demand and fire demand by the seasonal and daily peak multipliers. Other assumptions, definitions, and equations are unchanged from the correct solution.

Typical average annual daily demand is generally taken to be about 165 gal/person day. During summer months, the demand may increase by 20% to 30% and daily maximums may add another 50% to 100%. Fire demand contributes to increased demand with 500 gal/min as the acceptable minimum.

Assume that the following are representative.

parameter	value
Q_m	165 gal/person day
f_s	25% or 1.25
f_p	75% or 1.75

$$Q = (Q_m P + Q_f) f_s f_p$$
$$= \left(\begin{array}{c} \left(165 \ \dfrac{\text{gal}}{\text{person-day}}\right)(3200 \text{ people}) \\ \times \left(\dfrac{1 \text{ day}}{1440 \text{ min}}\right) + 500 \text{ gal/min} \end{array} \right)$$
$$\times (1.25)(1.75)$$
$$= 1896 \text{ gal/min} \quad (1900 \text{ gal/min})$$

(C) This incorrect solution does not include the seasonal and daily peak multipliers. Other assumptions, definitions, and equations are unchanged from the correct solution.

The average annual daily demand is generally taken to be about 160 to 165 gal/person day.

$$Q_f = 1020\sqrt{P}\left(1 - 0.01\sqrt{P}\right)$$
$$= 1020\sqrt{3.2}\left(1 - \left(0.01\sqrt{3.2}\right)\right)$$
$$= 1792 \text{ gal/min}$$
$$Q = Q_m P + Q_f$$
$$= \left(165 \ \frac{\text{gal}}{\text{person-day}}\right)(3200 \text{ people})\left(\frac{1 \text{ day}}{1440 \text{ min}}\right)$$
$$+ 1792 \text{ gal/min}$$
$$= 2159 \text{ gal/min} \quad (2200 \text{ gal/min})$$

SOLUTION 86

The population growth follows an arithmetic trend as illustrated by the relatively uniform change in population from decade to decade.

year	population	population change
1920	20,800	–
1930	23,400	2600
1940	25,100	1700
1950	27,900	2800
1960	29,800	1900
1970	32,600	2800
1980	35,200	2600
1990	37,700	2500
2000	40,100	2400
		19,300

n number of decades
ΔP population change for each decade

ΔP_m average population change

$$\Delta P_m = \frac{\sum \Delta P}{n} = \frac{19{,}300 \text{ people}}{8 \text{ decades}}$$
$$= 2413 \text{ people/decade}$$

Two decades are covered from 2000 to 2020.

P_f future population (in 2020)

$$P_f = 40{,}100 \text{ people} + (2 \text{ decades})\left(2413 \ \frac{\text{people}}{\text{decade}}\right)$$
$$= 44{,}926 \text{ people}$$

Beginning in the decade from 1970 to 1980, the per capita water demand has declined each decade. The declining trend seems to be flattening out to approximately 140 gal/capita day. Assume the demand remains constant at about 140 gal/capita day.

Q_f future daily demand   gal/day

$$Q_f = (44{,}926 \text{ people})\left(140 \ \frac{\text{gal}}{\text{capita day}}\right)$$
$$= 6.3 \times 10^6 \text{ gal/day}$$

The answer is (B).

Why Other Options Are Wrong

(A) This incorrect solution uses the average of the historical per capita water use to calculate future demand instead of the trend of the last few decades. Other assumptions, definitions, and equations are unchanged from the correct solution.

year	population	population change	per capita water use (gal/day)
1920	20,800	–	83
1930	23,400	2600	89
1940	25,100	1700	94
1950	27,900	2800	97
1960	29,800	1900	137
1970	32,600	2800	159
1980	35,200	2600	148
1990	37,700	2500	144
2000	40,100	2400	142
		19,300	1093

$$\Delta P_m = \frac{\sum \Delta P}{n} = \frac{19{,}300 \text{ people}}{8 \text{ decades}}$$
$$= 2413 \text{ people/decade}$$

$$P_f = 40{,}100 \text{ people} + (2 \text{ decades})\left(2413 \ \frac{\text{people}}{\text{decade}}\right)$$
$$= 44{,}926 \text{ people}$$

q_m average water demand   gal/capita day
q per capita water use   gal/day

$$q_m = \frac{\sum q}{n} = \frac{1093 \ \frac{\text{gal}}{\text{capita day}}}{9}$$
$$= 121 \text{ gal/capita day}$$

$$Q_f = (44{,}926 \text{ people})\left(121 \ \frac{\text{gal}}{\text{capita day}}\right)$$
$$= 5.4 \times 10^6 \text{ gal/day}$$

(C) This incorrect solution confuses years with decades, and projects the demand over a 20 yr period instead of two decades. Other assumptions, definitions, and equations are unchanged from the correct solution.

year	population	population change
1920	20,800	–
1930	23,400	2600
1940	25,100	1700
1950	27,900	2800
1960	29,800	1900
1970	32,600	2800
1980	35,200	2600
1990	37,700	2500
2000	40,100	2400
		19,300

$$\Delta P_m = \frac{\sum \Delta P}{n} = \frac{19{,}300 \text{ people}}{8 \text{ decades}}$$
$$= 2413 \text{ people/decade}$$

Twenty years are covered from 2000 to 2020.

$$P_f = 40{,}100 \text{ people} + (20 \text{ yr})\left(2413 \ \frac{\text{people}}{\text{decade}}\right)$$
$$= 88{,}360 \text{ people}$$

Assume the demand remains constant at about 140 gal/capita day.

$$Q_f = (88{,}360 \text{ people})\left(140 \ \frac{\text{gal}}{\text{capita day}}\right)$$
$$= 1.2 \times 10^7 \text{ gal/day}$$

(D) This incorrect solution uses the historical average population and the historical average per capita demand adjusted for the number of decades to calculate demand in 2020. Other definitions are the same as used in the correct solution.

year	population	per capita water use (gal/day)
1920	20,800	83
1930	23,400	89
1940	25,100	94
1950	27,900	97
1960	29,800	137
1970	32,600	159
1980	35,200	148
1990	37,700	144
2000	40,100	142
	272,600	1093

P population

$$Q_f = \frac{(\sum P)(\sum q)(n+2)}{n}$$

$$= \frac{(272,600)\left(1093 \ \dfrac{\text{gal}}{\text{capita day}}\right)(9+2)}{9}$$

$$= 3.6 \times 10^8 \ \text{gal/day}$$

SOLUTION 87

time period	average demand (gal/min)
0000–0200	6900
0200–0400	6500
0400–0600	8100
0600–0800	12,100
0800–1000	14,300
1000–1200	15,600
1200–1400	13,800
1400–1600	12,500
1600–1800	9900
1800–2000	8900
2000–2200	8300
2200–2400	7200
	124,100

n number of time periods
q_m average time period demand   gal/min
Q_m average daily demand   gal/min

$$Q_m = \frac{\sum q_m}{n}$$

$$= \frac{124,100 \ \dfrac{\text{gal}}{\text{min}}}{12}$$

$$= 10,342 \ \text{gal/min}$$

The reservoir is filling (net flow into the reservoir) during those time periods when the average time period demand is less than the average daily demand.

time period	average demand (gal/min)	net flow
0000–0200	6900	in
0200–0400	6500	in
0400–0600	8100	in
0600–0800	12,100	out
0800–1000	14,300	out
1000–1200	15,600	out
1200–1400	13,800	out
1400–1600	12,500	out
1600–1800	9900	in
1800–2000	8900	in
2000–2200	8300	in
2200–2400	7200	in

Sometime between 0500 and 0700 and between 1500 and 1700, the net flow changes direction. The midpoint of the time interval was used because the average demand for the 2 hr time interval was given. Assume that the flow-time relationship is linear during any 2 hr interval.

For the period from 0500 to 0700, apply linear interpolation to find that the change occurs at

$$0500 + \frac{(2 \text{ hr})(10,342 - 8100)}{12,100 - 8100} = 0500 + 1.12 \text{ hr}$$
$$= 0607 \quad (6{:}10 \text{ a.m.})$$

For the period from 1500 to 1700, apply linear interpolation to find that the change occurs at

$$1700 - \frac{(2 \text{ hr})(10,342 - 9900)}{12,500 - 9900} = 1700 - 0.34 \text{ hr}$$
$$= 1640 \quad (4{:}40 \text{ p.m.})$$

The net flow into the reservoir occurs from approximately 4:40 p.m. to 6:10 a.m.

The answer is (D).

Why Other Options Are Wrong

(A) This incorrect solution gives the time period during which the net flow is out of the reservoir and does not interpolate to define the time that the net flow direction changes. Other assumptions, definitions, and equations are unchanged from the correct solution.

time period	average demand (gal/min)
0000–0200	6900
0200–0400	6500
0400–0600	8100
0600–0800	12,100
0800–1000	14,300
1000–1200	15,600
1200–1400	13,800
1400–1600	12,500
1600–1800	9900
1800–2000	8900
2000–2200	8300
2200–2400	7200
	124,100

time period	average demand (gal/min)
0000–0200	6900
0200–0400	6500
0400–0600	8100
0600–0800	12,100
0800–1000	14,300
1000–1200	15,600
1200–1400	13,800
1400–1600	12,500
1600–1800	9900
1800–2000	8900
2000–2200	8300
2200–2400	7200
	124,100

$$Q_m = \frac{\sum q_m}{n}$$
$$= \frac{124{,}100 \ \frac{\text{gal}}{\text{min}}}{12}$$
$$= 10{,}342 \ \text{gal/min}$$

$$Q_m = \frac{\sum q_m}{n}$$
$$= \frac{124{,}100 \ \frac{\text{gal}}{\text{min}}}{12}$$
$$= 10{,}342 \ \text{gal/min}$$

The reservoir is filling (net flow into the reservoir) during those time periods when the average time period demand is greater than the average daily demand.

time period	average demand (gal/min)	net flow
0000–0200	6900	out
0200–0400	6500	out
0400–0600	8100	out
0600–0800	12,100	in
0800–1000	14,300	in
1000–1200	15,600	in
1200–1400	13,800	in
1400–1600	12,500	in
1600–1800	9900	out
1800–2000	8900	out
2000–2200	8300	out
2200–2400	7200	out

time period	average demand (gal/min)	net flow
0000–0200	6900	in
0200–0400	6500	in
0400–0600	8100	in
0600–0800	12,100	out
0800–1000	14,300	out
1000–1200	15,600	out
1200–1400	13,800	out
1400–1600	12,500	out
1600–1800	9900	in
1800–2000	8900	in
2000–2200	8300	in
2200–2400	7200	in

The net flow into the reservoir occurs from approximately 0600 to 1600.   (6:00 a.m. to 4:00 p.m.)

(B) This incorrect solution does not interpolate to define the time that the net flow direction changes. Other assumptions, definitions, and equations are unchanged from the correct solution.

The net flow into the reservoir occurs from approximately 1600 to 0600.   (4:00 p.m. to 6:00 a.m.)

(C) This incorrect solution makes an error in the linear interpolation calculation between 1600 and 1800. Other assumptions, definitions, and equations are unchanged from the correct solution.

time period	average demand (gal/min)
0000–0200	6900
0200–0400	6500
0400–0600	8100
0600–0800	12,100
0800–1000	14,300
1000–1200	15,600
1200–1400	13,800
1400–1600	12,500
1600–1800	9900
1800–2000	8900
2000–2200	8300
2200–2400	7200
	124,100

$$Q_m = \frac{\sum q_m}{n} = \frac{124{,}100 \ \frac{\text{gal}}{\text{min}}}{12}$$

$$= 10{,}342 \ \text{gal/min}$$

The reservoir is filling (net flow into the reservoir) during those time periods when the average time period demand is less than the average daily demand.

time period	average demand (gal/min)	net flow
0000–0200	6900	in
0200–0400	6500	in
0400–0600	8100	in
0600–0800	12,100	out
0800–1000	14,300	out
1000–1200	15,600	out
1200–1400	13,800	out
1400–1600	12,500	out
1600–1800	9900	in
1800–2000	8900	in
2000–2200	8300	in
2200–2400	7200	in

Sometime between 0600 and 0800 and between 1600 and 1800, the net flow changes direction.

For the period from 0600 to 0800, apply linear interpolation to find that the change occurs at

$$0600 + \frac{(2 \ \text{hr})(10{,}324 - 8100)}{12{,}100 - 8100} = 0600 + 1.11 \ \text{hr}$$

$$= 0707 \quad (7{:}10 \ \text{a.m.})$$

For the period from 1600 to 1800, apply linear interpolation to find that the change occurs at

$$1600 + \frac{(2 \ \text{hr})(10{,}324 - 9900)}{12{,}500 - 9900} = 1600 + 0.33 \ \text{hr}$$

$$= 1620 \quad (4{:}20 \ \text{p.m.})$$

The net flow into the reservoir occurs from approximately 4:20 p.m. to 7:10 a.m.

SOLUTION 88

The average annual demand is 90,000 ac-ft/yr. This is the slope of the average flow line plotted on the following illustration.

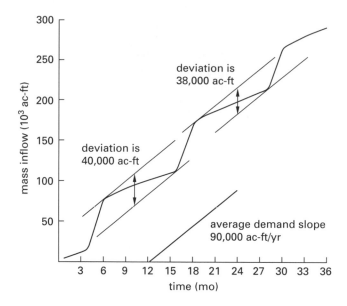

Lines parallel to the average annual demand slope are drawn tangent to the maximum points of deviation for each fill-and-empty cycle of the reservoir. The parallel line pair with the maximum deviation defines the reservoir capacity.

V required capacity   ac-ft

$$V = 40{,}000 \ \text{ac-ft}$$

The answer is (A).

Why Other Options Are Wrong

(B) This incorrect solution adds the volume defined by the deviation for each fill-and-empty cycle. The illustration is unchanged from the correct solution.

$$V = 40{,}000 \ \text{ac-ft} + 38{,}000 \ \text{ac-ft} = 78{,}000 \ \text{ac-ft}$$

(C) This incorrect solution assumes that the average annual demand and the reservoir capacity are equal.

The average annual demand is 90,000 ac-ft/yr. This is also the minimum reservoir capacity.

$$V = 90{,}000 \ \text{ac-ft}$$

(D) This incorrect solution assumes that the average annual demand multiplied by the number of years of record is the reservoir capacity.

The average annual demand is 90,000 ac-ft/yr for a period of 3 yr.

$$V = \left(90{,}000 \; \frac{\text{ac-ft}}{\text{yr}}\right)(3 \text{ yr}) = 270{,}000 \text{ ac-ft}$$

SOLUTION 89

Q	total flow rate	m³/d
Q_B	bypass flow rate	m³/d
Q_T	treatment flow rate	m³/d
TH	initial total hardness	mg/L as $CaCO_3$
TH_D	desired total hardness	mg/L as $CaCO_3$
TH_T	treated water total hardness	mg/L as $CaCO_3$

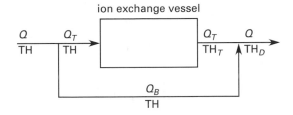

$$Q_T TH_T + Q_B TH = Q TH_D$$

For ion exchange, assume total hardness removal is 100% so that the treated water total hardness is zero.

$$Q_T(0) + Q_B TH = Q TH_D$$

$$Q_B = \frac{TH_D Q}{TH}$$

$$= \frac{\left(100 \; \frac{\text{mg}}{\text{L}} \text{ as } CaCO_3\right)\left(130\,000 \; \frac{\text{m}^3}{\text{d}}\right)}{382 \; \frac{\text{mg}}{\text{L}} \text{ as } CaCO_3}$$

$$= 34\,031 \text{ m}^3/\text{d}$$

$$Q_T = Q - Q_B$$

$$= 130\,000 \; \frac{\text{m}^3}{\text{d}} - 34\,031 \; \frac{\text{m}^3}{\text{d}}$$

$$= 95\,969 \text{ m}^3/\text{d}$$

TH_R total hardness removed   kg/d

$$TH_R = TH Q_T$$

$$= \left(382 \; \frac{\text{mg}}{\text{L}} \text{ as } CaCO_3\right)\left(95\,969 \; \frac{\text{m}^3}{\text{d}}\right)$$

$$\times \left((10)^{-6} \; \frac{\text{kg}}{\text{mg}}\right)\left(1000 \; \frac{\text{L}}{\text{m}^3}\right)$$

$$= 36\,660 \text{ kg/d}$$

C_R exchange resin capacity   kg/m³·resin

V_R exchange resin volume   m³/d

$$V_R = \frac{TH_R}{C_R} = \frac{36\,660 \; \frac{\text{kg}}{\text{d}}}{95 \; \frac{\text{kg}}{\text{m}^3\cdot\text{resin}}}$$

$$= 386 \text{ m}^3\cdot\text{resin/d}$$

n number of exchange vessels   –
V_C exchange resin volume per vessel   m³

$$n = \frac{V_R}{V_C} = \frac{\left(386 \; \frac{\text{m}^3\cdot\text{resin}}{\text{d}}\right)\left(1 \; \frac{\text{d}}{\text{regeneration}}\right)}{4 \; \frac{\text{m}^3\cdot\text{resin}}{\text{vessel}\cdot\text{regeneration}}}$$

$$= 96.5 \text{ vessels} \quad (97 \text{ vessels})$$

The answer is (C).

Why Other Options Are Wrong

(A) This incorrect solution bases the resin use on the bypass flow instead of the treatment flow. Other assumptions, definitions, and equations are the same as used in the correct solution.

$$Q_T = \frac{TH_D Q}{TH}$$

$$= \frac{\left(100 \; \frac{\text{mg}}{\text{L}} \text{ as } CaCO_3\right)\left(130\,000 \; \frac{\text{m}^3}{\text{d}}\right)}{382 \; \frac{\text{mg}}{\text{L}} \text{ as } CaCO_3}$$

$$= 34\,031 \text{ m}^3/\text{d}$$

$$TH_R = TH Q_T$$

$$= \left(382 \; \frac{\text{mg}}{\text{L}} \text{ as } CaCO_3\right)\left(34\,031 \; \frac{\text{m}^3}{\text{d}}\right)$$

$$\times \left((10)^{-6} \; \frac{\text{kg}}{\text{mg}}\right)\left(1000 \; \frac{\text{L}}{\text{m}^3}\right)$$

$$= 13\,000 \text{ kg/d}$$

$$V_R = \frac{TH_R}{C_R} = \frac{13\,000 \; \frac{\text{kg}}{\text{d}}}{95 \; \frac{\text{kg}}{\text{m}^3\cdot\text{resin}}}$$

$$= 137 \text{ m}^3\cdot\text{resin/d}$$

$$n = \frac{V_R}{V_C}$$

$$= \frac{\left(137 \; \frac{\text{m}^3\cdot\text{resin}}{\text{d}}\right)\left(1 \; \frac{\text{d}}{\text{regeneration}}\right)}{4 \; \frac{\text{m}^3\cdot\text{resin}}{\text{vessel}\cdot\text{regeneration}}}$$

$$= 34.2 \text{ vessels} \quad (34 \text{ vessels})$$

(B) This incorrect solution improperly calculated the total hardness removed. Other assumptions, definitions, and equations are the same as used in the correct solution.

$$Q_B = \frac{TH_D Q}{TH}$$

$$= \frac{\left(100 \ \frac{mg}{L} \text{ as } CaCO_3\right)\left(130\,000 \ \frac{m^3}{d}\right)}{382 \ \frac{mg}{L} \text{ as } CaCO_3}$$

$$= 34\,031 \ m^3/d$$

$$Q_T = Q - Q_B = 130\,000 \ \frac{m^3}{d} - 34\,031 \ \frac{m^3}{d}$$

$$= 95\,969 \ m^3/d$$

$$TH_R = (TH - TH_D) \, Q_T$$

$$= \left(382 \ \frac{mg}{L} \text{ as } CaCO_3 - 100 \ \frac{mg}{L} \text{ as } CaCO_3\right)$$

$$\times \left(95\,969 \ \frac{m^3}{d}\right)\left((10)^{-6} \ \frac{kg}{mg}\right)\left(1000 \ \frac{L}{m^3}\right)$$

$$= 27\,063 \ kg/d$$

$$V_R = \frac{TH_R}{C_R} = \frac{27\,063 \ \frac{kg}{d}}{95 \ \frac{kg}{m^3 \cdot resin}}$$

$$= 285 \ m^3 \cdot resin/d$$

$$n = \frac{V_R}{V_C} = \frac{\left(285 \ \frac{m^3 \cdot resin}{d}\right)\left(1 \ \frac{d}{regeneration}\right)}{4 \ \frac{m^3 \cdot resin}{vessel \cdot regeneration}}$$

$$= 71.2 \ vessels \quad (71 \ vessels)$$

(D) This incorrect solution bases the resin use on the total flow instead of correcting for bypass flow to maintain the desired total hardness concentration. Other assumptions, definitions, and equations are the same as used in the correct solution.

$$TH_R = TH Q$$

$$= \left(382 \ \frac{mg}{L} \text{ as } CaCO_3\right)\left(130\,000 \ \frac{m^3}{d}\right)$$

$$\times \left((10)^{-6} \ \frac{kg}{mg}\right)\left(1000 \ \frac{L}{m^3}\right)$$

$$= 49\,660 \ kg/d$$

$$V_R = \frac{TH_R}{C_R} = \frac{49\,660 \ \frac{kg}{d}}{95 \ \frac{kg}{m^3 \cdot resin}}$$

$$= 523 \ m^3 \cdot resin/d$$

$$n = \frac{V_R}{V_C}$$

$$= \frac{\left(523 \ \frac{m^3 \cdot resin}{d}\right)\left(1 \ \frac{d}{regeneration}\right)}{4 \ \frac{m^3 \cdot resin}{vessel \cdot regeneration}}$$

$$= 130.8 \ vessels \quad (130 \ vessels)$$

SOLUTION 90

Q flow rate $\quad$ ft³/sec
t residence time $\quad$ sec
V mixing basin volume $\quad$ ft³

$$V = Qt$$

$$= \left(5.0 \times 10^6 \ \frac{gal}{day}\right)\left(0.134 \ \frac{ft^3}{gal}\right)$$

$$\times (2 \ min)\left(\frac{1 \ day}{1440 \ min}\right)$$

$$= 931 \ ft^3$$

G velocity gradient $\quad$ sec⁻¹
μ dynamic viscosity $\quad$ lbf-sec/ft²
P mixing power $\quad$ ft-lbf/sec

For water at 60°F, the dynamic viscosity is 2.359×10^{-5} lbf sec/ft².

$$P = G^2 V \mu$$

$$= (700 \ sec^{-1})^2 \, (931 \ ft^3)\left(2.359 \times 10^{-5} \ \frac{lbf \cdot sec}{ft^2}\right)$$

$$= 10{,}762 \ ft \cdot lbf/sec$$

E motor efficiency $\quad$ fraction
P_a mixing power corrected
$\quad$ for motor efficiency $\quad$ hp

$$P_a = \frac{P}{E} = \frac{\left(10{,}762 \ \frac{ft \cdot lbf}{sec}\right)(1 \ hp)}{\left(550 \ \frac{ft \cdot lbf}{sec}\right)(0.88)} = 22.2 \ hp$$

The standard motor size just greater than 22.2 hp is 25 hp.

The answer is (D).

Why Other Options Are Wrong

(A) This incorrect solution multiplies instead of divides the mixing power by the motor efficiency and selects the standard motor size closest to the required power instead of sizing up. Other assumptions, definitions and equations are unchanged from the correct solution.

$$V = Qt$$
$$= \left(5.0 \times 10^6 \; \frac{\text{gal}}{\text{day}}\right) \left(0.134 \; \frac{\text{ft}^3}{\text{gal}}\right)$$
$$\times (2 \; \text{min}) \left(\frac{1 \; \text{day}}{1440 \; \text{min}}\right)$$
$$= 931 \; \text{ft}^3$$
$$P = G^2 V \mu$$
$$= \left(700 \; \text{sec}^{-1}\right)^2 \left(931 \; \text{ft}^3\right) \left(2.359 \times 10^{-5} \; \frac{\text{lbf-sec}}{\text{ft}^2}\right)$$
$$= 10{,}762 \; \text{ft-lbf/sec}$$
$$P_a = PE$$
$$= \left(10{,}762 \; \frac{\text{ft-lbf}}{\text{sec}}\right) (0.88) \left(\frac{1 \; \text{hp}}{550 \; \frac{\text{ft-lbf}}{\text{sec}}}\right)$$
$$= 17.2 \; \text{hp}$$

The standard motor size closest to 17.2 hp is 15 hp.

(B) This incorrect solution fails to include the correction for motor efficiency. Other assumptions, definitions and equations are unchanged from the correct solution.

$$V = Qt$$
$$= \left(5.0 \times 10^6 \; \frac{\text{gal}}{\text{day}}\right) \left(0.134 \; \frac{\text{ft}^3}{\text{gal}}\right)$$
$$\times (2 \; \text{min}) \left(\frac{1 \; \text{day}}{1440 \; \text{min}}\right)$$
$$= 931 \; \text{ft}^3$$
$$P = G^2 V \mu$$
$$= \left(700 \; \text{sec}^{-1}\right)^2 \left(931 \; \text{ft}^3\right) \left(2.359 \times 10^{-5} \; \frac{\text{lbf-sec}}{\text{ft}^2}\right)$$
$$\times \left(\frac{1 \; \text{hp}}{550 \; \frac{\text{ft-lbf}}{\text{sec}}}\right)$$
$$= 19.6 \; \text{hp}$$

The standard motor size just greater than 19.6 hp is 20 hp.

(C) This incorrect solution ignores the standard motor size, rounding the required horsepower to the nearest whole number. Other assumptions, definitions and equations are unchanged from the correct solution.

$$V = Qt$$
$$= \left(5.0 \times 10^6 \; \frac{\text{gal}}{\text{day}}\right) \left(0.134 \; \frac{\text{ft}^3}{\text{gal}}\right)$$
$$\times (2 \; \text{min}) \left(\frac{1 \; \text{day}}{1440 \; \text{min}}\right)$$
$$= 931 \; \text{ft}^3$$

$$P = G^2 V \mu$$
$$= \left(700 \; \text{sec}^{-1}\right)^2 \left(931 \; \text{ft}^3\right) \left(2.359 \times 10^{-5} \; \frac{\text{lbf-sec}}{\text{ft}^2}\right)$$
$$= 10{,}762 \; \text{ft-lbf/sec}$$
$$P_a = \frac{P}{E} = \frac{\left(10{,}762 \; \frac{\text{ft-lbf}}{\text{sec}}\right)}{(0.88)} \left(\frac{1 \; \text{hp}}{550 \; \frac{\text{ft-lbf}}{\text{sec}}}\right)$$
$$= 22.2 \; \text{hp} \quad (22 \; \text{hp})$$

SOLUTION 91

G	average velocity gradient	s^{-1}
G_t	average time-velocity gradient	–
t	hydraulic residence time	s

$$t = \frac{G_t}{G} = \frac{4.5 \times 10^4}{\dfrac{40}{\text{s}}} = 1125 \; \text{s}$$

Q	flow rate	m^3/d
V	mixing basin volume	m^3

$$V = Qt = \left(18\,000 \; \frac{\text{m}^3}{\text{d}}\right) (1125 \; \text{s}) \left(\frac{1 \; \text{d}}{86\,400 \; \text{s}}\right)$$
$$= 234 \; \text{m}^3$$

Assume for horizontally mounted paddles that the most efficient section for the paddles to rotate within will be a square. The length of each section will equal the depth.

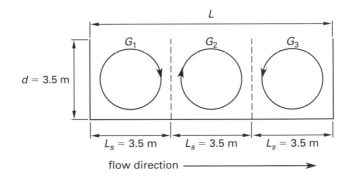

L	basin total length	m
L_s	section length	m

$$L = \sum L_s = 3.5 \; \text{m} + 3.5 \; \text{m} + 3.5 \; \text{m} = 10.5 \; \text{m}$$

d	basin depth	m
w	basin width	m

$$w = \frac{V}{Ld} = \frac{234 \; \text{m}^3}{(10.5 \; \text{m})(3.5 \; \text{m})}$$
$$= 6.4 \; \text{m} \quad (6.5 \; \text{m})$$

The answer is (C).

Why Other Options Are Wrong

(A) This incorrect solution chooses a basin length to width ratio of 4 to 1 as typical. Other assumptions, definitions, and equations are unchanged from the correct solution.

$$t = \frac{G_t}{G} = \frac{4.5 \times 10^4}{\dfrac{40}{s}} = 1125 \text{ s}$$

$$V = Qt = \left(18\,000 \ \frac{m^3}{d}\right)(1125 \text{ s})\left(\frac{1 \text{ d}}{86\,400 \text{ s}}\right)$$
$$= 234 \text{ m}^3$$

Assume for flocculation that a length to width ratio of 4 to 1 is typical.

$$L = 4w$$
$$V = Lwd = (4w)\,wd = 4w^2 d$$
$$w = \sqrt{\frac{V}{4d}} = \sqrt{\frac{234 \text{ m}^3}{(4)(3.5 \text{ m})}}$$
$$= 4.1 \text{ m} \quad (4.0 \text{ m})$$
$$L = (4)(4.0 \text{ m}) = 16 \text{ m}$$

(B) This incorrect solution divides the total volume by three and then sets the length per section and the width equal to each other. Other assumptions, definitions, and equations are unchanged from the correct solution.

$$t = \frac{G_t}{G} = \frac{4.5 \times 10^4}{\dfrac{40}{s}} = 1125 \text{ s}$$

$$V = Qt = \left(18\,000 \ \frac{m^3}{d}\right)(1125 \text{ s})\left(\frac{1 \text{ d}}{86\,400 \text{ s}}\right)$$
$$= 234 \text{ m}^3$$

V_s volume per section   m^3

$$V_s = \frac{234 \text{ m}^3}{3} = 78 \text{ m}^3$$

Assume $L_s = w$.

$$V_s = L_s wd = w^2 d$$
$$w = L_s = \sqrt{\frac{V_s}{d}} = \sqrt{\frac{78 \text{ m}^3}{(3.5 \text{ m})}}$$
$$= 4.7 \text{ m} \quad (5.0 \text{ m})$$
$$L = \sum L_s$$
$$= 5.0 \text{ m} + 5.0 \text{ m} + 5.0 \text{ m}$$
$$= 15 \text{ m}$$

(D) This incorrect solution assumes a hydraulic residence time of 30 min as typical for flocculation. The figure and other assumptions, definitions, and equations are unchanged from the correct solution.

Assume that the typical hydraulic residence time for flocculation is 30 min.

$$V = Qt = \left(18\,000 \ \frac{m^3}{d}\right)(30 \text{ min})\left(\frac{1 \text{ d}}{1440 \text{ min}}\right)$$
$$= 375 \text{ m}^3$$

Assume for horizontally mounted paddles that the most efficient section for the paddles to rotate within will be a square. The length of each section will equal the depth.

$$L = \sum L_s = 3.5 \text{ m} + 3.5 \text{ m} + 3.5 \text{ m}$$
$$= 10.5 \text{ m}$$
$$w = \frac{V}{Ld} = \frac{375 \text{ m}^3}{(10.5 \text{ m})(3.5 \text{ m})}$$
$$= 10.2 \text{ m} \quad (10 \text{ m})$$

SOLUTION 92

d	basin depth	m
L	basin length per section	m
V	mixing basin volume per section	m^3
w	basin width	m

$$V = Lwd = (4.0 \text{ m})(4.0 \text{ m})(8 \text{ m}) = 128 \text{ m}^3$$

D	paddle wheel diameter	m
l_c	paddle-side wall clearance	m

Assume a typically acceptable paddle-wall clearance of 0.3 m.

$$D = d - 2l_c = 4.0 \text{ m} - (2)(0.3 \text{ m}) = 3.4 \text{ m}$$

f_s	paddle slip factor	–
v_p	paddle tip velocity	m/s
ω	paddle rotational speed	rev/min

Assume a typical paddle slip factor of 0.75. Note that πD has units of m/rev.

$$v_p = \omega \pi D f_s$$
$$= \left(3 \ \frac{rev}{min}\right)\pi\left(3.4 \ \frac{m}{rev}\right)(0.75)\left(\frac{1 \text{ min}}{60 \text{ s}}\right)$$
$$= 0.40 \text{ m/s}$$

A_p	paddle surface area facing rotation	m^2
C_D	drag coefficient	–
G	velocity gradient	sec^{-1}
μ	dynamic viscosity	0.001 002 kg/m·s
ρ_w	water density	1000 kg/m^3

Assume that the drag coefficient for flat paddles is 1.8.

$$A_p = \frac{2G^2 V \mu}{C_D \rho_w \mathrm{v}_p^3}$$

$$= \frac{\left(45 \,\frac{1}{s}\right)^2 (2)(128 \text{ m}^3)\left(0.001\,002 \,\frac{\text{kg}}{\text{m}\cdot\text{s}}\right)}{(1.8)\left(1000 \,\frac{\text{kg}}{\text{m}^3}\right)\left(0.40 \,\frac{\text{m}}{\text{s}}\right)^3}$$

$$= 4.5 \text{ m}^2$$

The answer is (D).

Why Other Options Are Wrong

(A) This incorrect solution fails to cube the velocity term in the paddle area equation. Other assumptions, definitions, and equations are unchanged from the correct solution.

$$V = Lwd = (4.0 \text{ m})\,(4.0 \text{ m})\,(8 \text{ m})$$
$$= 128 \text{ m}^3$$

Assume an acceptable paddle-wall clearance of 0.3 m.

$$D = d - 2l_c = 4.0 \text{ m} - (2)\,(0.3 \text{ m})$$
$$= 3.4 \text{ m}$$

Assume a typical paddle slip factor of 0.75. Note that πD has units of m/rev.

$$\mathrm{v}_p = \omega \pi D f_s$$

$$= \left(3 \,\frac{\text{rev}}{\text{min}}\right) \pi \left(3.4 \,\frac{\text{m}}{\text{rev}}\right)(0.75)\left(\frac{1 \text{ min}}{60 \text{ s}}\right)$$

$$= 0.40 \text{ m/s}$$

Assume that the drag coefficient for flat paddles is 1.8.

$$A_p = \frac{2G^2 V \mu}{C_D \rho_w \mathrm{v}_p}$$

$$= \frac{\left(45 \,\frac{1}{s}\right)^2 (2)(128 \text{ m}^3)\left(0.001002 \,\frac{\text{kg}}{\text{m}\cdot\text{s}}\right)}{(1.8)\left(1000 \,\frac{\text{kg}}{\text{m}^3}\right)\left(0.40 \,\frac{\text{m}}{\text{s}}\right)}$$

$$= 0.72 \text{ m}^4/\text{s}^2$$

The units do not work.

(B) This incorrect solution fails to include the slip factor when calculating the paddle tip velocity. Other assumptions, definitions, and equations are unchanged from the correct solution.

$$V = Lwd = (4.0 \text{ m})\,(4.0 \text{ m})\,(8 \text{ m})$$
$$= 128 \text{ m}^3$$

Assume an acceptable paddle-wall clearance of 0.3 m.

$$D = d - 2l_c$$
$$= 4.0 \text{ m} - (2)\,(0.3 \text{ m})$$
$$= 3.4 \text{ m}$$

Assume a typical paddle slip factor of 0.75. Note that πD has units of m/rev.

$$\mathrm{v}_p = \omega \pi D$$

$$= \left(3 \,\frac{\text{rev}}{\text{min}}\right) \pi \left(3.4 \,\frac{\text{m}}{\text{rev}}\right)\left(\frac{1 \text{ min}}{60 \text{ s}}\right)$$

$$= 0.53 \text{ m/s}$$

Assume that the drag coefficient for flat paddles is 1.8.

$$A_p = \frac{2G^2 V \mu}{C_D \rho_w \mathrm{v}_p^3}$$

$$= \frac{\left(45 \,\frac{1}{s}\right)^2 (2)(128 \text{ m}^3)\left(0.001\,002 \,\frac{\text{kg}}{\text{m}\cdot\text{s}}\right)}{(1.8)\left(1000 \,\frac{\text{kg}}{\text{m}^3}\right)\left(0.53 \,\frac{\text{m}}{\text{s}}\right)^3}$$

$$= 1.9 \text{ m}^2$$

(C) This incorrect solution does not include the wall clearance when calculating the paddle diameter. Other assumptions, definitions, and equations are unchanged from the correct solution.

$$V = Lwd$$
$$= (4.0 \text{ m})\,(4.0 \text{ m})\,(8 \text{ m})$$
$$= 128 \text{ m}^3$$

Assume the paddle diameter to be equal to the basin depth.

$$D = 4.0 \text{ m}$$

Assume a typical paddle slip factor of 0.75. Note that πD has units of m/rev.

$$\mathrm{v}_p = \omega \pi D f_s$$

$$= \left(3 \,\frac{\text{rev}}{\text{min}}\right) \pi \left(4.0 \,\frac{\text{m}}{\text{rev}}\right)(0.75)\left(\frac{1 \text{ min}}{60 \text{ s}}\right)$$

$$= 0.47 \text{ m/s}$$

Assume that the drag coefficient for flat paddles is 1.8.

$$A_p = \frac{2G^2 V \mu}{C_D \rho_w \mathrm{v}_p^3}$$

$$= \frac{\left(45 \,\frac{1}{s}\right)^2 (2)\,(128 \text{ m}^3)\left(0.001\,002 \,\frac{\text{kg}}{\text{m}\cdot\text{s}}\right)}{(1.8)\left(1000 \,\frac{\text{kg}}{\text{m}^3}\right)\left(0.47 \,\frac{\text{m}}{\text{s}}\right)^3}$$

$$= 2.8 \text{ m}^2$$

SOLUTION 93

The overflow rate is equal to the rise velocity of the water in the sedimentation basin. Removal will result for those particles with a settling velocity that is greater than the overflow rate.

A_s settling zone surface area ft^2
q_o overflow rate ft^3/ft^2-hr
Q flow rate   gal/day

$$q_o = \frac{Q}{A_s} = \frac{\left(2.7 \times 10^6 \, \frac{gal}{day}\right)\left(0.134 \, \frac{ft^3}{gal}\right)}{(5700 \, ft^2)\left(24 \, \frac{hr}{day}\right)}$$

$$= 2.6 \, ft^3/ft^2\text{-}hr$$

$E\%$ removal efficiency   %
v_s particle settling velocity   in/sec

For calculating removal efficiency, the overflow rate and particle-settling velocity must have the same units.

$$E\% = \frac{v_s}{q_o} \times 100\%$$

$$= \frac{\left(0.008 \, \frac{in}{sec}\right) \times 100\% \left(3600 \, \frac{sec}{hr}\right)}{\left(2.6 \, \frac{ft^3}{ft^2\text{-}hr}\right)\left(12 \, \frac{in}{ft}\right)}$$

$$= 92\%$$

The answer is (D).

Why Other Options Are Wrong

(A) This incorrect solution makes an error when converting from hours to seconds in the efficiency equation. Other assumptions, definitions, and equations are unchanged from the correct solution.

$$q_o = \frac{Q}{A_s} = \frac{\left(2.7 \times 10^6 \, \frac{gal}{day}\right)\left(0.134 \, \frac{ft^3}{gal}\right)}{(5700 \, ft^2)\left(24 \, \frac{hr}{day}\right)}$$

$$= 2.6 \, ft^3/ft^2\text{-}hr$$

$$E\% = \frac{v_s}{q_o} \times 100\% = \frac{\left(0.008 \, \frac{in}{sec}\right) \times 100\% \left(60 \, \frac{sec}{hr}\right)}{\left(2.6 \, \frac{ft^3}{ft^2\text{-}hr}\right)\left(12 \, \frac{in}{ft}\right)}$$

$$= 1.5\%$$

(B) This incorrect solution divides the surface area by the flow rate to calculate efficiency. Other assumptions, definitions, and equations are unchanged from the correct solution.

$$E\% = \frac{A_s}{Q} \times 100\%$$

$$= \frac{(5700 \, ft^2) \times 100\% \left(7.46 \, \frac{gal}{ft^3}\right)}{\left(2.7 \times 10^6 \, \frac{gal}{day}\right)\left(\frac{1 \, day}{24 \, hr}\right)}$$

$$= 38\% \, hr/ft$$

The units do not work.

(C) This incorrect solution uses the wrong conversion from ft^3 to gal in the overflow rate calculation. Other assumptions, definitions, and equations are unchanged from the correct solution.

$$q_o = \frac{Q}{A_s} = \frac{\left(2.7 \times 10^6 \, \frac{gal}{day}\right)\left(0.264 \, \frac{ft^3}{gal}\right)}{(5700 \, ft^2)\left(24 \, \frac{hr}{day}\right)}$$

$$= 5.2 \, ft^3/ft^2\text{-}hr$$

$$E\% = \frac{v_s}{q_o} \times 100\%$$

$$= \frac{\left(0.008 \, \frac{in}{sec}\right) \times 100\% \left(3600 \, \frac{sec}{hr}\right)}{\left(5.2 \, \frac{ft^3}{ft^2\text{-}hr}\right)\left(12 \, \frac{in}{ft}\right)}$$

$$= 46\%$$

SOLUTION 94

f_P permeate recovery fraction   –
Q required feed water flow rate m^3/d
Q_o desired fresh water flow rate m^3/d

$$Q = \frac{Q_o}{f_p} = \frac{30\,000 \, \frac{m^3}{d}}{0.77} = 38\,961 \, m^3/d$$

A_m required membrane area m^2
G membrane flux rate $m^3/m^2 \cdot d$

$$A_m = \frac{Q}{G} = \frac{38\,961 \, \frac{m^3}{d}}{0.93 \, \frac{m^3}{m^2 \cdot d}} = 41\,894 \, m^2$$

P_m membrane packing density m^2/m^3
V_m membrane volume m^3

$$V_m = \frac{A_m}{P_m} = \frac{41\,894 \, m^2}{800 \, \frac{m^2}{m^3}} = 52.4 \, m^3$$

n_M number of membrane modules
V_M membrane module volume m^3

$$n_M = \frac{V_m}{V_M} = \frac{52.4 \text{ m}^3}{0.028 \dfrac{\text{m}^3}{\text{module}}} = 1871 \text{ module}$$

n_R total number of pressure vessels
n_P number of modules per pressure vessel

$$n_R = \frac{n_M}{n_P} = \frac{1871 \text{ module}}{12 \dfrac{\text{module}}{\text{vessel}}}$$

$$= 156 \text{ vessels} \quad (160 \text{ vessels})$$

The answer is (C).

Why Other Options Are Wrong

(A) This incorrect solution multiplies by instead of divides by the permeate recovery fraction when calculating the required feed water flow rate. Other assumptions, definitions, and equations are unchanged from the correct solution.

$$Q = Q_o f_p = \left(30\,000 \; \frac{\text{m}^3}{\text{d}}\right)(0.77) = 23\,100 \text{ m}^3/\text{d}$$

$$A_m = \frac{Q}{G} = \frac{23\,100 \dfrac{\text{m}^3}{\text{d}}}{0.93 \dfrac{\text{m}^3}{\text{m}^2 \cdot \text{d}}} = 24\,839 \text{ m}^2$$

$$V_m = \frac{A_m}{P_m} = \frac{24\,839 \text{ m}^2}{800 \dfrac{\text{m}^2}{\text{m}^3}} = 31 \text{ m}^3$$

$$n_M = \frac{V_m}{V_M} = \frac{31 \text{ m}^3}{0.028 \dfrac{\text{m}^3}{\text{module}}} = 1107 \text{ module}$$

$$n_R = \frac{n_M}{n_P} = \frac{1107 \text{ module}}{12 \dfrac{\text{module}}{\text{vessel}}} = 92 \text{ vessels}$$

(B) This incorrect solution fails to correct the desired fresh water flow rate for permeate recovery. Other assumptions, definitions, and equations are unchanged from the correct solution.

$$A_m = \frac{Q}{G} = \frac{30\,000 \dfrac{\text{m}^3}{\text{d}}}{0.93 \dfrac{\text{m}^3}{\text{m}^2 \cdot \text{d}}} = 32\,258 \text{ m}^2$$

$$V_m = \frac{A_m}{P_m} = \frac{32\,258 \text{ m}^2}{800 \dfrac{\text{m}^2}{\text{m}^3}} = 40.3 \text{ m}^3$$

$$n_M = \frac{V_m}{V_M} = \frac{40.3 \text{ m}^3}{0.028 \dfrac{\text{m}^3}{\text{module}}} = 1439 \text{ module}$$

$$n_R = \frac{n_M}{n_P} = \frac{1439 \text{ module}}{12 \dfrac{\text{module}}{\text{vessel}}} = 120 \text{ vessels}$$

(D) This incorrect solution calculates the required number of modules instead of the required number of pressure vessels. Other assumptions, definitions, and equations are unchanged from the correct solution.

$$Q = \frac{Q_o}{f_p} = \frac{30\,000 \dfrac{\text{m}^3}{\text{d}}}{0.77} = 38\,961 \text{ m}^3/\text{d}$$

$$A_m = \frac{Q}{G} = \frac{38\,961 \dfrac{\text{m}^3}{\text{d}}}{0.93 \dfrac{\text{m}^3}{\text{m}^2 \cdot \text{d}}} = 41\,894 \text{ m}^2$$

$$V_m = \frac{41\,894 \text{ m}^2}{800 \dfrac{\text{m}^2}{\text{m}^3}} = 52.4 \text{ m}^3$$

$$n_M = \frac{V_m}{V_M} = \frac{52.4 \text{ m}^3}{0.028 \dfrac{\text{m}^3}{\text{module}}}$$

$$= 1871 \text{ module} \quad (1900 \text{ module})$$

SOLUTION 95

Assume that the Carmen-Kozeny equations apply.

d_g geometric mean particle diameter
          between sieve sizes   m
d_{12} mesh opening for a #12 sieve   m
d_{16} mesh opening for a #16 sieve   m

For a 12×16 mesh, assume that 100% of the media passes a #12 sieve and is retained on a #16 sieve. The mesh opening for a #12 sieve is 0.0017 m and for a #16 sieve is 0.001 18 m.

$$d_g = \sqrt{d_{12}d_{16}} = \sqrt{(0.0017 \text{ m})(0.001\,18 \text{ m})}$$
$$= 0.0014 \text{ m}$$

ϕ particle shape factor   –
μ water dynamic viscosity kg/m·s
Re   Reynolds number   –
ρ_w water density, kg/m^3
v_s clean filter filtering velocity $\text{m}^3/\text{m}^2 \cdot \text{s}$

Assume that the particle shape factor for average sand is 0.75, the water density is 998.23 kg/m^3, and the water dynamic viscosity is 0.001 002 kg/m·s.

$$\text{Re} = \frac{\phi \rho_w v_s d_g}{\mu}$$

$$= \frac{(0.75)\left(998.23 \dfrac{\text{kg}}{\text{m}^3}\right)\left(0.003 \dfrac{\text{m}^3}{\text{m}^2 \cdot \text{s}}\right)(0.0014 \text{ m})}{0.001\,002 \dfrac{\text{kg}}{\text{m·s}}}$$

$$= 3.14$$

α media porosity   –
f friction factor   –

Assume that a typical porosity for granular media is 0.40.

$$f = \frac{(150)(1-\alpha)}{\text{Re}} + 1.75$$
$$= \frac{(150)(1-0.40)}{3.14} + 1.75$$
$$= 30.4$$

g gravitational acceleration   9.81 m/s^2
h head loss   m
L filter bed depth   m

$$h = \frac{f(1-\alpha)Lv_s^2}{\phi\alpha^3 d_g g}$$

$$= \frac{\left(\begin{array}{c}(30.4)(1-0.40)(0.75\ \text{m}) \\ \times\left(0.003\ \dfrac{\text{m}^3}{\text{m}^2\cdot\text{s}}\right)^2\left(100\ \dfrac{\text{cm}}{\text{m}}\right)\end{array}\right)}{(0.75)(0.40)^3(0.0014\ \text{m})\left(9.81\ \dfrac{\text{m}}{\text{s}^2}\right)}$$

$$= 19\ \text{cm}$$

The answer is (C).

Why Other Options Are Wrong

(A) This incorrect solution makes an error in the friction factor calculation and fails to cube the porosity term in the head loss equation. Other assumptions, definitions, and equations are unchanged from the correct solution.

For a 12 × 16 mesh, assume that 100% of the media passes a #12 sieve and is retained on a #16 sieve. The mesh opening for a #12 sieve is 0.0017 m and for a #16 sieve is 0.001 18 m.

$$d_g = \sqrt{d_{12}d_{16}} = \sqrt{(0.0017\ \text{m})(0.001\,18\ \text{m})}$$
$$= 0.0014\ \text{m}$$

Assume that the particle shape factor for average sand is 0.75, the water density is 998.23 kg/m^3, and the water dynamic viscosity is 0.001 002 kg/m s.

$$\text{Re} = \frac{\phi\rho_w v_s d_g}{\mu}$$
$$= \frac{(0.75)\left(998.23\ \dfrac{\text{kg}}{\text{m}^3}\right)\left(0.003\ \dfrac{\text{m}^3}{\text{m}^2\cdot\text{s}}\right)(0.0014\ \text{m})}{0.001\,002\ \dfrac{\text{kg}}{\text{m}\cdot\text{s}}}$$
$$= 3.14$$

Assume that a typical porosity for granular media is 0.40.

$$f = \frac{(150)(1-\alpha)+1.75}{\text{Re}}$$
$$= \frac{(150)(1-0.40)+1.75}{3.14}$$
$$= 29.2$$

$$h = \frac{f(1-\alpha)Lv_s^2}{\phi\alpha d_g g}$$

$$= \frac{\left(\begin{array}{c}(29.2)(1-0.40)(0.75\ \text{m}) \\ \times\left(0.003\ \dfrac{\text{m}^3}{\text{m}^2\cdot\text{s}}\right)^2\left(100\ \dfrac{\text{cm}}{\text{m}}\right)\end{array}\right)}{(0.75)(0.40)(0.0014\ \text{m})\left(9.81\ \dfrac{\text{m}}{\text{s}^2}\right)}$$

$$= 2.9\ \text{cm}$$

(B) This incorrect solution squares instead of cubes the porosity term in the head loss equation. Other assumptions, definitions, and equations are unchanged from the correct solution.

For a 12 × 16 mesh, assume that 100% of the media passes a #12 sieve and is retained on a #16 sieve. The mesh opening for a #12 sieve is 0.0017 m and for a #16 sieve is 0.001 18 m.

$$d_g = \sqrt{d_{12}d_{16}} = \sqrt{(0.0017\ \text{m})(0.001\,18\ \text{m})}$$
$$= 0.0014\ \text{m}$$

Assume that the particle shape factor for average sand is 0.75, the water density is 998.23 kg/m^3, and the water dynamic viscosity is 0.001 002 kg/m·s.

$$\text{Re} = \frac{\phi\rho_w v_s d_g}{\mu}$$
$$= \frac{(0.75)\left(998.23\ \dfrac{\text{kg}}{\text{m}^3}\right)\left(0.003\ \dfrac{\text{m}^3}{\text{m}^2\cdot\text{s}}\right)(0.0014\ \text{m})}{0.001\,002\ \dfrac{\text{kg}}{\text{m}\cdot\text{s}}}$$
$$= 3.14$$

$$f = \frac{(150)(1-\alpha)}{\text{Re}} + 1.75$$
$$= \frac{(150)(1-0.40)}{3.14} + 1.75$$
$$= 30.4$$

$$h = \frac{f(1-\alpha)Lv_s^2}{\phi\alpha^2 d_g g}$$

$$= \frac{\left(\begin{array}{c}(30.4)(1-0.40)(0.75\ \text{m}) \\ \times\left(0.003\ \dfrac{\text{m}^3}{\text{m}^2\cdot\text{s}}\right)^2\left(100\ \dfrac{\text{cm}}{\text{m}}\right)\end{array}\right)}{(0.75)(0.40)^2(0.0014\ \text{m})\left(9.81\ \dfrac{\text{m}}{\text{s}^2}\right)}$$

$$= 7.5\ \text{cm}$$

(D) This incorrect solution uses the #12 mesh opening as the particle diameter instead of using the geometric average between the two adjacent sieves and makes an error in the friction factor calculation. Other assumptions, definitions, and equations are unchanged from the correct solution.

For a 12×16 mesh, assume that the particle diameter is equal to the meh opening for a #12 sieve and is 0.0017 m.

d particle diameter   m

Assume that the particle shape factor for average sand is 0.75, the water density is 998.23 kg/m^3, and the water dynamic viscosity is 0.001 002 kg/m·s.

$$\text{Re} = \frac{\phi \rho_w v_s d_g}{\mu}$$

$$= \frac{(0.75)\left(998.23 \; \frac{\text{kg}}{\text{m}^3}\right)\left(0.003 \; \frac{\text{m}^3}{\text{m}^2 \cdot \text{s}}\right)(0.0017 \; \text{m})}{0.001 \, 002 \; \frac{\text{kg}}{\text{m} \cdot \text{s}}}$$

$$= 3.81$$

Assume that a typical porosity for granular media is 0.40.

$$f = \frac{(150)(1-\alpha) 1.75}{\text{Re}}$$

$$= \frac{(150)(1-0.40)(1.75)}{3.81}$$

$$= 41.3$$

$$h = \frac{f(1-\alpha) L v_s^2}{\phi \alpha^3 d_g g}$$

$$= \frac{\left(\begin{array}{c}(41.3)(1-0.40)(0.75 \; \text{m}) \\ \times \left(0.003 \; \frac{\text{m}^3}{\text{m}^2 \cdot \text{s}}\right)^2 \left(100 \; \frac{\text{cm}}{\text{m}}\right)\end{array}\right)}{(0.75)(0.40)^3(0.0014 \; \text{m})\left(9.81 \; \frac{\text{m}}{\text{s}^2}\right)}$$

$$= 25 \; \text{cm}$$

SOLUTION 96

For trial 1, use an overflow rate of 0.02 m/min and integrate the area to the left of the curve bounded by the overflow rate and the corresponding mass fraction remaining. Each grid element of the integrated area represents 0.0001 m/min.

v settling velocity   m/min
x fraction remaining   –
$v_i x_i$ integrated area   m/min

number of grid elements from illustration $= 54$

$$v_i x_i = \left(0.0001 \; \frac{\text{m}}{\text{min}}\right)(54)$$

$$= 0.0054 \; \text{m/min}$$

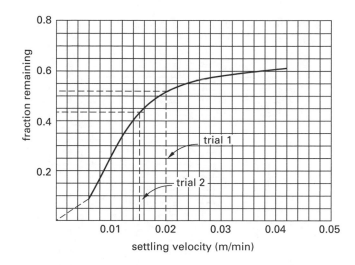

E fractional efficiency   –
v_c overflow rate   m/min
x_c mass fraction remaining corresponding
        to the overflow rate   –

$$E = (1 - x_c) + \frac{v_i x_i}{v_c}$$

$$= (1 - 0.52) + \frac{0.0054 \; \frac{\text{m}}{\text{min}}}{0.02 \; \frac{\text{m}}{\text{min}}}$$

$$= 0.75$$

$$0.75 < 0.80$$

Trial 2 is required.

For trial 2,

$$v_c = 0.015 \; \text{m/min}$$

number of grid elements from illustration $= 37.5$

$$v_i x_i = \left(0.0001 \; \frac{\text{m}}{\text{min}}\right)(37.5)$$

$$= 0.003 \, 75 \; \text{m/min}$$

$$E = (1 - x_c) + \frac{v_i x_i}{v_c}$$

$$= (1 - 0.42) + \frac{0.003 \, 75 \; \frac{\text{m}}{\text{min}}}{0.015 \; \frac{\text{m}}{\text{min}}}$$

$$= 0.83$$

Close enough. Trial 3 is not needed.

The answer is (C).

Why Other Options Are Wrong

(A) This incorrect solution uses the curve to find an overflow rate corresponding to a mass fraction remaining of 0.2. Other assumptions, definitions, and equations are unchanged from the correct solution.

Assume the fraction remaining that corresponds to an efficiency of 80% is 0.20.

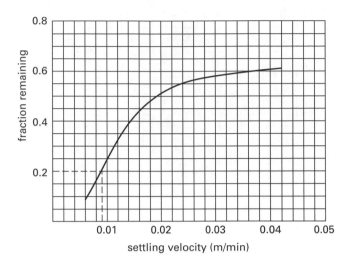

From the illustration, $v_c = 0.0085$ m/min.

(B) This incorrect solution integrates the area under the curve bounded by the overflow rate. Other assumptions, definitions, and equations are unchanged from the correct solution.

For trial 1,

$$v_c = 0.02 \text{ m/min}$$

Integrate the area under the curve bounded by the overflow rate.

number of grid elements from illustration = 48.5

$$v_i x_i = \left(0.0001 \ \frac{\text{m}}{\text{min}}\right)(48.5) = 0.004\,85 \text{ m/min}$$

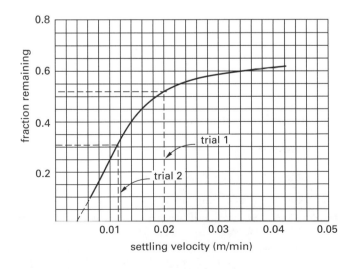

$$E = (1 - x_c) + \frac{v_i x_i}{v_c}$$

$$= (1 - 0.52) + \frac{0.00485 \ \frac{\text{m}}{\text{min}}}{0.02 \ \frac{\text{m}}{\text{min}}}$$

$$= 0.72$$
$$0.72 < 0.80$$

Trial 2 is required. For trial 2,

$$v_c = 0.011 \text{ m/min}$$

number of grid elements from illustration = 10.5

$$v_i x_i = \left(0.0001 \ \frac{\text{m}}{\text{min}}\right)(10.5)$$
$$= 0.001\,05 \text{ m/min}$$
$$E = (1 - x_c) + \frac{v_i x_i}{v_c}$$

$$= (1 - 0.3) + \frac{0.001\,05 \ \frac{\text{m}}{\text{min}}}{0.011 \ \frac{\text{m}}{\text{min}}}$$

$$= 0.80 \quad [\text{OK}]$$

(D) This incorrect solution makes an error in the fractional efficiency equation. The illustration and other assumptions, definitions, and equations are unchanged from the correct solution.

For trial 1,
$$v_c = 0.02 \text{ m/min}$$

number of grid elements from illustration = 54

$$v_i x_i = (54)\left(0.0001 \ \frac{\text{m}}{\text{min}}\right) = 0.0054 \text{ m/min}$$
$$E = -x_c + \frac{v_i x_i}{v_c}$$

$$= 0.52 + \frac{0.0054 \ \frac{\text{m}}{\text{min}}}{0.02 \ \frac{\text{m}}{\text{min}}}$$

$$= 0.79$$

Close enough. Trial 2 is not needed.

SOLUTION 97

ion	concentration (mg/L)	molecular weight (mg/mmol)	molarity (mmol/L)
Ca^{2+}	187	40	4.7
Mg^{2+}	49	24	2.0
HCO_3^-	618	61	10.1

Assume that the maximum hardness that can be precipitated is 40 mg/L as $CaCO_3$. The molecular weight of $CaCO_3$ is 100 mg/mmol.

CH HCO_3^- initial   mmol/L
CH_I HCO_3^- removed   mmol/L
CH_R HCO_3^- remaining   mmol/L
MgTH Mg^{2+} hardness initial   mmol/L
$MgTH_I$ Mg^{2+} hardness to be removed   mmol/L
$MgTH_O$ Mg^{2+} hardness precipitated   mmol/L
$MgTH_R$ Mg^{2+} hardness remaining   mmol/L
$NaOH_1$ NaOH added for Ca^{2+} as
       carbonate hardness removal   mmol/L
$NaOH_2$ NaOH added for Mg^{2+} as
       carbonate hardness removal   mmol/L
$NaOH_3$ NaOH added for Mg^{2+} as
       non-carbonate removal   mmol/L
TH_D desired total hardness   mmol/L
TH_F final total hardness   mmol/L
TH_R residual total hardness   mmol/L

$$TH_D = \left(100 \ \frac{mg}{L} \ \text{as } CaCO_3\right)\left(\frac{1 \ mmol}{100 \ mg} \ \text{as } CaCO_3\right)$$
$$= 1.0 \ mmol/L$$

$$TH_R = \left(40 \ \frac{mg}{L} \ \text{as } CaCO_3\right)\left(\frac{1 \ mmol}{100 \ mg} \ \text{as } CaCO_3\right)$$
$$= 0.4 \ mmol/L$$

$$TH_F = TH_D - TH_R = 1.0 \ \frac{mmol}{L} - 0.4 \ \frac{mmol}{L}$$
$$= 0.6 \ mmol/L$$

$$MgTH_I = MgTH = TH_F = 2.0 \ \frac{mmol}{L} - 0.6 \ \frac{mmol}{L}$$
$$= 1.4 \ mmol/L$$

$$(1)\left(\overset{Ca^{2+}}{4.7 \ \frac{mmol}{L}}\right) + (2)\left(\overset{2HCO_3^-}{4.7 \ \frac{mmol}{L}}\right) + (2)\left(\overset{2NaOH}{4.7 \ \frac{mmol}{L}}\right)$$
$$\rightarrow CaCO_3\downarrow + 2Na^+ + CO_3^- + H_2O$$

$$CH_R = CH - CH_I = 10.1 \ \frac{mmol}{L} - (2)\left(4.7 \ \frac{mmol}{L}\right)$$
$$= 0.7 \ mmol/L$$

$$NaOH_1 = (2)\left(4.7 \ \frac{mmol}{L}\right) = 9.4 \ mmol/L$$

$$\left(\frac{1}{2}\right)\left(\overset{Mg^{2+}}{0.7 \ \frac{mmol}{L}}\right) + \left(\frac{2}{2}\right)\left(\overset{2HCO_3^-}{0.7 \ \frac{mmol}{L}}\right) + \left(\frac{4}{2}\right)\left(\overset{4NaOH}{0.7 \ \frac{mmol}{L}}\right)$$
$$\rightarrow Mg(OH)_2\downarrow + 4Na^+ + 2CO_3^- + 2H_2O$$

$$MgTH_R = MgTH_I - MgTH_O$$
$$= 1.4 \ \frac{mmol}{L} - \left(\frac{1}{2}\right)\left(0.7 \ \frac{mmol}{L}\right)$$
$$= 1.05 \ mmol/L$$

$$NaOH_2 = \left(\frac{4}{2}\right)\left(0.7 \ \frac{mmol}{L}\right) = 1.4 \ mmol/L$$

$$(1)\left(\overset{Mg^{2+}}{1.05 \ \frac{mmol}{L}}\right) + (2)\left(\overset{2NaOH}{1.05 \ \frac{mmol}{L}}\right)$$
$$\rightarrow Mg(OH)_2\downarrow + 2Na^+$$

$$NaOH_3 = (2)\left(1.05 \ \frac{mmol}{L}\right) = 2.1 \ mmol/L$$

f NaOH fractional purity
MW   molecular weight of NaOH   mg/mmol
NaOH   NaOH dose   tonne/mo
Q flow rate m^3/d

$$NaOH = (NaOH_1 + NaOH_2 + NaOH_3) \ MW \frac{Q}{f}$$

$$= \left(9.4 \ \frac{mmol}{L} + 1.4 \ \frac{mmol}{L} + 2.1 \ \frac{mmol}{L}\right)$$
$$\times \left(40 \ \frac{mg}{mmol}\right) \times \frac{100\%}{83\%}\left(30\,000 \ \frac{m^3}{d}\right)$$
$$\times \left(30 \ \frac{d}{mo}\right)\left((10)^{-6} \ \frac{tonne \cdot L}{m^3 \cdot mg}\right)$$
$$= 560 \ tonne/mo$$

The answer is (B).

Why Other Options Are Wrong

(A) This incorrect solution does not correct for reagent purity. Other assumptions, definitions, and equations are the same as used in the correct solution.

ion	concentration (mg/L)	molecular weight (mg/mmol)	molarity (mmol/L)
Ca^{2+}	187	40	4.7
Mg^{2+}	49	24	2.0
HCO_3^-	618	61	10.1

Assume that the maximum hardness that can be precipitated is 40 mg/L as $CaCO_3$. The molecular weight of $CaCO_3$ is 100 mg/mmol.

$$TH_D = \left(100 \ \frac{mg}{L} \ \text{as } CaCO_3\right)$$
$$\times \left(\frac{1 \ mmol}{100 \ mg} \ \text{as } CaCO_3\right)$$
$$= 1.0 \ mmol/L$$

$$TH_R = \left(40 \ \frac{mg}{L} \ \text{as } CaCO_3\right)$$
$$\times \left(\frac{1 \ mmol}{100 \ mg} \ \text{as } CaCO_3\right)$$
$$= 0.4 \ mmol/L$$

$$TH_F = TH_D - TH_R$$
$$= 1.0 \ \frac{mmol}{L} - 0.4 \ \frac{mmol}{L}$$
$$= 0.6 \ mmol/L$$

$$MgTH_I = MgTH = TH_F$$

$$= 2.0 \ \frac{mmol}{L} - 0.6 \ \frac{mmol}{L}$$

$$= 1.4 \ mmol/L$$

ion	concentration (mg/L)	molecular weight (mg/mmol)	molarity (mmol/L)
Ca^{2+}	187	40	4.7
Mg^{2+}	49	24	2.0
HCO_3^-	618	61	10.1

The molecular weight of $CaCO_3$ is 100 mg/mmol.

$$(1) \overset{Ca^{2+}}{\left(4.7 \ \frac{mmol}{L}\right)} + (2) \overset{2HCO_3^-}{\left(4.7 \ \frac{mmol}{L}\right)} + (2) \overset{2NaOH}{\left(4.7 \ \frac{mmol}{L}\right)}$$

$$\rightarrow CaCO_3 \downarrow + 2Na^+ + CO_3^- + H_2O$$

$$(1) \overset{Ca^{2+}}{\left(4.7 \ \frac{mmol}{L}\right)} + (2) \overset{2HCO_3^-}{\left(4.7 \ \frac{mmol}{L}\right)} + (2) \overset{2NaOH}{\left(4.7 \ \frac{mmol}{L}\right)}$$

$$\rightarrow CaCO_3 \downarrow + 2Na^+ + CO_3^- + H_2O$$

$$CH_R = CH - CH_I$$

$$= 10.1 \ \frac{mmol}{L} - (2) \left(4.7 \ \frac{mmol}{L}\right)$$

$$= 0.7 \ mmol/L$$

$$NaOH_1 = (2) \left(4.7 \ \frac{mmol}{L}\right) = 9.4 \ mmol/L$$

$$CH_R = CH - CH_I$$

$$= 10.1 \ \frac{mmol}{L} - (2) \left(4.7 \ \frac{mmol}{L}\right)$$

$$= 0.7 \ mmol/L$$

$$NaOH_1 = (2) \left(4.7 \ \frac{mmol}{L}\right)$$

$$= 9.4 \ mmol/L$$

$$\left(\frac{1}{2}\right) \overset{Mg^{2+}}{\left(0.7 \ \frac{mmol}{L}\right)} + \left(\frac{2}{2}\right) \overset{2HCO_3^-}{\left(0.7 \ \frac{mmol}{L}\right)} + \left(\frac{4}{2}\right) \overset{4NaOH}{\left(0.7 \ \frac{mmol}{L}\right)}$$

$$\rightarrow Mg(OH)_2 \downarrow + 4Na+ + 2CO_3^- + 2H_2O$$

$$\left(\frac{1}{2}\right) \overset{Mg^{2+}}{\left(0.7 \ \frac{mmol}{L}\right)} + \left(\frac{2}{2}\right) \overset{2HCO_3^-}{\left(0.7 \ \frac{mmol}{L}\right)} + \left(\frac{4}{2}\right) \overset{4NaOH}{\left(0.7 \ \frac{mmol}{L}\right)}$$

$$\rightarrow Mg(OH)_2 \downarrow + 4Na+ + 2CO_3^- + 2H_2O$$

$$MgTH_R = MgTH_I - MgTH_O$$

$$= 1.4 \ \frac{mmol}{L} - \left(\frac{1}{2}\right) \left(0.7 \ \frac{mmol}{L}\right)$$

$$= 1.05 \ mmol/L$$

$$NaOH_2 = \left(\frac{4}{2}\right) \left(0.7 \ \frac{mmol}{L}\right) = 1.4 \ mmol/L$$

$$MgTH_R = MgTH_I - MgTH_O$$

$$= 2.0 \ \frac{mmol}{L} - \left(\frac{1}{2}\right) \left(0.7 \ \frac{mmol}{L}\right)$$

$$= 1.65 \ mmol/L$$

$$NaOH_2 = \left(\frac{4}{2}\right) \left(0.7 \ \frac{mmol}{L}\right)$$

$$= 1.4 \ mmol/L$$

$$(1) \overset{Mg^{2+}}{\left(1.05 \ \frac{mmol}{L}\right)} + (2) \overset{2NaOH}{\left(1.05 \ \frac{mmol}{L}\right)}$$

$$\rightarrow Mg(OH)_2 \downarrow + 2Na^+$$

$$(1) \overset{Mg^{2+}}{\left(1.65 \ \frac{mmol}{L}\right)} + (2) \overset{2NaOH}{\left(1.65 \ \frac{mmol}{L}\right)}$$

$$\rightarrow Mg(OH)_2 \downarrow + 2Na^+$$

$$NaOH_3 = (2) \left(1.05 \ \frac{mmol}{L}\right) = 2.1 \ mmol/L$$

$$NaOH = (NaOH_1 + NaOH_2 + NaOH_3) \ MW f Q$$

$$= \left(9.4 \ \frac{mmol}{L} + 1.4 \ \frac{mmol}{L} + 2.1 \ \frac{mmol}{L}\right)$$

$$\times \left(40 \ \frac{mg}{mmol}\right) \left(30\,000 \ \frac{m^3}{d}\right)$$

$$\times \left(30 \ \frac{d}{mo}\right) \left((10)^{-6} \ \frac{tonne \cdot L}{m^3 \cdot mg}\right)$$

$$= 464 \ tonne/mo \quad (460 \ tonne/mo)$$

$$NaOH_3 = (2) \left(1.65 \ \frac{mmol}{L}\right) = 3.3 \ mmol/L$$

$$NaOH = (NaOH_1 + NaOH_2 + NaOH_3) \ MW \frac{Q}{f}$$

$$= \left(9.4 \ \frac{mmol}{L} + 1.4 \ \frac{mmol}{L} + 3.3 \ \frac{mmol}{L}\right)$$

$$\times \left(40 \ \frac{mg}{mmol}\right) \left(\frac{100\%}{83\%}\right) \left(30,000 \ \frac{m^3}{d}\right)$$

$$\times \left(30 \ \frac{d}{mo}\right) \left((10)^{-6} \ \frac{tonne \cdot L}{m^3 \cdot mg}\right)$$

$$= 612 \ tonne/mo \quad (610 \ tonne/mo)$$

(C) This incorrect solution does not include in the calculation the hardness that is to remain in solution. Other assumptions, definitions, and equations are the same as used in the correct solution.

(D) This incorrect solution assumes that the bicarbonate is unlimited. Other assumptions, definitions, and equations are the same as used in the correct solution.

ion	concentration (mg/L)	molecular weight (mg/mmol)	molarity (mmol/L)
Ca^{2+}	187	40	4.7
Mg^{2+}	49	24	2.0

Assume that the maximum hardness that can be precipitated is 40 mg/L as $CaCO_3$. The molecular weight of $CaCO_3$ is 100 mg/mmol.

$$TH_D = \left(100 \ \frac{mg}{L} \ as \ CaCO_3\right)$$
$$\times \left(\frac{1 \ mmol}{100 \ mg} \ as \ CaCO_3\right)$$
$$= 1.0 \ mmol/L$$

$$TH_R = \left(40 \ \frac{mg}{L} \ as \ CaCO_3\right)$$
$$\times \left(\frac{1 \ mmol}{100 \ mg} \ as \ CaCO_3\right)$$
$$= 0.4 \ mmol/L$$

$$TH_F = TH_D - TH_R$$
$$= 1.0 \ \frac{mmol}{L} - 0.4 \ \frac{mmol}{L}$$
$$= 0.6 \ mmol/L$$

$$MgTH_I = MgTH = TH_F$$
$$= 2.0 \ \frac{mmol}{L} - 0.6 \ \frac{mmol}{L}$$
$$= 1.4 \ mmol/L$$

$$\overset{Ca^{2+}}{(1) \left(4.7 \ \frac{mmol}{L}\right)} + \overset{2HCO_3^-}{(2) \left(4.7 \ \frac{mmol}{L}\right)} + \overset{2NaOH}{(2) \left(4.7 \ \frac{mmol}{L}\right)}$$

$$\rightarrow CaCO_3 \downarrow + 2Na^+ + CO_3^- + H_2O$$

$$NaOH_1 = (2) \left(4.7 \ \frac{mmol}{L}\right) = 9.4 \ mmol/L$$

$$\overset{Mg^{2+}}{\left(1.4 \ \frac{mmol}{L}\right)} + \overset{2HCO_3^-}{(2) \left(1.4 \ \frac{mmol}{L}\right)} + \overset{4NaOH \rightarrow}{(4) \left(1.4 \ \frac{mmol}{L}\right)}$$

$$Mg(OH)_2 \downarrow + 4Na+ + 2CO_3^- + 2H_2O$$

$$NaOH_2 = (4) \left(1.4 \ \frac{mmol}{L}\right) = 5.6 \ mmol/L$$

$$NaOH = (NaOH_1 + NaOH_2 + NaOH_3) \ MW \frac{Q}{f}$$
$$= \left(9.4 \ \frac{mmol}{L} + 5.6 \ \frac{mmol}{L}\right) \left(40 \ \frac{mg}{mmol}\right)$$
$$\times \left(\frac{100\%}{83\%}\right) \left(30\,000 \ \frac{m^3}{d}\right)$$
$$\times \left(30 \ \frac{d}{mo}\right) \left((10)^{-6} \ \frac{tonne \cdot L}{m^3 \cdot mg}\right)$$
$$= 651 \ tonne/mo \quad (650 \ tonne/mo)$$

SOLUTION 98

The information given, such as the isotherm intercept and slope, suggests using the Freundlich isotherm equation. The alternative would be the Langmuir isotherm, but it includes empirical constants and not the isotherm intercept and slope terms.

C_f	DBP concentration remaining in solution	mg/L
C_o	DBP concentration before adsorption	mg/L
k	isotherm intercept	mg/g
M	powdered activated carbon (PAC) dose	g/L
$1/n$	isotherm slope	–

$$\frac{C_o - C_f}{M} = k C_f^{1/n}$$
$$M = \frac{C_o - C_f}{k C_f^{1/n}}$$
$$= \frac{\left(138 \ \frac{\mu g}{L}\right) \left(\frac{1 \ mg}{(10)^3 \ \mu g}\right) - \left(5 \ \frac{\mu g}{L}\right) \left(\frac{1 \ mg}{(10)^3 \ \mu g}\right)}{\left(21 \ \frac{mg}{g}\right) \left(\left(5 \ \frac{\mu g}{L}\right) \left(\frac{1 \ mg}{(10)^3 \ \mu g}\right)\right)^{0.54}}$$
$$= 0.11 \ g/L \ water \ treated \quad (110 \ mg/L)$$

The answer is (A).

Why Other Options Are Wrong

(B) This incorrect solution reverses the slope and intercept values in the isotherm equation. Other definitions and equations are the same as used in the correct solution.

n	isotherm intercept	mg/g
k	isotherm slope	–

$$M = \frac{C_o - C_f}{k C_f^{1/n}}$$
$$= \frac{\left(138 \ \frac{\mu g}{L}\right) \left(\frac{1 \ mg}{(10)^3 \ \mu g}\right) - \left(5 \ \frac{\mu g}{L}\right) \left(\frac{1 \ mg}{(10)^3 \ \mu g}\right)}{(0.54) \left(\left(5 \ \frac{\mu g}{L}\right) \left(\frac{1 \ mg}{(10)^3 \ \mu g}\right)\right)^{1/21 \ mg/g}}$$
$$= 0.32 \ g/L \ water \ treated \quad (320 \ mg/L)$$

(C) This incorrect solution uses the isotherm equation in the wrong form and wrongly defines the dose term in the equation. Other definitions and equations are the same as used in the correct solution.

X/M	PAC dose	g/L

$$\frac{X}{M} = kC_f^{1/n} = \left(21 \; \frac{mg}{g}\right)\left(\left(5 \; \frac{\mu g}{L}\right)\left(\frac{1 \; mg}{(10)^3 \; \mu g}\right)\right)^{0.54}$$
$$= 1.2 \; g/L \quad (1200 \; mg/L)$$

(D) This incorrect solution uses concentration units of $\mu g/L$ instead of mg/L. Other definitions and equations are the same as used in the correct solution.

C_f DBP concentration remaining in solution $\quad \mu g/L$

C_o DBP concentration before adsorption $\quad \mu g/L$

$$M = \frac{C_o - C_f}{kC_f^{1/n}}$$
$$= \frac{\left(138 \; \frac{\mu g}{L}\right) - \left(5 \; \frac{\mu g}{L}\right)}{\left(21 \; \frac{mg}{g}\right)\left(5 \; \frac{\mu g}{L}\right)^{0.54}}$$
$$= 2.7 \; g \cdot \mu g/mg \cdot L$$

The units do not work.

Assume units of g/L for the following calculation.

$$M = \left(2.7 \; \frac{g}{L}\right)\left(\frac{(10)^3 \; mg}{1 \; g}\right) = 2700 \; mg/L$$

SOLUTION 99

The Safe Drinking Water Act (SDWA) regulates all public drinking water systems in the United States. Public drinking water systems include all those serving at least 15 connections or serving at least 25 people for 60 days of each year. The purpose of the SDWA is to protect public health from exposure to both naturally occurring and man-made contaminants in the water supply. To satisfy this purpose, the SDWA includes regulation of waste disposal by injection into the groundwater.

The answer is (C).

Why Other Options Are Wrong

(A) This choice is incorrect because the SDWA does regulate all public drinking water systems in the United States.

(B) This choice is incorrect because the SDWA does regulate both naturally occurring and man-made contaminants in drinking water.

(D) This choice is incorrect because the SDWA does regulate injection of wastes through injection wells. The regulation occurs under the Underground Injection Control (UIC) program.

SOLUTION 100

At an initial pH of 9.7, the dominant alkalinity species will be carbonate.

For hydroxide alkalinity,

$$pH = 9.7$$
$$pH + pOH = 14$$
$$pOH = 14 - 9.7 = 4.3$$

$[OH^-]$ hydroxide concentration mol/L

$$pOH = -\log[OH^-]$$
$$[OH^-] = (10)^{-4.3} \; mol/L$$

EW calcium carbonate equivalent weight $\quad 50\,000$ mg/eq
OH^-_{alk} hydroxide alkalinity $\quad$ mg/L as $CaCO_3$
$|V|$ valence $\quad$ eq/mol

$$OH^-_{alk} = [OH^-]|V|EW$$
$$= \left((10)^{-4.3} \; \frac{mol}{L}\right)\left(1 \; \frac{eq}{mol}\right)$$
$$\times \left(50\,000 \; \frac{mg \; CaCO_3}{eq}\right)$$
$$= 2.5 \; mg/L \; as \; CaCO_3$$

When 0.03 N sulfuric acid is used as the titrant, 1.0 mL of acid will neutralize 1.5 mg of alkalinity as $CaCO_3$.

For carbonate alkalinity,

$CO_3^{2-}_{alk}$ carbonate alkalinity $\quad$ mg/L as $CaCO_3$
V_{sample} sample volume $\quad$ mL
$V_{titrant}$ titrant volume added $\quad$ mL

Titrating from pH 9.7 to pH 8.3 used 6 mL of titrant,

$$\left(\frac{1}{2}\right)CO_3^{2-}_{alk} = \left(\frac{1.5 \; mg \; as \; CaCO_3}{1 \; mL \; titrant}\right)\left(\frac{V_{titrant}}{V_{sample}}\right)$$
$$- OH^-_{alk}$$
$$= \frac{\left(\begin{array}{c}(1.5 \; mg \; as \; CaCO_3)(6 \; mL) \\ \times \left((10)^3 \; \frac{mL}{L}\right)\end{array}\right)}{(1 \; mL)(250 \; mL \; sample)}$$
$$- 2.5 \; \frac{mg}{L} \; as \; CaCO_3$$
$$= 33.5 \; mg/L \; as \; CaCO_3$$
$$CO_3^{2-}_{alk} = (2)\left(33.5 \; \frac{mg}{L} \; as \; CaCO_3\right)$$
$$= 67 \; mg/L \; as \; CaCO_3$$

For total alkalinity,

T_{alk} total alkalinity   mg/L as $CaCO_3$

Titrating from pH 9.7 to pH 4.5 used 18 mL of titrant,

$$T_{alk} = \left(\frac{1.5 \text{ mg as } CaCO_3}{1 \text{ mL titrant}} \right) \left(\frac{V_{titrant}}{V_{sample}} \right)$$

$$= \frac{(1.5 \text{ mg as } CaCO_3)(18 \text{ mL}) \left((10)^3 \frac{\text{mL}}{\text{L}} \right)}{(1 \text{ mL})(250 \text{ mL sample})}$$

$$= 108 \text{ mg/L as } CaCO_3$$

For bicarbonate alkalinity,

$HCO_3{}^-{}_{alk}$ bicarbonate alkalinity   mg/L as $CaCO_3$

$$HCO_3^-{}_{alk} = T_{alk} - OH_{alk}^- - CO_3^{2-}{}_{alk}$$

$$= 108 \frac{\text{mg}}{\text{L}} \text{ as } CaCO_3 - 2.5 \frac{\text{mg}}{\text{L}} \text{ as } CaCO_3$$

$$- 67 \frac{\text{mg}}{\text{L}} \text{ as } CaCO_3$$

$$= 38.5 \text{ mg/L as } CaCO_3$$

Check. Carbonate alkalinity dominates at 67 mg/L as $CaCO_3$.

The answer is (B).

Why Other Options Are Wrong

(A) This incorrect solution does not account for using the non-standard 0.03 N sulfuric acid titrant instead of the standard 0.02 N sulfuric acid titrant. Other assumptions, definitions, and equations are the same as used in the correct solution.

At an initial pH of 9.7, the dominant alkalinity species will be carbonate.

For hydroxide alkalinity,

$$pH = 9.7$$
$$pH + pOH = 14$$
$$pOH = 14 - 9.7 = 4.3$$
$$pOH = -\log[OH^-]$$
$$[OH^-] = (10)^{-4.3} \text{ mol/L}$$
$$OH_{alk}^- = [OH^-](|V|)\,EW$$
$$= \left((10)^{-4.3} \frac{\text{mol}}{\text{L}} \right) \left(1 \frac{\text{eq}}{\text{mol}} \right)$$
$$\times \left(50\,000 \frac{\text{mg } CaCO_3}{\text{eq}} \right)$$
$$= 2.5 \text{ mg/L as } CaCO_3$$

For carbonate alkalinity,

Assume 1.0 mL of acid will neutralize 1.0 mg of alkalinity as $CaCO_3$.

Titrating from pH 9.7 to pH 8.3 used 6 mL of titrant,

$$\left(\frac{1}{2} \right) CO_3^{2-}{}_{alk} = \left(\frac{1.0 \text{ mg as } CaCO_3}{1 \text{ mL titrant}} \right) \left(\frac{V_{titrant}}{V_{sample}} \right)$$

$$- OH^-{}_{alk}$$

$$= \frac{\left(\begin{array}{c} (1.0 \text{ mg as } CaCO_3)(6 \text{ mL}) \\ \times \left((10)^3 \frac{\text{mL}}{\text{L}} \right) \end{array} \right)}{(1 \text{ mL})(250 \text{ mL sample})}$$

$$- 2.5 \frac{\text{mg}}{\text{L}} \text{ as } CaCO_3$$

$$= 21.5 \text{ mg/L as } CaCO_3$$

$$CO_3^{2-}{}_{alk} = (2) \left(21.5 \frac{\text{mg}}{\text{L}} \text{ as } CaCO_3 \right)$$

$$= 43 \text{ mg/L as } CaCO_3$$

For total alkalinity,

T_{alk} total alkalinity   mg/L as $CaCO_3$

Titrating from pH 9.7 to pH 4.5 used 18 mL of titrant,

$$T_{alk} = \left(\frac{1.0 \text{ mg as } CaCO_3}{1 \text{ mL titrant}} \right) \left(\frac{V_{titrant}}{V_{sample}} \right)$$

$$= \frac{(1.0 \text{ mg as } CaCO_3)(18 \text{ mL}) \left((10)^3 \frac{\text{mL}}{\text{L}} \right)}{(1 \text{ mL})(250 \text{ mL sample})}$$

$$= 72 \text{ mg/L as } CaCO_3$$

For bicarbonate alkalinity,

$HCO_3{}^-{}_{alk}$ bicarbonate alkalinity   mg/L as $CaCO_3$

$$HCO_3^-{}_{alk} = T_{alk} - OH_{alk}^- - CO_3^{2-}{}_{alk}$$

$$= 72 \frac{\text{mg}}{\text{L}} \text{ as } CaCO_3 - 2.5 \frac{\text{mg}}{\text{L}} \text{ as } CaCO_3$$

$$- 43 \frac{\text{mg}}{\text{L}} \text{ as } CaCO_3$$

$$= 26.5 \text{ mg/L as } CaCO_3$$

Check. Carbonate alkalinity dominates at 43 mg/L as $CaCO_3$.

(C) This incorrect solution ignores the hydroxide alkalinity. Other assumptions, definitions, and equations are the same as used in the correct solution.

At an initial pH of 9.7, the dominant alkalinity species will be carbonate.

When 0.03 N sulfuric acid is used as the titrant, 1.0 mL of acid will neutralize 1.5 mg of alkalinity as $CaCO_3$.

For carbonate alkalinity,

Titrating from pH 9.7 to pH 8.3 used 6 mL of titrant.

$$\left(\frac{1}{2}\right)CO_3{}^{2-}{}_{alk} = \left(\frac{1.5 \text{ mg as } CaCO_3}{1 \text{ mL titrant}}\right)\left(\frac{V_{titrant}}{V_{sample}}\right)$$

$$= \frac{(1.5 \text{ mg as } CaCO_3)(6 \text{ mL})\left((10)^3 \frac{\text{mL}}{\text{L}}\right)}{(1 \text{ mL})(250 \text{ mL sample})}$$

$$= 36 \text{ mg/L as } CaCO_3$$

$$CO_3{}^{2-}{}_{alk} = (2)\left(36 \frac{\text{mg}}{\text{L}} \text{ as } CaCO_3\right)$$

$$= 72 \text{ mg/L as } CaCO_3$$

For total alkalinity,

T_{alk} total alkalinity   mg/L as $CaCO_3$

Titrating from pH 9.7 to pH 4.5 used 18 mL of titrant.

$$T_{alk} = \left(\frac{1.5 \text{ mg as } CaCO_3}{1 \text{ mL titrant}}\right)\left(\frac{V_{titrant}}{V_{sample}}\right)$$

$$= \frac{(1.5 \text{ mg as } CaCO_3)(18 \text{ mL})\left((10)^3 \frac{\text{mL}}{\text{L}}\right)}{(1 \text{ mL})(250 \text{ mL sample})}$$

$$= 108 \text{ mg/L as } CaCO_3$$

For bicarbonate alkalinity,

$HCO_3{}^-_{alk}$ bicarbonate alkalinity   mg/L as $CaCO_3$

$$HCO_3{}^-{}_{alk} = T_{alk} - CO_3{}^{2-}{}_{alk}$$

$$= 108 \frac{\text{mg}}{\text{L}} \text{ as } CaCO_3 - 72 \frac{\text{mg}}{\text{L}} \text{ as } CaCO_3$$

$$= 36 \text{ mg/L as } CaCO_3$$

Check. Carbonate alkalinity dominates at 72 mg/L as $CaCO_3$.

(D) This incorrect solution assumes that all the carbonate alkalinity is titrated above pH 8.3 instead of one-half of it being titrated. Other assumptions, definitions, and equations are the same as used in the correct solution.

For hydroxide alkalinity,

$$pH = 9.7$$

$$pH + pOH = 14$$

$$pOH = 14 - 9.7 = 4.3$$

$$pOH = -\log[OH^-]$$

$$[OH^-] = (10)^{-4.3} \text{ mol/L}$$

$$OH^-alk = (|V|)EW$$

$$= \left((10)^{-4.3} \frac{\text{mole}}{\text{L}}\right)\left(1 \frac{\text{eq}}{\text{mol}}\right)$$

$$\times \left(50\,000 \text{ mg} \frac{CaCO_3}{\text{eq}}\right)$$

$$= 2.5 \text{ mg/L as } CaCO_3$$

When 0.03 N sulfuric acid is used as the titrant, 1.0 mL of acid will neutralize 1.5 mg of alkalinity as $CaCO_3$

For carbonate alkalinity,

Titrating from pH 9.7 to pH 8.3 used 6 mL of titrant.

$$CO_3{}^{2}{}^-{}_{alk} = \left(\frac{1.5 \text{ mg as } CaCO_3}{1 \text{ mL titrant}}\right)\left(\frac{V_{titrant}}{V_{sample}}\right) - OH^-alk$$

$$= \frac{(1.5 \text{ mg as } CaCO_3)(6 \text{ mL})\left((10)^3 \frac{\text{mL}}{\text{L}}\right)}{(1 \text{ mL})(250 \text{ mL sample})}$$

$$- 2.5 \frac{\text{mg}}{\text{L as } CaCO_3}$$

$$= 33.5 \text{ mg/L as } CaCO_3$$

For total alkalinity,

Titrating for pH 9.7 to pH 4.5 used 18 mL of titrant.

$$T_{alk} = \left(\frac{1.5 \text{ mg as } CaCO_3}{1 \text{ mL titrant}}\right)\left(\frac{V_{titrant}}{V_{sample}}\right)$$

$$= \frac{(1.5 \text{ mg as } CaCO_3)(18 \text{ mL})\left((10)^3 \frac{\text{mL}}{\text{L}}\right)}{(1 \text{ mL})(250 \text{ mL sample})}$$

$$= 108 \text{ mg/L as } CaCO_3$$

For bicarbonate alkalinity,

$$HCO_3{}^-{}_{alk} = T_{alk} - OH^-{}_{alk} - CO_3{}^{2}{}^-{}_{alk}$$

$$= 108 \frac{\text{mg}}{\text{L as } CaCO_3} - 2.5 \frac{\text{mg}}{\text{L as } CaCO_3}$$

$$- 33.5 \frac{\text{mg}}{\text{L as } CaCO_3}$$

$$= 72 \frac{\text{mg}}{\text{L as } CaCO_3}$$

Carbonate alkalinity dominates at 72 mg/L as $CaCO_3$.

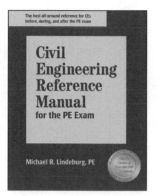

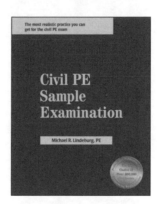
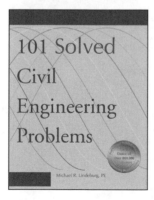